COPPER IN SOILS
AND PLANTS

COPPER IN SOILS AND PLANTS

Proceedings of the Golden Jubilee International Symposium on 'Copper in Soils and Plants' held at Murdoch University, Perth, Western Australia on May 7-9, 1981 under the sponsorship of the Australian Academy of Technological Sciences.

Edited by

J.F. Loneragan
A.D. Robson
R.D. Graham

Academic Press

A Subsidiary of Harcourt Brace Jovanovich, Publishers

Sydney   New York   London   Toronto   San Francisco

Academic Press Rapid Manuscript Reproduction

Printed in Australia

National Library of Australia Cataloguing-in-Publication Data
Copper in Soils and Plants

    Bibliography
    Includes Index
    ISBN 0 12 455520 9 ✓

    1. Copper. 2. Soils — Copper content. 3. Plants,
Effects of copper on.
    I. Loneragan, J.F. II. Robson, A.D.
    III. Graham, R.D.

631.4'16

Library of Congress Catalog Card Number: 80-70774

CONTENTS

Copper in Plants

Copper in Soil - Plant Systems

12. Copper Concentrations in Plants and their Relationship to Soil Properties

S.C. Jarvis

13. Diagnosis of Copper Deficiency and Toxicity

A.D. Robson and D.J. Reuter

14. Distribution and Correction of Copper Deficiency in Crops and Pastures

J.W. Gartrell

15. Conclusion: Copper in Soils and Plants —An Overview

A.D. Robson

Abstracts of Posters

Index

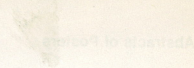

Abstracts of Posters

CONTRIBUTORS

Alston, A.M., Waite Agricultural Research Institute, University of Adelaide, Glen Osmond, South Australia, 5064.

Barrow, N.J., Division of Land Resource Management, Commonwealth Scientific and Industrial Research Organisation, Private Bag, Post Office, Wembley, Western Australia. 6014.

Best, Errol K., Queensland Wheat Research Institute, Toowoomba, Queensland, 4350

Bevege, D.I., Department of Forestry, Indooroopilly, Queensland. 4068

Beveridge, S.J., Faculty of Medicine, University of Newcastle, New South Wales, 2300

Bussler, Wolfgang, Department of Plant Nutrition, Faculty of International Agricultural Development, Technical University, Berlin, 33

Fitch, Alanah, Department of Agronomy, University of Illinois, Urbana, U.S.A. 61801

Gartrell, J.W., Department of Agriculture, Jarrah Road, South Perth, Western Australia. 6151

Gilkes, R.J., Department of Soil Science and Plant Nutrition, University of Western Australia, Nedlands, Western Australia. 6009

Graham, Robin D., Department of Agronomy, Waite Agricultural Research Institute, University of Adelaide, Glen Osmond, South Australia. 5064

Grundon, Noel J., Queensland Wheat Research Institute, Toowoomba, Queensland. 4350

Harry, Susan P., Department of Agronomy, Waite Agricultural Research Institute, Glen Osmond, South Australia. 5064

Hill, J., Commonwealth Scientific and Industrial Research Organisation, Division of Forest Research, St. Lucia, Queensland. 4067

James, Robert O., Emulsion Research Laboratory, Eastman Kodak Company, Rochester, New York, U.S.A. 14615

Jarvis, S.C., Division of Soils and Plant Nutrition, Grassland Research Institute, Hurley, United Kingdom. SL6 5LR

Kubota, Joe, Soil Conservation Service, USDA, U.S. Plant, Soil and Nutrition Laboratory, Ithaca, New York, U.S.A. 14853

Loneragan, J.F., School of Environmental and Life Sciences, Murdoch University, Murdoch, Western Australia. 6150

McBride, Murray B., Department of Agronomy, Cornell University, Ithaca, New York, U.S.A. 14853

Merry, Richard H., Division of Soils, Commonwealth Scientific and Industrial Research Organisation, Glen Osmond, South Australia. 5064

Parker, A.J., School of Mathematical and Physical Sciences, Murdoch University, Murdoch, Western Australia. 6150

Pondrilei, K. Simon, Department of Primary Industry, Konedobu, Papua New Guinea

Reddy, G.D., Waite Agricultural Research Institute, University of Adelaide, Glen Osmond, South Australia. 5064

Reuter, D.J., Department of Agriculture and Fisheries, Adelaide, South Australia. 5000

Robson, A.D., Department of Soil Science and Plant Nutrition, The University of Western Australia, Nedlands, Western Australia. 6009

Simmons, W.J., Department of Soil Science and Plant Nutrition, The University of Western Australia, Nedlands, Western Australia. 6009

Snowball, K., Department of Soil Science and Plant Nutrition, The University of Western Australia, Nedlands, Western Australia. 6009

Stevenson, F.J., Department of Agronomy, University of Illinois, Urbana, U.S.A. 61801

Tiller, Kevin G., Division of Soils, Commonwealth Scientific and Industrial Research Organisation, Glen Osmond, South Australia. 5064

Turvey, N.D., A.P.M. Forests Pty Ltd., Traralgon, Victoria. 3844

Velayutham, K.S., Department of Primary Industry, Konedobu, Papua New Guinea

Walker, Colin D., U.S. Plant, Soil and Nutrition Laboratory, Tower Road, Ithaca, New York, U.S.A. 14853

Walker, S., John Innes Institute, Colney Lane, Norwich, Norfolk, United Kingdom. NR4 7UH

Walker, W.R., Department of Chemistry, University of Newcastle, New South Wales. 2308

Webb, John, School of Mathematical and Physical Sciences, Murdoch University, Murdoch, Western Australia. 6150

Welch, Ross M., U.S. Plant, Soil and Nutrition Laboratory, Tower Road, Ithaca, New York, U.S.A. 14853

Woolhouse, H.W., John Innes Institute, Colney Lane, Norwich, Norfolk, United Kingdom. NR4 7UH

PREFACE

Fifty years ago, the first definitive evidence for the essentiality of copper for the growth of higher plants was published. The burst of research activity which followed established copper deficiency as a limiting condition for plant growth and animal production on numerous soils in all continents.

The last ten years has seen a renaissance of that early interest in copper in agricultural systems. It has also seen a growing awareness of copper toxicity to plants and animals of pollutants from mine and sewage wastes. Furthermore, changing technology has exacerbated problems of deficiency and toxicity. The resulting research activity has expanded our understanding of the behaviour of copper in many processes. It therefore seemed timely for workers in relevant fields to meet and review this new information and to try and relate it to the behaviour of copper in soils and plant systems.

The Golden Jubilee of the recognition of copper as an essential element for plant growth provided an appropriate time for such a review. The scheduling in 1981 of the "Fourth International Symposium on Trace Element Metabolism in Man and Animals" in Western Australia, where so much research on copper has been done, provided an appropriate venue.

We wish to record our thanks to the Australian Academy of Technological Sciences for sponsoring the Symposium, to the organisations (p. xv) which provided financial support, and to members of the Organising Committee (p. xv). Our special thanks go to Dee Cahill and to Derek Plaskett for assistance in preparing the proceedings of the Symposium for publication.

We also wish to pay a tribute to the late Professor Alan Posner whose untimely death in the early stages of planning the Symposium deprived us of a valued friend and colleague. Alan Posner was an outstanding soil scientist whose work had included the study of processes involving the behaviour of copper in soils. We will miss his lively mind at the Symposium discussions but in the work reviewed we will remember his many contributions to this subject and to soil science generally.

ACKNOWLEDGEMENTS

SPONSOR
Australian Academy of Technological Sciences

ORGANISING COMMITTEE
N.J. Barrow, Commonwealth Scientific and Industrial Research Organisation, Western Australia
J.W. Gartrell, Department of Agriculture, Western Australia
R.D. Graham, Waite Agricultural Research Institute, South Australia
J.F. Loneragan (Chairman), Murdoch University, Western Australia
The late A.M. Posner, University of Western Australia
A.D. Robson, University of Western Australia
P. Southern, CSBP & Farmers Ltd., Western Australia

FINANCIAL SUPPORTERS
Australian Academy of Technological Sciences
Australian Wheat Industry Research Council
Australian Wool Corporation
CSBP & Farmers Limited
Mt Isa Mining Holdings Limited
Murdoch University
Wheat Industry Research Committee of Western Australia

INTRODUCTION: THE CHEMISTRY OF COPPER

A.J. Parker

School of Mathematical and Physical Sciences
Murdoch University
Perth, Western Australia

COPPER MINERALS

Copper is widely distributed geographically and geologically. Deposits have been formed throughout geological time. However, copper ore bodies are rare and occupy only a very small fraction of known rock formations. The average amount of copper in the earth's crust is about 70ppm (Hodgson, 1963). Copper frequently occurs as the metal and in the Lake Superior region of the USA masses weighing several tonnes have been found. More frequently copper occurs as a primary sulfide mineral, such as bornite (Cu_5FeS_4) or chalcopyrite ($CuFeS_2$). The secondary copper sulfide minerals such as chalcocite (Cu_2S) and covellite (CuS) are forms of copper which have been leached and precipitated. Oxidised copper minerals are formed by weathering of sulfides and occur as oxides, hydroxides, carbonates, chlorides and silicates (Butts, 1964). Copper and sulfur form sulfides in which copper is in the +1 rather than +2 oxidation state. The so-called "cupric sulfide", CuS, is more usefully thought of as copper(I) disulfide, $Cu^+.S_2^=.Cu^+$ (Vaughan and Craig, 1978). Copper(II) and sulfide ions are not compatible since copper (II) is reduced and sulfide is oxidised when these species are in contact in solution, in melts and in the gas phase. Most copper sulfide minerals possess semi-conductor properties. In the natural state copper sulfide crystals contain impurities and defects, so they are good conductors (Vaughan and Craig, 1978). Whether it is long term weathering or short term leaching with bacteria, ferric sulfate or

COPPER IN SOILS AND PLANTS

ferric chloride, the oxidation of copper sulfide is an electrochemical or corrosion process (Parker *et al.*, 1981). Many
sulfide ore bodies are gigantic electrochemical cells which
have cathodic (reductive) sites at the oxygen-rich ground
level and anodic sites (oxidation) deeper down near the water
table. Thus weathering and oxidised zones are sometimes
found at depth in copper ore bodies. Familiarity with semi-
conductors and electrochemical properties (Vaughan and Craig,
1978) is recommended for those interested in naturally occurring copper, be they mineralogists, extractive metallurgists
or soil scientists.

ATOMIC AND PHYSICAL PROPERTIES (Cotton and Wilkinson, 1972)

Copper of atomic number 29 is the first element of sub-
group 1B (the first transition series) of the periodic table.
Copper has very little relation to the main group I elements,
such as sodium and potassium, and is noticeably different
from silver and gold. The electronic structure of the free
copper atom is $1s^2$, $2s^2$, $2p^6$, $3s^2$, $3p^6$, $3d^{10}$, $4s^1$. The
single 4s electron is outside a filled 3d shell. The filled
$3d^{10}$ shell is much less effective than is a noble gas shell
(e.g. $3p^6$ of potassium) in shielding this electron from the
positive nuclear charge. Thus the 4s electron is rather
difficult to remove from copper. Copper's first ionization
potential (7.72 eV) reflects this greater energy for electron
removal than does the first ionization potential of the alkali metals (e.g. K = 4.34 eV for example). Like all elements
of the first transition series (e.g. Cr, Mn, Fe, Co, Ni) and
unlike Li, Na, K, Rb, of Group I, two electrons are removed
relatively easily from copper atoms (20.29 eV) in the gas
phase to form Cu^{2+} (copper II). This ion is quite stable in
water, but unlike any other elements in the transition series,
the second ionization potential of copper is sufficiently
higher than the first and the effects of the ions environment is such as to allow a variety of stable copper(I)
species to also exist. Whether copper exists as copper(I)
rather than copper(II) depends very much on the physical environment (i.e. crystal lattice, melt, gas phase, solution),
the solvent, the concentration and what ligands (bases) are
present. Thus copper(I) is stable in water containing excess
halide ions, acetonitrile, pyridine, or cyanide ions, in an-
hydrous acetonitrile, ammonia, or pyridine, and as solid
$CuCl$, $CuCN$, CuI, $CuSCN$, Cu_2O, Cu_2S, etc. However, in water,
copper(I) disproportionates (eq.1) as the sulfate, nitrate,

or perchlorate, when attempts are made to prepare 10^{-1}-10^{-2}M
Cu_2SO_4, $CuNO_3$ or $CuClO_4$ solutions in water. The hydrated
Cu^+ ion is not a stable species in water at the concentration
normally used by chemists.

An interesting aspect of copper in wet soils (Lindsay,
1979), as distinct from the aqueous copper salt solutions
normally encountered by chemists, is that in very dilute solution (10^{-6}M - 10^{-7}M) copper(I) is moderately stable with
respect to its disproportionation into copper metal and Cu^{2+}.
The equilibrium constant K for (1) is $10^6 M^{-1}$ at 25° in water,
so that in "chemical solutions" (10^{-2} - 10^{-3}M), very little
copper(I) is present.

$$2Cu+ \overset{K}{\rightleftharpoons} Cu + Cu^{2+} \tag{1}$$

However, since the average concentration of dissolved copper
in copper-rich moist soils is 10^{-6} - 10^{-7}M, that copper could
be present as 1 x 10^{-7}M Cu^{2+} and 3 x 10^{-7}M Cu^+, for example.
Hydrated Cu^+ would be the predominant copper species. This
mathematical curiosity arises of course because of the
squared $[Cu^+]$ term in K = $[Cu^{2+}]$ $[Cu^+]^2$.

The isotopes of copper are Cu^{63} (69%) and Cu^{65} (31%).
The highly active Cu^{64}, of half life 12.8 hours, is the most
suitable for tracer work. Decay of Cu^{64} leads to Zn^{64} or
Ni^{64} by electron or positron emission respectively (Fronsbein and Firminhac, 1954).

CHEMISTRY OF COPPER IONS IN DIFFERENT ENVIRONMENTS

Standard reduction potentials at 25° for various forms of
copper(I) and copper(II) in aqueous solutions, sometimes containing other bases, are in Table 1. The Table shows the
electron accepting tendencies of these copper species under
the conditions specified. The more positive the reduction
potential, the more readily electrons are accepted by the
copper containing species to the left of the reduction
equation. Equally of course, Table 1 shows the ability of
copper metal or copper(I) (i.e. species to the right of the
reduction equation) to *donate* electrons to suitable electron
acceptors in various environments. The more negative the
reduction potential, the more readily does the copper or the
copper(I) species on the right donate electrons (i.e. the
more readily is the copper oxidised). For reasons associated
with the poor shielding of the copper nucleus by its d^{10}
electrons, as discussed above for the gas phase processes,

TABLE I.   Electron Acceptor Tendencies[a] (i.e. Reduction
           Potentials (nhe)) of Copper(I) and Copper(II)
           Systems in Water at 25°

Reduction Equation	E^O volt nhe
$Cu_2S + 2e^- \rightarrow 2Cu + S^{2-}$	-0.95
$CuS + 2e^- \rightarrow Cu + S^{2-}$	-0.76
$Cu_2O + H_2O + 2e^- \rightarrow 2Cu + 2OH^-$	-0.36
$CuSCN + e^- \rightarrow Cu + SCN^-$	-0.27
$CuI + e^- \rightarrow Cu + I^-$	-0.19
$Cu(NH_3)_2^+ + e^- \rightarrow Cu + 2NH_3$	-0.11
$Cu(NH_3)_4^{2+} + 2e^- \rightarrow Cu + 4NH_3$	-0.05
$CuBr + e^- \rightarrow Cu + Br^-$	$+0.03$
$Cu(NH_3)_4^{2+} + e^- \rightarrow Cu(NH_3)_2^+ + 2NH_3$	$+0.06$
$CuCl + e^- \rightarrow Cu + Cl^-$	$+0.12$
$Cu^{2+} + e^- \rightarrow Cu^+_{aq}$	$+0.17$
$Cu(An)_3^+ + e^- \rightarrow Cu^0 + 3CH_3CN$	$+0.18$
$CuS + e^- \rightarrow \frac{1}{2}Cu_2S + \frac{1}{2}S^{2-}$	$+0.19$
$Cu^{2+} + 2e^- \rightarrow Cu^0$	$+0.34$
$Cu+ + e^- \rightarrow Cu^0$	$+0.52$
$Cu^{2+} + Cl^- + e^- \rightarrow CuCl$	$+0.57$
$Cu^{2+} + I^- + e^- \rightarrow CuI$	$+0.88$

[a]The more positive the reduction potential, the more readily
electrons are accepted by the copper containing species to
the left of the reduction equation to form the species shown
to the right of the equation.

copper is a noble metal from which an electron is donated
with difficulty to species dissolved in water. This is despite the strong solvation of the relatively small Cu^+ and
Cu^{2+} ions, which form on oxidation of copper. In other than
pure water however, as shown in Tables 1-3, the electron
acceptor tendencies of copper ions and the electron donor
properties of copper metal, depend very much on the environment of the copper ions, i.e. solid or liquid phase, nature
of solvent, nature of copper-bases, etc. (Cotton and Wilkinson, 1972).

The term copper-base refers to any electron pair donor,
e.g. NH_3, H_2O, Cl^-, $(NH_2)_2C = S$, amino acid, hydroxy species,
thiol, etc., which forms complexes with Cu^+ or Cu^{2+}. One
uses the expression "copper-base" rather than simply "base"
because relative base strengths towards copper ions (copper-
basicity) often have little relation to base strengths towards hydrogen ions (proton-basicity). Thus the pK_b of
bases in water (Table 2) is not necessarily an indication of
the relative strengths with which the same bases form complexes with Cu^+ or Cu^{2+}(Ahrland *et al.*, 1958; Edwards, 1964;
Sillen and Martell, 1964). For example, Cl^-(aq) or various
RO^-(aq) species are stronger bases towards H_3O^+(aq) than are
I^-(aq) and RS^-(aq) species, but I^-(aq) and RS^-(aq) are much
stronger bases towards Cu^+(aq) than are RO^-(aq) or Cl^-(aq)
(Tables 1 and 2). Ammonia is a very much stronger base towards H_3O^+ than is acetonitrile in water (Table 2) and is
also a very much stronger base towards Cu^{2+} than is acetonitrile (Table 1). However, ammonia in water is only
stronger by a factor of 10^6 as a base towards Cu^+ than is
acetonitrile (Table 2). Indeed the term copper-base is itself inadequate, because one should compare bases in terms of
their copper(II)-base strengths and their copper(I)-base
strengths. The two properties are often different.

As a general rule, bases whose basic centre is an oxygen
or nitrogen atom, or especially a chelating centre involving
both these atoms, are strong copper(II)-bases, whereas those
bases whose basic centre is a halide ion, a sulfur, phosphorus, arsenic, or selenium atom, or multiply-bonded nitrogen or carbon, are strong copper(I)-bases (Sillen and
Martell, 1964; Cotton and Wilkinson, 1972). Examples are
in Tables 1-3.

The disproportionation equilibrium of Cu^+ (eq.1) well
illustrates the electron acceptor properties of Cu^{2+} and
Cu^+ and the electron donor properties of Cu^+ and Cu^0 in
various environments. Values of the equilibrium constant, K,
for Cu^+ disproportionation differ by 10^{50} through different
chemical environments in solution, as shown in Tables 2 and 3

TABLE II. *Electron Acceptor Properties of Cu^{2+} and Cu^+, in different environments, Proton and Copper(I)-basicities and Free Energies of Transfer of Cu^+ from water to water containing 2 Molar Bases at 25°*

Solvent[e]	log K[b,f]	$\Delta G_{tr} Cu^+$[c,f] kJ mol^{-1}	log β_2[d]	pKa[h]
Water	-6.2	0	-	-
KCN	-[i]	-193	24.0	9.3
$Na_2S_2O_3$	-[i]	-87	11.69	1.7
Me_2NCHS	-[i]	-81	-	-
KI	-[i]	-59	8.9	-9.5
KSCN	-[i]	-	11.0	-1.8[g]
NH_3	-3.4	-57	10.9	9.2
KBr	-8.4	-41	5.9	-9
Pyridine	-6.6	-41	6.6[g]	5.2
KCl	-6.0	-34	4.88	-8
CO[a]	-3.0	-30	-	-
$CH_2=CH_2$[a]	-2.0	-25	-	-
CH_3CN	-3.6	-24	4.3[g]	<1

[a] *At 5 atmospheres pressure.*

[b] *K is the equilibrium constant for eq.1, $2Cu^+ \rightleftharpoons Cu + Cu^{2+}$.*

[c] *The free energy of transfer of Cu^+ from pure water to the solvent S indicated is based on the assumption of negligible liquid junction potential in the cell $Cu/CuClO_4.S/sce$.*

[d] *β_2 is the stability constant for the equilibrium, $Cu^+ + 2\bar{X} \rightleftharpoons CuX_2^-$ in water at 25° from Sillen and Martell (1964), where X is the base shown. It indicates copper(I) basicity.*

[e] *The "solvents" are 2M solutions of these bases in water.*

[f] *Data from R.I. Tilley, Ph.D. Thesis, Australian National University, 1974.*

[g] *Sillen and Martell (1971).*   [h] *pKa is the acid dissociation constant in water of the conjugate acid of the specified base, from Sillen and Martell (1964). It indicates proton basicity.*
[i] *Copper(II) ions are reduced to copper(I) species in these solutions.*

(Tilley, 1974; Pritam Singh, 1980). Tables 2 and 3 contain single ion free energies of transfer of Cu^+ and Cu^{2+} from water to other solvents at 25°. In Table 2 the "solvents" are water containing 2M copper(I) and copper(II)-bases (Parker *et al.*, 1977). In Table 3 transfer is from water to other pure solvents and the anion which accompanies the copper ions (NO_3^-, SO_4^{2-}, ClO_4^-) is a very much weaker copperbases than are the solvents.

The concept of the free energy of transfer of single ions (ΔG_{tr} ion) between different solvents is a most useful one (Parker, 1969; Popovych, 1970). It allows rapid and easy prediction of many thermodynamic properties of ions in different solvents. For example, solubility products, redox potentials, rate constants (k), stability constants of complexes and acidities in non-aqueous solvents, S, can be calculated from ΔG_{tr} ion, for the various species involved in the equilibrium or rate process, and from the equilibrium constant K^W, potential E^W, or rate constant k^W, in water (Parker, 1969). Equations (2)-(4) are used.

$$\ln K^S = \ln K^W - \frac{\Delta G^O_{tr}}{RT} \qquad (2)$$

$$E_s^O = E_w^O - \frac{\Delta G^O_{tr}}{NF} \qquad (3)$$

$$\ln k^S = \ln k^W + \frac{\Delta G_{tr}}{RT} \text{ (reactants)} - \frac{\Delta G_{tr}}{RT} \text{ (transition} \qquad (4)$$
$$\text{state)}$$

Thermodynamic properties of aqueous inorganic copper systems have been calculated by Duby (1977). Instability constants of copper complexes in water are in the Chemical Society's special publication 17 (Sillen and Martell, 1964) and solubilities of copper salts in water are in Seidell (1958), so that where ΔG_{tr} data is also available, chemistry in other environments can be predicted.

Despite their utility, single ion thermodynamic quantities cannot be measured through a thermodynamically acceptable procedure, so ΔG_{tr} ion values in Tables 2 and 3 are derived via the extrathermodynamic assumption that $\Delta G_{tr}Ph_4As^+ = \Delta G_{tr}Ph_4B^-$ (Popovych, 1970). Values of $\Delta G_{tr}Cu^{2+}$ for transfer from water to a variety of solvents are beginning to appear (Tilley, 1974; Pritam Singh, 1980).

I have laboured the non-aqueous thermodynamics of copper ions, because in plants, soils and biological materials the environment of the contained copper is likely to differ from that of hydrated copper ions in water. As noted, the aqueous chemistry of copper, e.g. the stability of Cu^{2+} and instability of Cu^+ and the solubility of their salts, may not be a useful guide to the chemistry of copper in situations of interest to this audience. I believe that a good understanding of stability constants of complexes (Sillen and Martell, 1964) and of non-aqueous solvation energies (Parker, 1976) is helpful to you.

AQUATED COPPER(II) ION

The most relevant copper species for the soil scientist is probably the aquated copper(II) ion, $Cu(H_2O)_6^{2+}$. Since natural ground waters have a pH in the range 5-8, and since $Cu(H_2O)_6^{2+}$ is a moderately strong acid of pK_a 6.8, its conjugate base, $CuOH(H_2O)_5^+$, is often the major species present in ground waters. It is not always helpful to speak of Cu^{2+} and $CuOH^+$ in water because the water molecules bound to copper(II) in $Cu(H_2O)_6^{2+}$ are copper-bases and the copper-water complexion has a well defined, but distorted, octahedral geometry. It also has a typical blue colour, which is not exhibited by the uncomplexed Cu^{2+} ion. Stronger copper-bases than water (e.g. NH_3, Cl^-) displace one or more water molecules if present in an aqueous solution of copper(II) salts and as noted $Cu(H_2O)_6^{2+}$ is the strongest acid of the aquated first transition series divalent cations. The acidities of the $M(H_2O)_6^{2+}$ species in $M(H_2O)_6^{2+} + H_2O \rightleftharpoons MOH(H_2O)_5^+ + H_3O^+$ decrease in the order of M, with pK_a values (Sillen and Martell, 1964) in parentheses, $Cu^{2+}(6.8) > Zn^{2+}(8.8) > Co^{2+}(8.9) > Fe^{2+}(9.5) > Ni^{2+}(10.6)$, $Mn^{2+}(10.6)$.

Water bound in the co-ordination shell of a cation is much more acidic than pure water, in water as solvent, because of the effect of the positive charge of the ion. The degree of acidification of a species $M(H_2O)_m^{n+}$ depends in part on the charge of the cation. Thus pK_a values of aquated cations decrease in the series $Tl^+(13.2) > Mg^{2+}(11.4) > Al^{3+}(5.1)$. However, the much greater acidity of $Cu(H_2O)_6^{2+}(6.8)$ over $Mg(H_2O)_4^{2+}(11.4)$, $Fe(H_2O)_6^{2+}(9.5)$ and $Ni(H_2O)_6^{2+}(10.6)$ suggests a specially strong interaction of OH with Cu^{2+} in $CuOH(H_2O)_5^+$.

Thus the $CuOH(H_2O)_5^+$ ion is an important species in ground water (pH 5-9) containing low concentrations of copper. Of course at higher pH, or in solutions containing more concentrated copper, solid species such as CuO and $Cu(OH)_2$ precipitate from a solution of $Cu(OH_2)_6^{2+}$.

Because of the strong affinity of copper(II) for hydro-oxyl groups, most copper(II) salts crystallise from water with water of crystallisation or containing OH groups. Species such as $CuSO_4.5H_2O$, $[Cu(NH_3)_4](OH)_2$, $CuCl_2.2H_2O$, $Na_2[Cu(OH)_4]$, $Na_5[CuCl(OH)_6(H_2O)].6H_2O$, $Sr[Cu(OH)_4].H_2O$ with colours ranging from pale blue to violet indicate the variety of copper(II) hydroxyl interactions.

COPPER(II) AS AN OXIDANT

Because there are many bases which complex strongly with copper(I), which is a d^{10} cation, but weakly with copper(II), a d^9 cation, such bases can be used to enhance the electron acceptor properties (i.e. oxidising strength) of copper(II) solutions. Such bases raise the reduction potential (increase the electron acceptor properties of Cu^{2+}) of the process: $Cu^{2+} + e^- \rightleftarrows Cu^+$ to more positive values by stabilising Cu^+ more than they stabilize Cu^{2+}. For example, the reduction potential is 0.17 volt for this system as non-basic sulphate in water, but is 0.88V in the presence of M iodide ions in water (Table 1). We have found many useful applications of such copper(I)-bases in the extractive metallurgy of copper, gold, and silver, especially when the bases are in non-aqueous solvents (Parker *et al.*, 1975). For example, the following reactions (4)-(12), in which Cu(II) accepts one electron from a suitable donor, proceed readily in the solvent system indicated. In water, no reaction occurs, or the reverse reaction is virtually complete.

$$Cu^{2+} + Cu^0 \rightarrow 2Cu^+, \text{ in } CH_3CN/H_2O \tag{4}$$

$$Cu^{2+} + Ag^0 \rightarrow Cu^+ + Ag^+, \text{ in } CH_3CN/H_2O \tag{5}$$

$$Cu^{2+} + CuS \rightarrow 2Cu^+ + S, \text{ in } CH_3CN/H_2O \tag{6}$$

$$2Cu^{2+} + Cu_2S \rightarrow 4Cu^+ + S, \text{ in } CH_3CN/H_2O \tag{7}$$

$$2CuO + SO_2 \rightarrow Cu_2SO_4, \text{ in } CH_3CN/H_2O \tag{8}$$

$$3Cu^{2+} + Cu_2SO_3CuSO_3 2H_2O \rightarrow 2Cu_2SO_4 + 2Cu^+ + 4H^+,$$
$$\text{in } CH_3CN/H_2O \tag{9}$$

$$CuCl_2 + Au + 2Cl^- \rightarrow AuCl_2^- + CuCl_2^-, \text{ in dry}$$
$$\text{dimethylsulfoxide} \tag{10}$$

$$3CuCl_2 + CuFeS_2 \rightarrow FeCl_2 + 4CuCl + 2S, \text{ in brine} \tag{11}$$

$$2CuCl_2 + Ag_2S + 4Cl^- \rightarrow 2AgCl_2^- + 2CuCl_2^- + S, \text{ in}$$
$$\text{dry dimethylsulfoxide} \tag{12}$$

Reactions (4)-(9) can be used to produce very pure copper or silver metal from impure materials as discussed below. Reaction (10) is reversed when water is added to the dry dimethylsulfoxide (DMSO) and this allows very pure gold to be prepared by dissolving the gold in jewellery, in anode slimes from copper refineries, in dental alloys, in nuggets, in battery sands, etc. with dry DMSO containing $CuCl_2$ and $CaCl_2$ (eq.10) (Parker, 1981). The solution is then filtered and water is added to reverse reaction (1) and precipitate pure gold. Reaction (12) can be used to dissolve silver from materials such as anode slimes, photographic residues, jewellery, coinage, etc. When water is added to the filtered solution pure silver chloride precipitates (Parker *et al.*, 1979).

As noted, (Tables 1 and 2), copper(I) forms extremely strong complexes with bases containing sulfur atoms, (e.g. cystein), with iodide ion and cyanide ion. Thus copper(II), which does not form strong complexes with such bases, is an extremely strong oxidant in the presence of such bases. Indeed these bases are oxidised by Cu^{2+} and the two species are not compatible in aqueous solution (eq.13 and 14).

$$2Cu^{2+} + 4I^- \rightarrow 2CuI + I_2 \tag{13}$$

$$2Cu^{2+} + 4RS^- \rightarrow 2CuSR + RSSR \tag{14}$$

TABLE III. *Electron Acceptor Properties of Cu^{2+} and Cu^+ in Pure Solvents and Free Energies of Transfer of Copper Ions from Water to Pure Solvents at 25°.[b] Assumption, $\Delta G_{tr} Ph_4 As^+ = \Delta G_{tr} Ph_4 B^-$. Molar Concentration Scale.*

Solvent	$\Delta G_{tr} Cu^+$ kJ mol^{-1}	$\Delta G_{tr} Cu^{2+}$ kJ mol^{-1}	log K[a]
Water	0	0	+ 6.2
Vacuum	+1000	+2944	- 159
Methanol	+ 7	+ 28	+ 3.6
Formamide	-	- 28	-
Dimethylformamide	-	- 25	-
Dimethylsulfoxide	- 50	- 42	- 0.8
Acetonitrile	- 50	+ 68	- 23
Propylene carbonate	-	+ 73	-
Dimethylformthiamide	- 109	+ 73	- 45
N-methylpyrrolidone	- 38	- 16	- 3.6
Ammonia	-	-	-
Ethylene diamine	-	-	+ 5.0

[a] *Expressed as disproportionation constants K for the reaction $2Cu^+ \rightleftharpoons Cu + Cu^{2+}$.*

[b] *Pritam Singh, Ph.D. Thesis, Murdoch University, Western Australia, 1980.*

SOLUBILITIES OF COPPER SALTS

Copper(I), like silver(I), forms many slightly water soluble "salts" with anions X$^-$, in which copper(I) is stable. Such salts possess some degree of covalent bonding. The CuX species is often a weak electrolyte (i.e. it is not dissociated into free ions). A solution such as CuCl in acetonitrile-water is an example. Some solubilities of copper salts and minerals are in Table 4. Copper(II) salts, CuX$_2$, are usually much more soluble in water than the corresponding copper(I) salts and are stronger electrolytes. The great stability of $Cu(SCN)_2^-$ (Table 2) and very low solubility of CuSCN (Table 4), despite the low proton-basicity of SCN^- (Table 2) is noteworthy in that it reflects the strong affinity of Cu^+, a d^{10} cation, for bases like SCN^- containing sulfur and multiply-bound carbon and nitrogen.

TABLE IV.   *Solubilities of Copper Salts and Minerals in Water at 25°.*

Cu(I) salt	log K_{so} [a]	Cu(II) salt	log K_{so} [b]
CuCl	− 4.49	$Cu(OH)_2$	− 19.36
CuBr	− 8.23	CuO	− 20.38
CuN_3	− 8.3	$Cu_2(OH)_2CO_3$	− 21.51
CuI	− 11.96	$Cu_3(OH)_2(CO_3)_2$	− 21.53
CuSCN	− 11.3 [c]	$Cu_4(OH)_6SO_4$	− 17.2
CuCN	− 19.5	$Cu_2(OH)_3Cl$	− 17.35
Cu_2S	− 49		
Cu_2O	− 14.8		

[a] *Sillen and Martell (1964).*
[b] *Baes and Mesmer (1976).*
[c] *Sillen and Martell (1971).*

Many of the copper(I) salts in Table 4, (e.g. CuCl) are much more soluble in organic solvents, or in mixtures of organic solvents with water, than they are in water, especially if the organic solvent (e.g. acetonitrile or pyridine) is a strong copper(I)-base.

This property can be used to purify CuCl, e.g. that generated in some copper hydrometallurgical processes. The CuCl is dissolved in an acetonitrile-water mixture, insoluble impurities are separated, then the acetonitrile is distilled. Pure CuCl crystallises out in the residual water (Preston *et al.*, 1980).

COPPER COMPLEXES AND THEIR APPLICATIONS

The first transition series of the Periodic Table, which includes vanadium, chromium, manganese, iron, cobalt and nickel, is noted for the variety of the complexes which form with bases. The complexes of copper(I) and copper(II) are no exception and a variety of co-ordination numbers and geometries are observed.

The principles and terminology of transition metal complexes are well described by Cotton and Wilkinson (1972). In the simplest complexes, the base acts as a single electron pair donor and the copper ion acts as an electron pair acceptor, e.g. CuCl. Chelating bases are so named because they have two (bidentate) or more (tri and quadridentate) electron donor sites (often on O, S or N atoms) forming a "claw" which grips the copper ion. They form particularly strong complexes and examples are shown in I-VI. Such complexes are of significance in soils, in plants and in biological systems.

I   II

III

IV

V

VI

Structures IV and V represent chelation by Kelex and Lix 65 respectively. These are two important industrial chelating bases, which are used on a very large scale for the solvent extraction of copper at copper mines (Ritcey and Ashbrook, 1979). Amino acids are good chelating bases for copper ions as shown in VI (McBride, 1981; Stevenson and Fick, 1981; Walker and Webb, 1981).

A type of complex of particular importance for copper(I) is where the base donates lone pairs of electrons to Cu^+ but also has vacant low energy orbitals e.g. π orbitals. The base then acts both as an electron donor and an acceptor in a process known as back bonding (Cotton and Wilkinson, 1972). The d^{10} electrons are accepted from Cu^+ by the base into empty π or π^* orbitals. Examples of such bases are carbon monoxide, isocyanides (RNCO), nitric oxide, pyridines, nitriles, alkenes, alkynes, cyanide ion and thiocyanate ion. (Cotton and Wilkinson, 1972). All form strong complexes with copper(I) (Table 2), (Sillen and Martell, 1964 and 1971).

Copper(II) complexes, when in crystals, have distorted geometries rather than regular tetrahedra, octahedra, etc. (Cotton and Wilkinson, 1972). One typical distortion is an elongation along one four-fold axis, so that four short copper to base bonds are in a plane with two long bonds perpendicular to and on opposite sides of the plane. Copper(II) salts range in color from the rare yellow or orange to the more familiar blue or green. Planar complexes of copper(II), e.g. $Cu(NH_3)_4^{2+}$, and distorted octahedral shapes, e.g. $Cu(H_2O)_6^{2+}$ are common, but most other geometries, including copper bonding, are described by Cotton and Wilkinson (1972) for copper complexes. The blue aquo copper(II) ion $Cu(H_2O)_6^{2+}$ is formed when most copper(II) salts are dissolved in water. Complexes in which basic nitrogen is bound to copper(II) are especially stable. They are often a much deeper blue than the aquo complex, as shown by the $Cu(NH_3)_4^{2+}$ion in water.

The complexes of copper with copper-bases find application in a number of ways. Because of their back bonding tendencies with copper(I), carbon monoxide, ethene and acetylene are removed from gas streams by bubbling them through solutions of CuCl (Cotton and Wilkinson, 1972). Copper(I) salts catalyse many reactions of alkenes, alkynes, aromatic compounds and diazo compounds by complexing with these species. The oxidative dimerization of acetylene is an example (Fieser and Fieser, 1969).

The copper(I)-acetonitrile complex, $Cu(CH_3CN)_3^+$, promises application in the extractive metallurgy of copper (Parker *et al.*, 1975; Parker, 1976). Cu_2SO_4 is a stable species in water containing acetonitrile, but in water without acetonitrile it immediately disproportionates (eq.1) to copper and $CuSO_4$. Solutions of Cu_2SO_4 in acetonitrile are made by reducing $CuSO_4$, e.g. with copper, sulfur dioxide, nickel, iron, or copper sulfides. $CuSO_4$ in acetonitrile-water is a very powerful oxidising agent (Parker *et al.*, 1977) (Table 3). Thus reaction (1) is reversed in acetonitrile-water and scrap copper can be dissolved with $CuSO_4$ to give Cu_2SO_4 solutions.

When acetonitrile is distilled from Cu_2SO_4 solutions, reaction (1) proceeds in the forward direction and very pure crystalline copper and an aqueous solution of $CuSO_4$ are formed, as the Cu_2SO_4 disproportionates. The acetonitrile distillate and the aqueous $CuSO_4$ are recycled to dissolve more scrap copper. This provides a method of refining scrap copper (Parker, 1973).

The concept of distilling acetonitrile from Cu_2SO_4 solutions is known as thermal disproportionation. It is an alternative to recovering copper from aqueous $CuSO_4$ by electrolysis of the solutions (Parker and Muir, 1976).

Complexes of copper with oxygen are important catalysts for many redox reactions, such as the oxidation of ascorbic acid and the conversion of alkenes into aldehydes. The $Cu(MeCN)_4^+$ oxidation (eq. 15-18) may serve as a model for such processes (Gray, 1969). Copper complexes with histidine,

$$Cu(MeCN)_4^+ + O_2 \rightarrow Cu(MeCN)_4^+O_2 \tag{15}$$

$$Cu(MeCN)_4O_2^+ + H^+ \rightarrow Cu^{2+} + HO_2 + 4MeCN \tag{16}$$

$$Cu^+ + HO_2 \rightarrow Cu^{2+} + HO_2^- \tag{17}$$

$$H^+ + HO_2^- \rightarrow H_2O_2 \tag{18}$$

ethylene diamine and 2,2'-bipyridine have also been used to model the carrying of oxygen in biological systems (Siegel and McCormick, 1971).

BIOCHEMICAL ASPECTS (Peisach *et al.*, 1966; Walker and Webb, 1981)

The topic is discussed in later papers at this conference and I simply note that the chemistry discussed in this paper, i.e. the redox properties of Cu^{2+}/Cu^+ in different chemical environments, the structure of copper complexes and the different interaction of various basic atoms with copper(I) versus copper(II) are fundamental chemical aspects of the biochemistry of copper.

Copper proteins exhibit electron transfer and oxidase activity. Ascorbic acid oxidase (8 g A of Cu per mole of protein) is widely distributed in plants and micro-organisms. As the name implies, it catalyses oxidation of ascorbic acid by oxygen. Cytochrome oxidase contains copper and heme in equal proportions and is the terminal electron acceptor of the mitochondrial oxidative pathway. Tryrosinases require copper to catalyse the formation of melanin pigments.

Hemocyanin is an important oxygen carrier (one O_2 per 2Cu atoms) in molluscs and arthropods. Other copper proteins include plastocyanin, superoxide dismutase, ribulose diphosphate oxygenase and plantacyanin.

PHYSICAL METHODS IN THE INVESTIGATION OF COPPER COMPOUNDS

Inorganic chemists have developed a variety of techniques for investigating the structure, bonding oxidation states and properties of copper compounds. Several of these are of relevance to soil scientists and are described by Cotton and Wilkinson (1972).

Magnetic effects. These relate to the number and distribution of electrons in a species. Copper(II) is a paramagnetic species, i.e. it is attracted into a magnetic field because it has unpaired electrons associated with its d^9 electronic configuration. Copper(I) is diamagnetic and is repelled out of a magnetic field because its electrons are paired in the d^{10} configuration. Thus the magnetic susceptibility of copper compounds can tell us something of the electronic structure of copper in the compound. Simple copper(II) complexes have magnetic moments usually in the range 1.75 - 2.20 Bohr magnetons.

Electron spin resonance (esr) - This technique is extremely sensitive and is related to magnetic effects. It is observed when molecules or ions containing one or more unpaired electrons are placed in a magnetic field and radiation is applied to the system. Radiation in the microwave region is swept through the appropriate frequency range and the frequency of maximum absorption is detected under a fixed magnetic field. Esr measurements can give information about the population of electrons in orbitals, and the degree of hybridization of the orbitals. For example, esr can distinguish between two Cu(II) ions in different environ-

ments in the one compound. Copper(I) and copper(II) when
present as type III copper (Walker and Webb, 1981) in proteins are not esr active and the technique is used to detect
such copper. "Type III copper" does *not* indicate that the
copper is in +3 oxidation state.

Electronic absorption spectra. The position, intensity
and width of the absorption bands, when copper(II) complexes
are irradiated with visible or near ultraviolet light, relate
to the interaction of inner d orbitals of copper with their
chemical environment. Different bases produce very different absorption spectra, and this can indicate the stereo-
chemistry of a copper(II) complex. Time does not permit
development of this topic, which is well covered in Cotton
and Wilkinson (1972), but it is relevant to the classification of type (I) "deep blue" copper proteins, which absorb
light at 600-900nm, and type (II) "non blue" copper proteins.
As noted, type III copper is a form of copper not detectable
by esr, but which has electronic absorption at 330-350nm
(Walker and Webb, 1981).

ANALYTICAL CHEMISTRY OF COPPER (Leininger, 1954; Elwell and
    Scholes, 1967; Young, 1971)

Much of the wet analytical chemistry of copper has been
replaced by instrumental methods such as anodic stripping,
inductively coupled plasma (ICP) spectrometry, atomic absorption, polarography and X-ray methods (Ewing, 1975). However,
a brief review of some of the older chemical methods based
on Leininger (1954) and Elwell and Scholes (1967) is of
interest, because it reveals much information about the
chemistry of copper which is frequently omitted from modern
inorganic texts, but is frequently useful in the laboratory.

Detection. Copper forms one of the least soluble sulfides and is precipitated as a black sulfide from 0.25 M HCl.
Many other sulfides do not precipitate from such acidic
solutions. The sulfide is soluble in hot dilute nitric acid
but is insoluble in ammonium polysulfide solution. This
separates it from arsenic, antimony, tin and molybdenum.
Copper(II) ions form a deep blue colour in the presence of
excess ammonia and for reasons discussed above, give iodine
and copper(I) iodide on addition of KI (eq.13). Many spot
tests for copper have been developed.

Separation. Solvent extraction of Cu^{2+} from water into water insoluble kerosene is now widely used both on a laboratory and industrial scale. Copper-bases, such as Kelex, (IV), and Lix 65, (V), which are water-insoluble but kerosene-soluble, are used for solvent extraction. Copper(I) or copper(II) may be separated from elements other than arsenic, antimony, tin, molybdenum, bismuth, lead and silver by its precipitation with hydrogen sulfide from 0.25 M HCl. Copper (I) is precipitated as white CuSCN with potassium thiocyanate. This separates it from most ions, except silver, lead, mercury, selenium, tellurium and gold. Excess sodium hydroxide precipitates $Cu(OH)_2$ and thus separates Cu^{2+} from molybdenum, tungsten, vanadium and arsenic, which do not precipitate.

Elemental copper is precipitated onto aluminium in 10% sulfuric acid and this separates it from iron. Copper is also deposited quantitatively in dense adherent deposits from acid solution by electrodeposition onto a platinum cathode. Electrodeposition is used extensively (e.g. anodic stripping) for analysis, but more noble elements like gold, silver, mercury, selenium, tellurium, arsenic, antimony, bismuth, tin, molybdenum and iron interfere.

Gravimetric and volumetric procedures. Copper is selectively precipitated from solution by a number of organic reagents, to give stable compounds of definite composition. Reagents include α-benzoin oxime from ammonia solutions, salicylaldoxime at pH 2.6 - 3.0, 8-hydroxyquinoline, cupferron, and anthranilic acid.

The iodometric titration of Cu^{2+} is empirical but is strongly favoured by many workers. Solutions of copper(II) salts generate iodine and copper(I) iodide in the presence of excess iodide ions (equation 13). The iodide is titrated with standard sodium thiosulfate, using starch indicator. Only a few elements (iron(III) and arsenic) interfere. The solution is boiled with nitric acid, prior to neutralization and addition of KI, to oxidise iron(II), arsenic and antimony. Iron(III) interference is prevented by complexing it with ammonium bifluoride. The precipitated cuprous iodide absorbs some iodine, but addition of ammonium thiocyanade minimises this problem.

Solutions of copper(I) salts can be titrated with standard $KMnO_4$ solution. Precipitated cuprous thiocyanate may be titrated with permanganate or potassium iodate. Molybdenum does not interfere.

Colorimetric procedures.   Sodium diethyldithiocarbamate is added to an ammoniacal solution of copper(II).   An orange-brown stable dispersion is formed.   The transmittance of the aqueous dispersion, or of the extract into carbon tetrachloride, chloroform, amyl alcohol or bromobenzene, is measured.   Many elements interfere, but the method applies down to 0.5ppm and was recommended (pre AA) for estimating copper in water and plant materials.   Zinc dibenzyldithiocarbamate is a source of the base which is less susceptible to interference from nickel.

Diphenylthiocarbazone is used to detect 10ppm copper as a red violet copper complex in carbon tetrachloride.   Palladium, gold, silver, mercury, bismuth and ferric ions interfere, but the method has been recommended.   Other reagents which give coloured species suitable for analysis, of about 0.05mg copper, include dithioxamide, concentrated HBr, ferricyanide irons in acetic acid, H_2S and 1:10 phenanthroline.

For further information about copper chemistry, I recommend Gmelins Handbuch der Anorganischen Chemie, Verlag Chemie, Weinheim Germany, Volume A, Part 2, and Volume B, Parts 1 to 4.

REFERENCES

Ahrland, S., Chatt, J. and Davies, N.R. (1958). *Quart. Review, 12*, 265.
Baes, C.F. and Mesmer, R.E. (1976). "The Hydrolysis of Cations", Wiley-Interscience, New York, N.Y.
Butts, A. (1954) ed.   "Copper the Metal its Alloys and Compounds", Reinhold Publishing Corp., New York, N.Y.
Cotton, F.A. and Wilkinson, G. (1972). "Advanced Inorganic Chemistry", 3rd ed. Interscience Publishers, New York, N.Y.
Duby, P. (1977). "The Thermodynamic Properties of Aqueous Inorganic Copper Systems", Int. Copper Res. Assoc. Monograph IV, New York, N.Y.
Edwards, J.O. (1964). "Inorganic Reaction Mechanisms", Benjamin, New York, N.Y.
Elwell, W.T. and Scholes, I.R. (1967). "Analyses of Copper and its Alloys", Pergamon Press, Oxford.
Ewing, G.W. (1975). "Instrumental Methods of Chemical Analysis", 4th ed. McGraw-Hill, New York, N.Y.
Fieser, L.M. and Fieser, M. (1969). "Reagents for Organic Synthesis", Wiley, New York, N.Y.

Fronsbein, J. and Firminhac, R.H. (1954). *Ch.21 in* "Copper the Metal its Alloys and Compounds", (ed. A. Butts), Reinhold Publishing Corp., New York, N.Y.

Gray, R.D. (1969) *J. Amer. Chem. Soc., 91,* 56.

Hodgson, J.F. (1963). Advances in Agronomy *15,* 119.

Leininger, E. (1954). *Ch.46 in* "Copper the Metal, its Alloys and Compounds" (ed. A. Butts), Reinhold Publishing Corp., New York, N.Y.

Lindsay, W.L. (1979). "Chemical Equilibria in Soils", Interscience Publishing, New York, N.Y.

McBride, M.B. (1981). Chapter 2, this Volume.

Parker, A.J. (1969) *Chem. Reviews, 69,* 1.

Parker, A.J. (1973) *Search, 4,* 426.

Parker, A.J. (1976) Electrochimica Acta, *21,* 1.

Parker, A.J. (1981) Unpublished work.

Parker, A.J., Clare, B.W. and Smith, R.P. (1979) Hydrometallurgy, *4,* 233.

Parker, A.J., Clarke, D.A. Couche, R.A., Miller, G., Tilley, R.I. and Waghorne, W.E. (1977). *Aust. J. Chem., 30,* 1661.

Parker, A.J. and Muir, D.M. (1976). *Min. Mag., 537.*

Parker, A.J., Paul, R.L. and Power, G.P. (1981). *Aust. J. Chem., 34,* 000.

Parker, A.J., Waghorne, W.E., Giles, D.E., Sharp, J.H., Alexander, R. and Muir, D.M. (1975). U.S. Pat. 3,865,744.

Peisach, J., Aisen, P. and Blumberg, W.E. (eds.) (1966). "The Biochemistry of Copper", Academic Press, New York, N.Y.

Popovych, O. (1970). *Crit. Rev. Anal. Chem., 1,* 73.

Preston, J.S., Muir, D.M. and Parker, A.J. (1980). Hydrometallurgy, *5,* 227.

Pritam Singh. (1980). Ph.D. Thesis, Murdoch University.

Ritcey, G.M. and Ashbrook, A.W. (1979). "Solvent Extraction Principles and Applications to Process Metallurgy", Elsevier, Amsterdam.

Seidell, A. (1958). "Solubilities of Inorganic and Metal-Organic Compounds", 4th ed. American Chemical Society, Washington D.C.

Siegel, H. and McCormick, D.B. (1971). *J. Amer. Chem. Soc., 93,* 2041.

Sillen, L.G. and Martell, A.E. (1964). "Stability Constants of Metal Ion Complexes", Special Publication No.17, Chemical Society, London.

Sillen, L.G. and Martell, A.E. (1971). "Stability Constants of Metal Ion Complexes", Special Publication No.25, Chemical Society, London.

Stevenson, F.J. and Fick, A. (1981). Chapter 4, this Volume.
Tilley, R.I. (1974). Ph.D. Thesis, Australian National
    University.
Vaughan, D.J. and Craig, J.R. (1978). "Mineral Chemistry of
    Metal Sulfides", Cambridge University Press, Cambridge.
Walker, C.D. and Webb, J. (1981). Chapter 9, this Volume.
Young, R.S. (1971). "Chemical Analysis in Extractive
    Metallurgy", C. Griffin & Co., London.

COPPER IN SOILS

FORMS AND DISTRIBUTION OF COPPER IN SOLID AND SOLUTION PHASES OF SOIL

Murray B. McBride

Department of Agronomy
Cornell University
Ithaca, New York

I. INTRODUCTION

Copper is probably the most versatile of all of the agronomically important trace elements in its ability to chemically interact with soil mineral and organic components as well as form precipitates with sulphide, carbonate, hydroxide and other anions. Thus, numerous forms of Cu are likely to exist in any given soil environment. The properties and occurrence of these forms under various soil conditions is the subject of the present chapter.

II. MINERAL FORMS OF COPPER

A. *Primary and Secondary Minerals*

Although the average Cu concentration in the earth's crust is about 70 ppm (Hodgson, 1963), the element is much more abundant in basaltic than granitic rocks (Krauskopf, 1972) and has a tendency to be excluded from carbonate rocks. The abundance of Cu in basaltic rocks may be explained by the greater occurrence of sulphides and ferromagnesian silicates in basalt than in granitic rocks. The most common primary minerals of Cu are simple and complex sulphides (Table 1), with strong covalent bonds formed between reduced copper (Cu^+) and sulphide (S^{2-}) anions. The

COPPER IN SOILS AND PLANTS

Copyright © 1981 by Academic Press Australia.
All rights of reproduction in any form reserved.
ISBN 0 12 455520 9

TABLE I.   *Common Copper Minerals*[a]

Group	Mineral	Cu oxidation state	Structure
Metallic	chalcite Cu	0	cubic close packing, Cu atoms in coordination number 12
Sulphides	high-chalcosite Cu_2S	+1	planar layers of Cu and S in 3-fold coordination
	covellite CuS	+1,+2	2/3 Cu(+1) and 1/3 Cu(+2), Cu(+2) in planar CuS_3 triangles, Cu(+1) in CuS_4 tetrahedra
	chalcopyrite $CuFeS_2$	+1	sphalerite-type structure, Cu in tetrahedral coordination with cubic close-packed S atoms
	bornite Cu_5FeS_4	+1	cubic antifluorite-type structure, Cu and Fe in tetrahedral coordination, with 18 Cu + Fe atoms randomly distributed among 24 sites
	enargite Cu_3AsS_4	+1	wurtzite-like structure, with Cu and As in tetrahedral coordination

TABLE I (continued)

Group	Mineral	Cu oxidation state	Structure
Sulphides (Cont'd.)	tetrahedrite $Cu_{12}(Sb,As)_4S_{13}$	+1, +2	framework structure with $Cu(+1)S_4$ tetrahedra. Large holes in framework contain 4 $Cu(+1)$ and 2 $Cu(+2)$ in 3-fold planar coordination, also 4 $Sb(+3)$ or $As(+3)$ in 3-fold pyramidal coordination
Oxides	cuprite Cu_2O	+1	O atoms at centre and vertices of cube, Cu in linear coordination between two O atoms
	tenorite CuO	+2	Cu coordinated to four O by planar dsp^2 bonding
Carbonates	malachite $Cu_2(OH)_2CO_3$	+2	Cu in distorted octahedra, two O and two OH in planar coordination, with remaining two axial positions occupied by distant O atoms for half the Cu and OH groups for the other half
	azurite $Cu_3(OH)_2(CO_3)_2$	+2	2/3 of Cu in rectangular coordination with two O and two OH, 1/3 of Cu in pyramidal coordination with three O and two OH

TABLE I (continued)

Group	Mineral	Cu oxidation state	Structure
Silicates	chrysocolla $CuSiO_3 \cdot 2H_2O$	+2	infinite SiO_3 chains (like pyroxene), linked by Cu coordinated to four O atoms
Sulphates	brochantite $Cu_4(OH)_6SO_4$	+2	octahedral Cu coordination: four OH groups in a plane and one O atom and one OH group at axial positions
Chlorides	atacamite $Cu_2(OH)_3Cl$	+2	centres of OH and Cl in cubic close packing, Cu in 6-fold coordination of two types: four OH plus two Cl, five OH plus one Cl

a Structural data from Povarennykh (1972).

copper sulphide structures are more covalent in nature than the silicate, oxide, or carbonate structures. However, copper is capable of isomorphous substitution in ferromagnesian silicates, an excellent example being Cu^{2+} replacement for Mn^{2+} in manganese garnets (Povarennyk, 1972). It is also likely that limited quantities of Cu^{2+} can substitute for Fe^{2+} or Mg^{2+} in pyroxenes and amphiboles, since Cu^{2+} forms analogous silicate structures [ie. shattuckite, $Cu_5(Si_2O_6)_2(OH)_2$; and plancheite, $Cu_7(Si_4O_{11})_2(OH)_2$]. Although the ionic radius of Cu^{2+} (0.73 Å) is similar to that of Fe^{2+} (0.78 Å) and Mg^{2+} (0.72 Å) in octahedral coordination, isomorphous substitution for Fe^{2+} and Mg^{2+} in silicates may be limited by the greater Pauling electronegativity of Cu^{2+} (2.0) when compared to Fe^{2+} (1.8) or Mg^{2+} (1.3). It is generally considered that, when substitution in an ionic crystal is possible, the element having the lower electronegativity will be preferred because of its ability to form a more ionic bond (Mason, 1966). This principle derives from the directional nature of covalent bonds that requires a specific crystal field symmetry to maximize Cu-anion bond energy, as well as the limited ability of silicate oxygen atoms to form covalent bonds.

While the predominant minerals of Cu in the earth's crust are sulphides (largely in the +1 oxidation state), the metallic form of Cu (Table 1) is also common in reducing environments. However, upon exposure to conditions at the earth's surface, Cu^+ and Cu^o are oxidized to the +2 oxidation state. The flooding of soils may cause Cu^+, or even Cu^o, to become thermodynamically more stable than Cu^{2+} (Lindsay, 1979), although this reduction process requires that the soil atmospheric level of O_2 be reduced to an extremely low concentration. In aerobic soil environments, the primary sulphide minerals are dissolved by weathering processes. The dissolution of chalcopyrite can be written:

$$4CuFeS_2 + 17O_2 + 10H_2O \rightarrow 4Cu^{2+} + 4Fe(OH)_3 + 8SO_4^{2-} + 8H^+$$

A number of secondary minerals of Cu^{2+}, including oxides, carbonates, silicates, sulphates, and chlorides can then form (Table 1), although all are relatively soluble. At low pH and under strong leaching conditions at the earth's surface, most of these minerals would not persist. It has generally been demonstrated that those Cu minerals which are expected to be the least soluble under conditions prevalent in surface soils (ie. tenorite, malachite) are much too soluble to control the very low observed activities of Cu^{2+}

in soil solution (Lindsay, 1979; McBride and Blasiak, 1979; Cavallaro and McBride, 1980). This result is clearly indicated in Figure 1, where the activity of Cu^{2+} in a soil as measured by an ion-selective electrode was well below the value given by the solubility line of CuO or $Cu_2(OH)_2CO_3$ at all pH values. Since the solubilities of malachite and azurite are usually calculated by assuming that the soil solution is in equilibrium with atmospheric CO_2 (partial pressure = $10^{-3.52}$ atm), the solubility lines of these carbonates would be shifted toward lower positions on Figure 1 if the soil atmosphere contained elevated concentrations of CO_2. Figure 2 demonstrates that malachite will become more stable than tenorite at CO_2 partial pressures near 0.1 atmosphere. However, since the results of Figure 1 were obtained on soil suspensions equilibrated with atmospheric CO_2 in the laboratory, the precipitation of Cu^{2+} as a pure

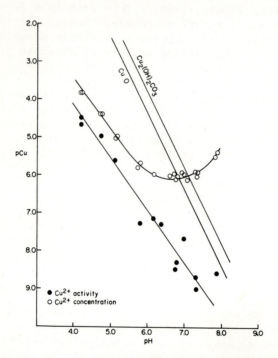

Fig. 1.   Total Cu concentration and Cu^{2+} activity in soil solution as a function of adjusted pH for an acid mineral soil amended with Cu^{2+} (80 ml of 0.015 $\underline{M}$ CaCl$_2$ solution containing 40 ppm Cu^{2+} added to 10 g soil).

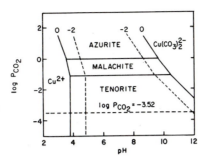

Fig. 2.   Predominance diagram for oxide and carbonate mineral phases of Cu(+2).   The broken horizontal line represents the atmospheric pressure of CO_2.   The broken lines labelled with -2 mark the limits of the P_{CO_2} and pH conditions which will cause precipitation of solid phases at a copper concentration of $10^{-2} \underline{M}$ (adapted from Schindler, 1967).

carbonate, oxide, or hydroxycarbonate could not have occurred. Soils of low native acidity that are amended with Cu^{2+} tend to have pCu- pH solubility lines with slopes approaching 2, while more acid soils have slopes between 1.0 and 1.5 (Cavallaro and McBride, 1980). These deviations from the theoretical slope of 2 expected for the precipitation of Cu^{2+} as $Cu(OH)_2$, CuO or $Cu_2(OH)_2CO_3$ suggest that simple precipitation phenomena are not generally responsible for the "adsorption" of Cu^{2+} added to non-alkaline soils. It is even less likely that pure secondary mineral phases of Cu^{2+} would exist in most soils without the addition of Cu^{2+} amendments, since soils contain on the average only 20-30 ppm of copper in total (Hodgson, 1963; Lindsay, 1979). This small amount could easily exist in adsorbed forms and as impurities in other minerals. Because of the extremely low concentrations of free Cu^{2+} in soil solutions, estimates of Cu^{2+} activity have rarely been made. However, a chelate equilibrium technique produced values of Cu^{2+} activity in calcareous and alkaline soils on the order of $10^{-12} \underline{M}$ (Sinha et al., 1978), demonstrating that soluble free Cu^{2+} in non-acid soils is several order of magnitude lower than the level predicted from solubility products of the least soluble pure solid phases. It has been suggested that $CuFe_2O_4$, an inverse ferrite structure (Cerovic et al., 1972), may determine the solubility of Cu^{2+} in soil solution (Lindsay,

1979). However, the tetrahedral coordination of Fe^{3+} required for this structure may inhibit its formation under soil conditions. In addition, considering that many mineral soils contain appreciable quantities of iron and aluminum oxide with high specific surface areas, one would expect Cu^{2+} to exist in chemisorbed or occluded forms rather than as a separate phase. Either situation may involve the formation of $Cu^{2+}-O-Fe^{3+}$ bonds, although the solubility of chemisorbed or occluded Cu^{2+} is likely to be lower than the solubility of a pure Cu^{2+} mineral.

B.   Occluded and Co-precipitated Copper

Since a relatively small fraction of the total copper in many soils is isotopically exchangeable (McLaren and Crawford, 1974; Kline and Rust, 1966), and chemical fractionation schemes generally show that the bulk of native copper in soil clays (aluminosilicate and oxide minerals) is only extractable by dissolution of the clays (McLaren and Crawford, 1973a; Shuman, 1979), it is likely that a significant fraction of soil copper is "buried" in various mineral structures. Calcareous soils, for example, have large fractions of non-diffusible Cu in a form that is soluble in dilute acid (Kline and Rust, 1966). Non-diffusible Cu cannot be adsorbed on surfaces, and therefore is probably present as an impurity within carbonate minerals. Although Cu^{2+} can adsorb on calcite (Heydemann, 1959), the ionic radius of Cu^{2+} (0.73 Å) may prevent it from readily substituting for Ca^{2+} in the calcite structure, unlike Mn^{2+}. However, Cu^{2+} may substitute more freely for Mg^{2+} (0.72 Å radius) in magnesium carbonate or Fe^{2+} (0.78 Å radius) in iron carbonate.

Non-calcareous soils contain Cu fractions which cannot be extracted by weak acid, but are solubilized by chemical treatments which dissolve iron and manganese oxides or layer silicate clays (McLaren and Crawford, 1973a). This "residual" Cu, which represents the bulk of the total in many soils, exists in a largely non-diffusible form (McLaren and Crawford, 1974; Kline and Rust, 1966). Non-diffusible Cu in soils may reside within oxide structures, since co-precipitation of Cu^{2+} in aluminum and iron hydroxides occurs readily (McBride, 1978; Nalovic et al., 1975). Such substitutions, however, inhibit crystallization and produce high surface area oxides. The charge deficit created by substitution of Al^{3+} or Fe^{3+} by Cu^{2+} must be balanced by a

structural defect or an adsorbed cation. Thus, the microcrystalline nature of oxides containing ionic impurities
probably permits the structural charge associated with Cu^{2+}
substitution to be located at or near surfaces, where it can
be balanced by surface ions. Studies of synthetic magnetite
(Fe_3O_4) and maghemite ($\gamma-Fe_2O_3$) have indicated that Cu^{2+} ion
impurities tend to accumulate near the particle surfaces
(Sidhu et al., 1980), but are partially ejected from the
structure upon alteration of maghemite to hematite, a
process requiring the conversion of tetrahedral Fe(+3) to
octahedral Fe(+3). The ideal hematite ($\alpha-Fe_2O_3$) structure
does not readily accommodate divalent ions such as Cu^{2+}. An
observed correlation between the quantities of Cu and Mn
dissolved by the treatment of soil with oxalate suggests
that Cu associates more strongly with manganese oxides than
iron oxides (McLaren and Crawford, 1973a). Associations
between copper and iron oxides have commonly been observed
in soils (McKenzie, 1975; Childs and Leslie, 1977), although
no evidence was found for copper-manganese associations in
the same soils. Thus, the data regarding Cu-Mn and Cu-Fe
associations often appear to be contradictory. However, the
abundance of iron relative to manganese may give Cu a better
opportunity to associate with iron-rich secondary mineral
phases in soils despite a very strong affinity of manganese
oxides for Cu.

Since Cu^{2+} is capable of isomorphous substitution in
octahedral positions of layer silicate clays, a portion of
the "residual" Cu that is released from soils by HF treatment may originate from crystalline silicate clays, although
crystalline oxides or primary mineral particles may contribute part of the HF-soluble Cu. After removal of exchangeable Cu^{2+}, and destroying organic matter and iron oxides,
Shuman (1979) found that the remaining clay fraction of a
number of soils often contained a relatively large percentage of the total soil copper.

III. ADSORBED FORMS OF COPPER

A. *Adsorption on Clays and Oxides*

Unlike many other metallic ions, Cu^{2+} can be "specifically" adsorbed by layer silicate clays, iron oxides,
aluminum oxides, manganese oxides, and organic matter.
Specific adsorption refers here to adsorption in the
presence of excess quantities of Ca^{2+} or some other elec-

trostatically bonded metal ion that is capable of preventing significant Cu^{2+} adsorption by simple ion exchange.

On layer silicates, Cu^{2+} adsorption is not anomalous if the pH is maintained at relatively low values (<5). Thus, Cu^{2+} and Ca^{2+} have approximately equal preference for exchange sites of montmorillonite at low pH (El-Sayed et al., 1970). Electron spin resonance (ESR) investigations of adsorbed Cu^{2+} on fully wetted smectites and kaolinites have shown the metal to tumble freely as the hexaquo $Cu(H_2O)_6^{2+}$ species at interlayer or external surface positions (McBride, 1976; Clementz et al., 1973), although $Cu(H_2O)_6^{2+}$ is held more rigidly in the interlayers of vermiculite. Thus the ions are readily exchanged from surfaces by other cations. However, as the pH is raised, the layer silicate surfaces apparently promote hydrolysis of Cu^{2+} with the concomitant release of protons (Farrah and Pickering, 1976 a, 1976 b), and the adsorbed copper becomes much less exchangeable. Hydrolysis may be driven by the closer association of weakly hydrated species such as $CuOH^+$ and $Cu_2(OH)_2^{2+}$ with layer silicate surfaces when compared with strongly hydrated Cu^{2+}. A progressive loss in ESR signal of Cu^{2+} adsorbed on smectite with increasing pH (McBride, unpublished data) suggests that Cu-hydroxy dimers or polymers nucleate on the planar surfaces. While $CuOH^+$ is paramagnetic and is expected to produce an ESR signal (unless line broadening arises from the lowered symmetry of $CuOH^+$ compared to $Cu(H_2O)_6^{2+}$), $Cu_2(OH)_2^{2+}$ may be diamagnetic as a result of Cu-Cu interaction and electron spin pairing. Although apparent "adsorption" of Cu^{2+} by layer silicates in excess of the CEC may occur at high pH as a result of $Cu(OH)_2$ precipitation (Bingham et al., 1964), layer silicate clays added to solutions that are undersaturated with respect to $Cu(OH)_2$ precipitation adsorb Cu^{2+} in a partially or fully hydrolyzed form (Farrah and Pickering, 1976a). In effect, layer silicate clays shift the apparent solubility product of $Cu(OH)_2$ to a lower value than that observed in the absence of clay.

At trace levels in soils, Cu^{2+} solubility may be controlled by specific interactions with hydrous oxide impurities in layer silicates rather than adsorption on the silicates (Jenne, 1968). Mossbauer and ESR spectroscopy reveal that even "pure" smectites used for laboratory studies contain ultrafine particles of iron oxides (Goodman, 1978; McBride, 1979). Amorphous Fe and Al hydroxides as well as crystalline oxyhydroxides readily adsorb Cu^{2+} despite the presence of excess alkali metal ions (Kinniburgh et al., 1976; Forbes et al., 1976). In fact, with the possible exception of Pb^{2+}, Cu^{2+} is the most strongly adsorbed of all the divalent transition and heavy metals on Fe and Al oxides and oxyhydroxides. Since the mechanism of adsorption, unlike the loose electrostatic association observed on permanent charge clays, appears to involve formation of a direct surface Cu-O-Al or Cu-O-Fe bond (McBride, 1978), the quantity of surface hydroxyl groups may determine the maximum level of chemisorption possible. However, surface OH groups coordinated to two Al or Fe ions are probably not active in chemisorption. Chemisorption of Cu^{2+} on a crystalline mineral like gibbsite is likely to occur only at edges, where hydroxyls coordinated to single Al^{3+} ions exist. In general, microcrystalline and amorphous oxides should have much greater capacities to chemisorb Cu^{2+} than crystalline oxides with low surface areas. Mossbauer spectroscopy has demonstrated that iron oxides (hematite and goethite) in some highly weathered soils are microcrystalline, with particle sizes on the order of 100 Å (Bigham et al., 1978). It is reasonable to expect, then, that oxides can play a significant role in specific adsorption of Cu^{2+} by most soils, as suggested by Jenne (1968). Relatively small quantities of free iron and aluminum oxides in soils may have disproportionately large active surface areas.

Like Fe and Al oxides, Mn oxides specifically adsorb Cu^{2+}, with the level of adsorption increasing as a function of pH (Murray et al., 1968; Murray, 1975). The affinity of synthetic Mn oxides for Cu^{2+} is even stronger than that of Fe or Al oxides (McKenzie, 1980). Although laboratory prepared manganese oxides may be quite different from manganese-rich oxide components in soils, a correlation has been found between the amount of specifically adsorbed Cu^{2+} in soils and "free" manganese oxide content (McLaren and Crawford, 1973b). Chemisorption at manganese oxide surfaces may be an important process despite the presence of greater quantities of Fe and Al oxides. Although Cu^{2+} is strongly adsorbed by manganese oxides, it does not show the anomalously high affinity that Mn^{2+} and Co^{2+} have for these oxides (McKenzie, 1980). There is some evidence that Mn^{2+} and Co^{2+} adsorption by manganese oxides involves oxidation of the metals to the +3 oxidation state. The Mn^{3+} and Co^{3+} ions are relatively easily stabilized in certain ligand fields, while the Cu^{3+} oxidation state is highly unstable.

It is reasonable to conclude that Cu^{2+} adsorption on manganese oxides occurs by a different mechanism from Mn^{2+} or Co^{2+} adsorption.

B. *Adsorption on Organic Matter*

Complexation by organic matter has long been recognized as an effective mechanism of Cu^{2+} retention in soils. The Cu^{2+} ion is directly bonded to two or more organic functional groups (mainly carboxylic, carbonyl and phenolic), so that the ion is immobilized in a rigid inner-sphere complex (McBride, 1978b). Trace levels of Cu^{2+} in peats may also be strongly bound in porphyrin complexes (Goodman and Cheshire, 1973). Copper is almost unique in its ability to form inner-sphere complexes with soil organics at low pH; other divalent transition metals such as Mn^{2+}, Fe^{2+}, Co^{2+}, and Ni^{2+} are probably bound by outer-sphere complexation with the metal retaining its inner hydration sphere (Bloom and McBride, 1979). However, as the pH of organic matter is raised, the ability of these other metals to form inner-sphere complexes is enhanced.

A comparison of adsorption by various soil constituents (Figure 3) suggests that organic matter and Mn oxides are the most likely materials to retain Cu^{2+} in a non-exchangeable form in soils. Copper adsorption maxima have been correlated with organic carbon and "free" manganese oxide contents of soils (McLaren and Crawford, 1973b), a result consistent with the adsorption data of Figure 3. Although

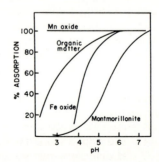

Fig. 3. *Specific adsorption of Cu^{2+} by soil constituents as a function of pH. (One gram sample equilibrated with 200 ml of 0.05 M $CaCl_2$ containing 5 ppm copper). (Adapted from McLaren and Crawford, 1973b).*

the isotopic exchangeability of Cu^{2+} adsorbed in organic
matter is lower than the exchangeability of Cu^{2+} adsorbed on
pure iron oxides, manganese oxides or layer silicates,
organically-bound Cu^{2+} in soils is probably the largest
contributor to the isotopically exchangeable (diffusible)
fraction (McLaren and Crawford, 1974; Kline and Rust, 1966).
The importance of organic forms of copper is demonstrated by
the fact that alkali extraction techniques designed to
remove most of the organic matter from mineral soils tend to
solubilize a greater fraction of total soil copper than any
other metal (Cheshire et al., 1977). Organic solids serve
the function of holding Cu^{2+} in a kinetically available but
thermodynamically stable (insoluble) form. While Cu^{2+}
freshly adsorbed on oxides is apparently accessible to
isotopic exchange, further adsorption of other metals on the
surfaces may occlude Cu^{2+} in soil oxide particles over long
periods of time. Exchangeable and weak acid – extractable
copper represent a small percentage of the total copper in
most soils (McLaren and Crawford, 1973 a), indicating that
little Cu^{2+} exists in unamended soils as exchange ions on
permanent charge sites of clays or as specifically adsorbed
forms on the external surfaces of layer silicates or oxides.
The bulk of the copper is complexed by organic matter,
occluded in oxides and substituted in primary or secondary
minerals.

IV.   SOLUTION FORMS OF COPPER

The association and hydrolysis constants of Cu^{2+} (see
Chapter 1) indicate that formation of significant concentrations of soluble nitrate, chloride, sulphate, or phosphate complexes is unlikely in most soil solutions (Sanders
and Bloomfield, 1980), although sulphate and chloride may
form complexes with Cu^{2+} in natural waters (Mann and
Deutscher, 1977) or saline soils (Mattigod and Sposito,
1977). Only hydroxy and carbonate complexes are expected to
commonly exist as significant species in soil solution. The
activities of the major soluble forms of Cu(+2) in equilibrium with tenorite are depicted as a function of pH in
Figure 4, assuming that the aqueous phase is in equilibrium
with atmospheric CO_2. Hydrolysis products of Cu^{2+} (CuOH$^+$ and
$Cu_2(OH)_2^{2+}$) are significant below pH 8, although the log K^o
of -8.0 used as the CuOH$^+$ formation constant in this diagram
may be an overestimate of the stability of CuOH$^+$ (Baes and
Mesmer, 1976). Similarly, several K^o values given in the

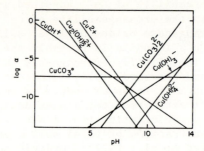

Fig. 4.   Activities of Cu(+2) species in equilibrium with CuO as a function of pH (25°C, ionic strength = 0, $logP_{CO_2}$ = -3.52) (Adapted from Schindler, 1967).

literature for $Cu(OH)_2^o$ and $Cu(OH)_3^-$ formation are probably too high. As the pH approaches about 8, the overall solubility of Cu(+2) decreases to a minimum, then increases above pH 8 as carbonate and anionic hydroxy complexes become important (Figure 4). Thus, it has been suggested that a major inorganic form of complexed Cu(+2) in neutral and alkaline soil solutions is $CuCO_3^o$ (Sanders and Bloomfield, 1980; Mattigod and Sposito, 1977). It has been estimated that the $(CuCO_3^o)/(Cu^{2+})$ ratio in some soil solutions at pH 7.0 is 4, assuming that about 2×10^{-3} M HCO_3^- is present in solution (Sanders and Bloomfield, 1980). At very high pH, $Cu(OH)_4^{2-}$ and $Cu(CO_3)_2^{2-}$ become the predominant soluble forms of Cu(+2) (Figure 4), although these species would be restricted to highly alkaline soils.

As predicted by Figure 4, the solubility of Cu(+2) in soils (as expressed by total dissolved copper rather than uncomplexed Cu^{2+}) approaches a minimum near pH 7 and increases above this pH (Figure 1). However, a number of studies have produced evidence that the natural Cu(+2) in soil solutions at higher pH does not exist as inorganic complexes, but is largely complexed with soluble organics (Hodgson et al., 1965, 1966). The total complexed Cu(+2) in solution generally increases above pH 7 because of the greater solubility of soil organics at higher pH (McBride and Blasiak, 1979), while the "free" soluble Cu^{2+} at higher pH is very much lower, commonly in the range of 10^{-8}-10^{-9} M (McBride and Blasiak, 1979; Hodgson et al., 1965). Total native Cu in soil solution is often about 10^{-6}-10^{-7} M

(Bradford et al., 1971), so that the ratio of total dissolved Cu to free (uncomplexed) Cu^{2+} can be greater than 100 (Hodgson et al., 1966).

Techniques such as electron spin resonance (ESR) and anodic stripping voltammetry (ASV) have allowed studies of the speciation of Cu to be made on solutions containing natural organics. The ESR spectra of Cu-fulvic acid complexes suggest that fulvic acids complex low levels of Cu^{2+} at three to four ligand positions, the donor groups being mainly carboxylate, carbonyl and phenolate (Bresnahan et al., 1978). The observed non-labile behaviour of Cu^{2+}-humic acid and fulvic acid complexes, revealed by the unavailability of the Cu(+2) for reduction by ASV (Shuman and Woodward, 1977; O'Shea and Mancy, 1976) is consistent with the strong inner-sphere multidentate complexation indicated by ESR. Since inorganic copper complexes are available for reduction, ASV has distinguished between inorganic and organic complexes in natural waters, indicating that organic complexes are dominant (Shuman and Woodward, 1977). Most other metals investigated (eg. Co^{2+}, Zn^{2+}, Ca^{2+}) have been shown to complex with organics in soil solutions and natural waters to a much smaller extent than Cu^{2+}.

Since numerous studies using different techniques have all produced the conclusions that organic complexes are generally more important in soil solutions than inorganic complexes, and that the ratio of complexed to uncomplexed Cu is very high, some consideration must be given to the effect of organic complexation on Cu adsorption processes. Chemical models predict that Cu^{2+} adsorption by mineral and organic surfaces should be reduced by the presence of soluble chelating agents such as citrate (Vuceta and Morgan, 1978). This prediction is verified by the data shown in Figures 5 and 6, with the adsorption of Cu^{2+} by montmorillonite inhibited to various extents by soluble organic complexing agents. Thus, EDTA and citrate compete very effectively with the exchange sites of the clay, maintaining virtually all of the added Cu^{2+} in solution as an organic complex (Figure 5). Fulvic acid is of intermediate effectiveness, while salicylate and phthalate are the least able to prevent adsorption of Cu^{2+} by the clay. The level of "free" Cu^{2+}, measured by an ion-selective electrode, controls the quantity of Cu^{2+} adsorbed by the clay, so that the relationship between adsorbed Cu^{2+} and Cu^{2+} activity is not greatly affected by the presence of organics (Figure 6). It is likely, then, that Cu^{2+} will not adsorb on exchange sites

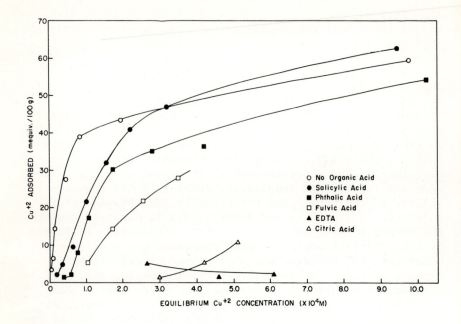

Fig. 5. Adsorption isotherms of Cu^{2+} on montmorillonite in the presence and absence of organic acids, expressed in terms of the equilibrium copper concentration (organic acid concentration equals $2 \times 10^{-3}\underline{M}$, pH $\simeq$ 5).

of permanent charge clays in soils because of its much stronger affinity for organics or other sites of specific adsorption. Complexes of Cu^{2+} with organic acids do not adsorb significantly on layer silicate clays because of their net negative charge (Bloomfield et al., 1976), although fulvic and humic acids added to Cu^{2+}-saturated clays appear to adsorb (Theng and Scharpenseel, 1975) as Cu^{2+} is extracted from exchange sites and precipitates as an insoluble Cu^{2+}-organic acid complex.

Unlike layer silicate clays, Fe or Al oxides may show enhanced as well as inhibited adsorption of Cu^{2+} in the presence of soluble organics. Because of the ability of oxides to adsorb organic acids, including fulvic and humic acids, by ligand exchange reactions (Parfitt et al., 1977a, 1977b), the presence of these organics on oxide surfaces may enhance adsorption of Cu^{2+} if only a portion of the func-

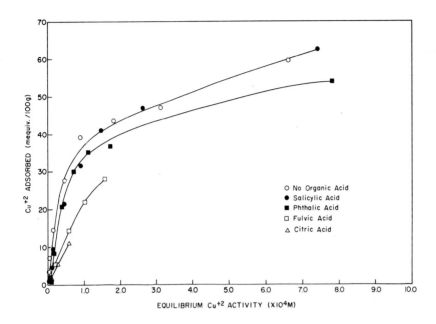

Fig. 6.   Adsorption isotherms of Cu^{2+} on montmorillonite in the presence and absence of organic acids, expressed in terms of the equilibrium Cu^{2+} activity.

tional groups are involved in bonds to the surface. The remaining groups are then free to bond with Cu^{2+}, so that an oxide-organic-metal "complex" is formed. Some organics such as glutamic acid (Davis and Leckie, 1978) and fulvic acid (Ufkes, 1979) can enhance Cu^{2+} adsorption at a given pH, while others have little effect (<u>eg</u>. salicylic acid) or an inhibitory effect (<u>eg</u>. picolinic acid) on adsorption. Organics may then act to mobilize or immobilize Cu^{2+} in soil solution, with the net effect depending on the nature of the adsorbing mineral surface as well as the type of organic. It is generally true, however, that small organic acid molecules solubilize Cu^{2+} by chelation and complexation reactions, thereby increasing the total dissolved Cu in soil solution above the concentration expected in the absence of organics.

V.   DISTRIBUTION OF COPPER IN SOIL PROFILES

Copper is one of the least mobile of the trace elements, with the result that many soil profiles show little
variation of total Cu concentration with depth (Hodgson,
1963). However, movement of Cu from highly leached surface
horizons into lower horizons has been observed in podzols.
In addition, organic-enriched surface horizons often contain
higher concentrations of Cu than the lower horizons as a
result of accumulation in the residues of plant tissue. Of
more interest than the total Cu is the "available" Cu as a
function of depth. Copper extractable by weak acid usually
varies with depth in a profile to a much greater extent than
total Cu (Swaine and Mitchell, 1960), and is commonly higher
in the organic-rich surface layer than in the lower horizons. In addition the Cu that is extractable by alkali or
EDTA generally decreases with depth in soil profiles
(Krahmer and Bergmann, 1978; Karim et al., 1976), while the
chemically less labile forms of Cu (eg. occluded Cu) increase with depth. Thus, organic matter in soil serves the
important role of maintaining Cu, released initially by
dissolution of minerals, in a complexed form that is available for uptake by plants yet is relatively resistant to
movement by acid leaching.

VI.   SUMMARY

It is evident that most of the Cu in soils is highly
insoluble and can only be extracted by strong chemical
treatments which dissolve various mineral structures or
solubilize organic matter. Nevertheless, a significant
"pool" of diffusible Cu exists, probably in organic complexes, which is in equilibrium with the very low level of
Cu in soil solution. Most of the soluble Cu in surface
soils is organically complexed, a fact which causes the
total Cu in soil solution to be greater than expected if
soluble organics were not present. Despite this effect of
soluble organics, measurement of the very low concentrations
of dissolved Cu in most soils is difficult, so that extracting agents are often used to remove Cu from the "pool" of
diffusible Cu. Critical to the design of extracting techniques is an understanding of the kinetics of Cu release
from various soil components. Until the rates of release

are understood, the relative importance to plants of the various forms of diffusible and non-diffusible Cu cannot be evaluated.

ACKNOWLEDGEMENTS

Research supported in part by NSF grants no. EAR-7923290 and EAR-7710215.

REFERENCES

Baes, C.F., and Mesmer, R.E. (1976). "The Hydrolysis of Cations". Wiley-Interscience, New York.

Bigham, J.M., Golden, D.C., Bowen, L.H., Buol, S.W., and Weed, S.B. (1978). *Soil Sci. Soc. Am. J.*, 42, 816.

Bingham, F.T., Page, A.L., and Sims, J.R. (1964). *Soil Sci. Soc. Am. Proc.* 28, 351.

Bloom, P.R., and McBride, M.B. (1979). *Soil Sci. Soc. Am. J.* 43, 687.

Bloomfield, C., Kelso, W.I., and Pruden, G. (1976). *J. Soil Sci.* 27, 16.

Bradford, G.R., Bair, F.L., and Hunsaker, V. (1971). *Soil Sci.* 112, 225.

Bresnahan, W.T., Grant, C.L., and Weber, J.H. (1978). *Anal. Chem.* 50, 1675.

Cavallaro, N., and McBride, M.B. (1980). *Soil Sci. Soc. Am. J.* 44, 729.

Cerovic, D., Momcilovic, I., and Kiss, S.J. (1972). *J. Materials Sci.* 7, 735.

Cheshire, M.V., Berrow, M.L., Goodman, B.A., and Mundie, C.M. (1977). *Geochim. Cosmochim. Acta* 41, 1131.

Childs, C.W., and Leslie, D.M. (1977). *Soil Sci.* 123, 369.

Clementz, D.M., Pinnavaia, T.J., and Mortland, M.M. (1973). *J. Phys. Chem.* 77, 196.

Davis, J.A., and Leckie, J.O. (1978). *Environ. Sci. Tech.* 12, 1309.

El-Sayed, M.H., Burau, R.G., and Babcock, K.L. (1970). *Soil Sci. Soc. Am. Proc.* 34, 397.

Farrah, H., and Pickering, W.F. (1976a). *J. Chem.* 29, 1167.

Farrah, H., and Pickering, W.F. (1976b). *Aust. J. Chem.* 29, 1177.

Forbes, E.A., Posner, A.M., and Quirk, J.P. (1976). *J. Soil Sci.* 27, 154.

Goodman, B.A. (1978). *Clay Minerals* 13, 351.

Goodman, B.A., and Cheshire, M.V. (1973). *Nature New Biology* 244, 158.

Heydemann, A.   (1959).   *Geochim. Cosmochim. Acta* 15, 305.

Hodgson, J.F.   (1963).   *Advances in Agronomy* 15, 119.

Hodgson, J.F., Geering, H.R., and Norvell, W.A. (1965).
Soil Sci. Soc. Am. Proc. 29, 665.

Hodgson, J.F., Lindsay, W.L., and Trierweiler, J.F. (1966).
Soil Sci. Soc. Am. Proc. 30, 723.

Jenne, E.A. (1968). *In* "Trace Inorganics in Water" (R.A.
Baker, Ed.), p. 337. *Adv. Chem. Ser.*, No. 73.

Karim, H., Sedberry, J.E., and Miller, B.J. (1976). Communi-
cations in Soil Science and Plant Analysis 7, 437.

Kinniburgh, D.G., Jackson, M.L., and Syers, J.K. (1976).
Soil Sci. Soc. Am. J. 40, 796.

Kline, J.R., and Rust, R.H. (1966). *Soil Sci. Soc. Am.
Proc.* 30, 188.

Krahmer, R., and Bergmann, W. (1978). Archiv. fur Ackerund
Pflanzenbau und Bodenkunde 22, 405.

Krauskopf, K.B. (1972). *In* "Micronutrients in Agriculture"
(J.J. Mortvedt, P.M. Giordano, and W.L. Lindsay, eds.),
p.7.   Soil Sci. Soc. Am., Madison, Wisconsin.

Lindsay, W.L. (1979). "Chemical Equilibria in Soils".
Wiley-Interscience, New York.

McBride, M.B.   (1976).   *Clays and Clay Minerals* 24, 88.

McBride, M.B.   (1978a). *Soil Sci. Soc. Am. J.* 42, 27.

McBride, M.B.   (1978b). *Soil Science* 126, 200.

McBride, M.B.   (1979).   *Clays and Clay Minerals* 27, 224.

McBride, M.B., and Blasiak, J.J. (1979). *Soil Sci. Soc. Am.
J.* 43, 866.

McKenzie, R.M. (1975). *Aust. J. Soil Res.* 13, 177.

McKenzie, R.M. (1980). *Aust. J. Soil Res.* 18, 61.

McLaren, R.G., and Crawford, D.V. (1973a). *J. Soil Sci.*
24, 172.

McLaren, R.G., and Crawford, D.V. (1973b). *J. Soil Sci.*
24, 443.

McLaren, R.G., and Crawford, D.V. (1974). *J. Soil Sci.*
25, 111.

Mann, A.W., and Deutscher, R.L. (1977). Chemical Geology
19, 253.

Mason, B. (1966). "Principles of Geochemistry". John
Wiley and Sons, New York.

Mattigod, S.V., and Sposito, G. (1977). *Soil Sci. Soc. Am.
J.* 41, 1092.

Murray, J.W. (1975). *Geochim. Cosmochim. Acta* 39, 505.

Murray, D.J., Healy, T.W., and Fuerstenau, D.W. (1968). *In*
"Adsorption from Aqueous Solution" (W.J. Weber and E.
Matijevic, eds.), p. 74. *Adv. Chem. Ser.*, No. 79.

Nalovic, L., Pedro, G., and Janot, C. (1975). *Proc. Int. Clay Conf.* (Mexico City), 601.

O'Shea, T.A., and Mancy, K.H. (1976). Analytical Chemistry 48, 1603.

Parfitt, R.L., Fraser, A.R., and Farmer, V.C. (1977a). *J. Soil Sci.* 28, 289.

Parfitt, R.L., Fraser, A.R., Russell, J.D., and Farmer, V.C. (1977b). *J. Soil Sci.* 28, 40.

Povarennykh, A.S. (1972). "Crystal Chemical Classification of Minerals". Plenum Press, New York.

Sanders, J.R., and Bloomfield, C. (1980). *J. Soil Sci.* 31, 53.

Schindler, P.W. (1967). *In* "Equilibrium Concepts in Natural Water Systems" (W. Stumm, ed.), p. 196. *Adv. Chem. Ser.*, No. 67.

Shuman, L.M. (1979). *Soil Science* 127, 10.

Shuman, M.S., and Woodward, G.P. (1977). *Environ. Sci. Tech.* 11, 809.

Sidhu, P.S., Gilkes, R.J., and Posner, A.M. (1980). *Soil Sci. Soc. Am. J.* 44, 135.

Sinha, M.K., Dhillon, S.K., Dhillon, K.S., and Dyanand, S. (1978). *Aust. J. Soil Res.* 16, 19.

Swaine, D.J., and Mitchell, R.L. (1960). *J. Soil Sci.* 11, 347.

Theng, B.K.G., and Scharpenseel, H.W. (1975). *Proc. Int. Clay Conf.* (Mexico City), 643.

Ufkes, E.A. (1979). M.S. Thesis, Cornell University.

Vuceta, J., and Morgan, J.J. (1978). *Environ. Sci. Tech.* 12, 1302.

3

COPPER REACTIONS WITH INORGANIC COMPONENTS OF SOILS INCLUDING UPTAKE BY OXIDE AND SILICATE MINERALS

Robert O. James[1]

CSIRO Physical Technology Unit
Ryde, New South Wales, Australia

N.J. Barrow

CSIRO Division of Land Resources Management
Wembley, Western Australia

I. INTRODUCTION

The reaction between Cu and soil, or soil constituents, is one example of a diverse group of processes which occur at surfaces and which are controlled or used by man. The processes include: the activation of oxide, silicate, and sulfide minerals toward various organic collectors in mineral processing; the dispersion of geo-chemical indicator elements from potential ore bodies; the general problem of reaction of trace metals with soils; and the similar problem of reaction between trace metals in large bodies of water and particles in suspension. As a result, information on the nature of the process tends to come from diverse sources. A common observation, however, is that adsorption of Cu ions increases as the pH increases. The increase may be very sharp so that one may change from little adsorption, to almost complete adsorption within little more than one pH unit (Fig. 1). Similar behaviour may also be observed for other metals (Fig. 1) - though the pH at which adsorption starts differs from element to element.

Present address: *Emulsion Research Laboratory,*
Eastman Kodak Company, Rochester, N.Y.,
U.S.A.

In recent years there have been considerable advances in our understanding of this behaviour and in this article we will attempt to summarise these. However before doing so it is necessary to briefly review: the characteristics of adsorbate - that is the metal ion species present in solution; the characteristics of the adsorbents; and the characteristics of the adsorption.

II.   THE METAL IONS PRESENT IN SOLUTION

This topic is dealt with in greater detail by other contributors and by Baes and Mesmer (1976) and Sylva and Davidson (1978).   It suffices to note that Cu ions tend to hydrolyse in solution to give, for example:

$$Cu^{++} + H_2O \rightleftharpoons CuOH^+ + H^+$$

and

$$CuOH^+ + H_2O \rightleftharpoons Cu(OH)_2 + H^+$$

Polymeric forms also occur, especially in concentrated solution and at high pH.   The pK of the first reaction is about 8 and this means that at pH 4 only about 0.01% of the Cu ions in dilute solution are present as $CuOH^+$.   However the proportion increases tenfold for each unit increase in

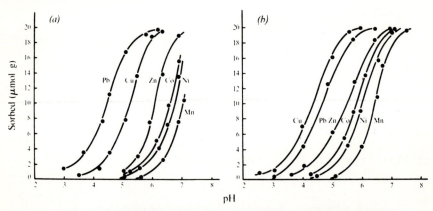

FIGURE 1.   *Adsorption of Cu, and other heavy metals on iron oxides as a function of pH, 20 μmol/g added. (a) Hematite   (b)   Goethite. (From McKenzie, 1980)*

pH. Thus adsorption is often observed to occur in a region in which the concentration of $CuOH^+$ ions, though small, increases rapidly with increasing pH (Fig. 1). In addition, Cu ions may form complexes with anions such as carbonate, phosphate and chloride (Libus and Tialowska 1975, Mann and Deutscher 1977).

In soil solutions the picture becomes even more complex because of the opportunity for the Cu ions to be complexed by a range of organic molecules. Because of this complexity, most studies on the process of adsorption have been done in the presence of simple electrolytes such as dilute solutions of nitrate (Table 1).

At high pH, Cu ions may be removed from solution by precipitation of the oxide, the hydroxide or the hydroxy carbonate. However, it has been shown several times by McBride and co-workers that the concentration of copper in soil is too low for such precipitates to be controlling the concentration (McBride 1978a, b; Cavallaro and McBride 1978; McBride and Blasiak 1979). Thus, removal of Cu from solution is not due to reaction with other ions in the solution phase - that is precipitation. Rather it is by reaction with atoms on the surface of the solid phase - that is adsorption.

III CHARACTERISTICS OF THE ADSORBENTS

A wide range of materials has been shown to be capable of adsorbing Cu ions from solution (Table 1). Included are oxides of iron, manganese and silicon. One of the important properties of these materials is that they may adsorb and desorb protons from solution - that is, the charge carried by the material is variable. At low pH the surface charge tends to be positive due to adsorption of protons; at high pH it tends to be negative due to loss of protons (or adsorption of hydroxyls). Thus there must be a pH at which the material carries no net charge. This point of zero charge is an important characteristic of the material. For manganese oxides it is at about pH 2 (McKenzie 1980) and thus in the middle range of pH, manganese oxides tend to bear a negative charge.
For iron oxides it is at about pH 8, and in the middle range of pH, the iron oxides therefore tend to be positively charged. Adsorption also occurs on clays and, as Farrah and Pickering (1979) show, these materials may also take protons from solution - that is they also have a

variable charge. This is in addition to the fixed charge which may be present as a result of isomorphous substitutions within the lattice. The charge carried by such materials at a given pH can be measured by, for example, finding the point of zero charge by acid-base titration and then calculating the protons taken up, or given off, from the amount of acid or alkali required to move to the desired pH.

In describing such systems one other measure is often used. Because the charge on the particles is balanced by ions from the solution, there is a separation of charge near the surface. As a result, electrostatic potentials are created. If the particles are then made to move in an electric field, the electrostatic potential near the surface can be measured. This electrokinetic potential, or zeta potential, is the potential at the hydrodynamic shear plane in the diffuse layer near the solid surface.

IV   CHARACTERISTICS OF ADSORPTION

Four main approaches have been used to investigate adsorption of metal ions, such as Cu, on colloid surfaces. These include: direct measurement of Cu remaining in solution or adhering to the surface under a range of conditions; measurement of the acid-base reactions accompanying adsorption (such as the release of protons); measurement of the electrokinetic potential before and after adsorption; and spectroscopic techniques such as UV-visible spectroscopy, electron spin resonance and X-ray photoelectron spectroscopy of Cu adsorbed on various solids. The first three methods have provided much information on the pH dependence of adsorption and formation of coatings at high metal concentrations and pH. However, they do not give direct information about the adsorbed state of Cu or other ions and this is of fundamental importance for developing predictive models. The more recent applications of spectroscopic methods must improve our view of the adsorbed state.

A.   *Direct Measurement of Adsorption*

Figure 1 illustrates typical results obtained for the adsorption of Cu, and of other metals, on iron oxides. All the metals show similar curves and the sequence is fairly consistent. Cu and Pb are adsorbed at low pH, Zn, Co, Ni and Mn follow. This is a similar order to that for the first hydrolysis constant - that is, adsorption seems to occur over a pH range in which the $MeOH^+$ ions are at a similar, although low, concentration. Adsorption also increases rapidly over a pH range in which the concentration of $MeOH^+$ ions increases rapidly. Such results have therefore often been taken to imply that the $MeOH^+$ ions are the adsorbed species. However other explanations are possible for the increase in adsorption with increase in pH. These will be given in detail later - but briefly we can argue that since adsorption displaces protons from the surface, a decreased concentration of protons will favour adsorption. We can also argue that an increase in pH makes the surface electrostatically more suitable for adsorption, because it decreases the positive charge - or increases the negative charge. For a positively charged surface, the repulsion of the positively-charged metal ions becomes therefore smaller with increases in pH.

Compared to its importance, relatively few studies have been made of adsorption of Cu in the presence of complexing ligands. Davis and Leckie (1978b) showed that adsorption of Cu was in some cases increased and in others decreased by the presence of complexing ligands. From their own work, and that of MacNaughton and James (1974), they concluded that ligands which are themselves adsorbed increase adsorption and ligands which are not adsorbed, such as chloride decrease adsorption. However the effect of a ligand also varies between metals. Adsorption of Hg and Ag is reduced in the presence of chloride (MacNaughton and James (1974), Forbes *et al.* 1974) but adsorption of Cu, Pb, Zn, and Cd is increased (Forbes 1973; Forbes *et al.* 1974).

B.   *Measurement of the Acid-Base Reactions Accompanying Adsorption*

As explained earlier, suspensions of metal oxides tend to exchange protons with the solution depending on the pH. One can therefore plot base (or acid) added against the resulting pH as in Fig. 2. This effect is modified in the presence of metal ions in such a way as to indicate that adsorption of the metal ions has released protons to the solution (Fig. 2). Although the effect has been widely observed, its magnitude is far from constant. Murray (1975) found that about 1.2 H^+ ions were released per Cu ion adsorbed and, as Fig. 3 shows, for Cu this ratio did not

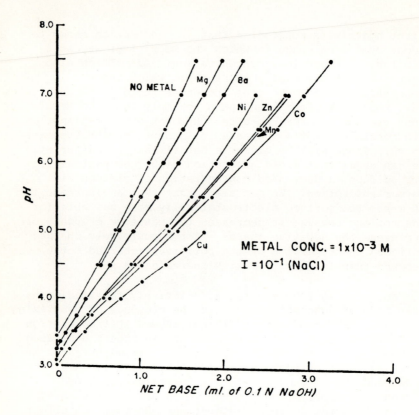

FIGURE 2.   The acid-base potentiometric titration curves
for manganese dioxide in 10^{-3} mol dm^{-3} ionic solutions of the
metals Cu, Co, Mn, Zn, Ni, Ba and Mg with 10^{-1} mol/dm^3
NaCl as ionic medium.   (From Murray, 1975)

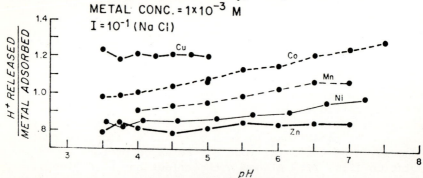

FIGURE 3. H^+ released/metal adsorbed ratios for 10^{-3}
mol/dm^3 Cu, Ni, Co, Mn and Zn at the manganese dioxide/aqueous
NaCl interface.   (From Murray, 1975)

change with pH - though it increased with some of the ions studied. In contrast, Schindler *et al.* (1976) reported that the number of H^+ ions released, per Cu ion adsorbed, increased from 1 to 2 as pH increased. McKenzie (1979, 1980) found that, for Cu adsorbed on the manganese oxide, birnessite, the ratio was about 1 and for another manganese oxide, cryptomelane, the ratio was about 1.5. The value formed by Forbes *et al.* (1976) for adsorption on goethite was about 2. Perhaps some of this diversity is due to experimental error. However the several studies involved a range of adsorbents and a range of conditions (Table 1) and it is likely that much of the diversity is due to real differences in the process.

Though the subject will be dealt with in more detail later, it is appropriate at this stage to ask why protons are released when trace metals are adsorbed. Two mechanisms may be envisaged. First when a positively charged ion is adsorbed on a surface, it is reasonable to expect that the disturbance in charge will be partly compensated by release of positive ions, such as protons, from the surface. This will be so whatever the detail of the adsorption reaction. Second if ions are adsorbed in a proportion different from that in the solution, an hydrolysis of the ions remaining in solution may be induced. For example, if $MeOH^+$ ions were adsorbed, their removal would be compensated by hydrolysis of Me^{++} ions thus releasing a proton. This could be described as a surface-induced hydrolysis.

C. Measurement of the Electrokinetic Potential

Measurements of the electrokinetic or zeta potential have been used to aid investigations into adsorption for a long time - though they have been used rather seldom for studying adsorption of Cu. Such measurements do not allow us to calculate directly the surface charge or the ionic adsorption density. However if the zeta potential of a colloid dispersion is measured in the presence and in the absence of Cu, then the difference between the calculated diffuse layer charges is the net amount of charge that crosses the shear plane on the addition of Cu. James (1971) compared the effects of Cu, Pb and Zn on the electrophoretic mobility of quartz particles. Figure (4) shows that, in the absence of the metals, the value was negative - reflecting the low point of zero charge of quartz. With increasing pH, the value decreased because of increased adsorption of hydroxide ions. In the presence of metals,

TABLE I. Some Direct Copper Adsorption Studies

Adsorbent, Spec. Surf. Area or Particle Size	Electrolyte	pH Range, ( ) indicates Cu hydroxide formation	Approx, Cu. Conc.	Reference
Quartz SiO_2 0.09 m^2/g	-	4 – 6	$10^{-5}-10^{-3}$ mol/e	Richardson & Hawkes (1956)
Amorph ferric hydroxide	-	4 – 7(8)	$10^{-6}-10^{-5}$ mol/dm	Hem & Skougstad (1960)
Fithian illite API standard clay #35 < 2μm diam 1000 ppm	Artificial river water and 0.7M	4 – 7(-12) 4 – 7(-12)	50 and 100 ppb	O'Connor & Kester (1975)
Hydrous manganese dioxide	0.1 N NaCl	3 – 5	$10^{-3} mol/dm^3$	Murray (1975)
SiO_2 Min-U-Sil	0.1 to 0.001 NaCl and various complexing ligands	4 – 7(8)	$10^{-6} mol/dm^3$	Vuceta (1976) Vuceta & Morgan (1978)
Goethite 2gm/dm^3/3	0.1 N KNO_3	3 – 6	$10^{-5} mol/dm^3$	Grimme (1968)
Goethite (16.2 m^2/dm^3) & quartz (82.5 m^2/dm^3)	0.1 N $NaNO_3$	4 – 7	$10^{-6} mol/dm^3$	Davis (1977); Davis & Leckie (1978a, b)

TABLE I.   *(Contd.)*

Goethite 89 m^2	0.075 N $NaNO_3$	4 – 6	$3.2x10^{-5}$	Forbes, Posner & Quirk (1974, 1976)
MnO_2 birnessite Fe_2O_3 hematite FeOOH goethite	0.1 N KNO_3	2 – 6 3 – 6 3 – 6	10^{-4}–10^{-3} M	McKenzie (1979, 1980)
SiO_2 "Silikagel H" $3.7x10^9$ cm^2 & "Aerosil 200" $1.6x10^9$ cm^2/kg	1 mol/dm^3	4 – 6.5		Schindler, Fürst Dick & Wolf (1976)
Bentonite < 2µm	variable	4 – 6	$4x10^{-6}$–$3x10^{-4}$	Steger (1973)
Georgia kaolinite, 12 m^2/g chlorite & illite (14 m^2/g) (49 m^2/g)	variable	5 – 6	100 ppm $1.6x10^{-3}$ mol/dm^3	Koppelman & Dillard (1977)
Kaolinite < 2µm	various complexing ligands	4 – 6	$1.6x10^{-4}$ mol/dm^3	Farrah & Pickering (1976a)
Illite & Montmorillonite	various complexing ligands	4 – 6	$1.6x10^{-4}$ mol/dm^3	Farrah & Pickering (1976b, 1977, 1979)
Georgia kaolinite CEC ∿ 1.2meq/100g	0.01 M NaCl & variable	4 – 6	10^{-6}–10^{-3} mol/dm^3	McBride (1978a)
Various soils	variable and also 0.1M $CaCl_2$	4 – 6	10^{-7}–10^{-4}	Cavallaro & McBride (1978)

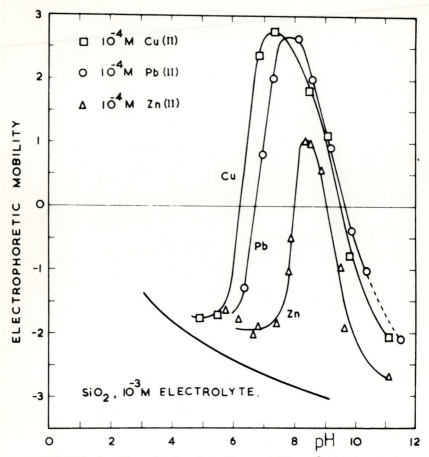

FIGURE 4.   *The electrophoretic mobility of quartz particles (0.05 g/dm^3) in 10^{-4} mol/dm^3 solution of Cu(II), Pb(II)
and Zn(II) nitrates as a function of pH.   (From James, 1971)*

this trend was sharply modified.   At a characteristic pH,
the decreasing trend was reversed and positive values were
obtained.   This reflects the adsorption of positively-charged
metal ions.   The sequence was Cu $>$ Pb $>$ Zn, and this
differed from the adsorption measurements of Schindler *et
al.* (1976) who found Pb $>$ Cu $>$ Cd.   At high pH, the
electrophoretic mobility again decreased and became
negative – that is the values passed through a point of
zero mobility and thus of zero charge.   These effects have
been explained qualitatively by James and Healy (1972).   In
addition to postulating the adsorption of positively
charged ions, they argued that, in the presence of excess

amounts of metal ions, adsorbed hydrolysed ions condensed to form polymeric hydrous oxides on the subtrate. The surface coatings mask the substrate properties and the particles behave as if they were Cu, Zn or Pb hydroxides. They therefore show their own characteristic points of zero charge.

D *Spectroscopic techniques*

The adsorbed forms of Cu may be deduced from the previously-mentioned approaches but these deductions may not give an unambiguous answer. It is desirable to have further information about the bonding arrangements at the surface. This is being attempted by spectroscopic investigation of the adsorbed ions. Koppelman and Dillard (1977) used x-ray photoelectron spectroscopy to study Cu adsorbed at pH 5 on to the clay mineral chlorite. They showed that adsorbed Cu hydroxides or oxides were not formed at this pH. There was a shift of the photoelectron energy spectrum for the adsorbed Cu which was unlike that for other divalent metals. From this, they concluded that the adsorbed form may be a hydrolysed species such as $CuOH^+$ rather than the aquo species $Cu(H_2O)_6^{2+}$.

McBride has used electron spin resonance to study Cu adsorbed on various minerals including kaolinite, hectorite and montmorillonite and he has summarised his findings in an earlier chapter in this book. Briefly he found that Cu was present as the octahedrally hydrated ion $CuH_2O)_6^{2+}$. However, he also found that as the pH was increased, there was evidence that hydroxy forms were present. Measurements in the visible spectrum have been used to study Cu ions in ion exchange resins but have not identified any large concentrations of ions other than the aquo species $Cu(H_2O)_6^{2+}$ (Nortia and Laitinen 1968).

Thus spectroscopic methods suggest that different bonds may be found under different conditions. Investigations using a range of colloids and a range of pH values will be needed to understand this complexity.

V. MODELS FOR ADSORPTION OF Cu AND OTHER TRACE METALS

Until relatively recently it has not been possible to describe simultaneously all the phenomena of adsorption in

ar:y more than a qualitative way. Where appropriate, some
of these descriptions have been given above. However the
increased availability of computers has made it possible to
solve the equations which govern several components and
several reactions at equilibrium. This leads to
quantitative descriptions of the phenomena – or models.
These may then be compared with the observations. However
a difficulty with this approach is often that a limited
range of observations is made. If only adsorption is
measured, a particular model may describe this very
closely. Yet this model may be inappropriate if, say,
values for proton release are also available. (Several
models may describe an elephant's trunk in isolation; only
one is appropriate if the rest of the elephant is visible.)
Ideally the model chosen should describe all of the phenomena
of adsorption – though models which are less than this
ideal may sometimes be useful in restricted circumstances.
For copper there do not appear to be any modelling studies
that meet this ideal. Most have used the increase in
adsorption with pH as the main criterion for assessment of
the model.

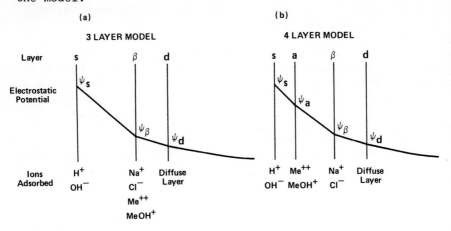

*FIGURE 5. Diagram representing the interface between
the solution and the surface on which Cu is adsorbed. The
change in the electrostatic potential ψ with distance is shown
as are the ions allocated to each layer.*

A basic assumption of many recent models of adsorption
is that the complex distribution of ions near a surface can
be simplified so that particular ions are allocated to
specified mean planes of adsorption. As an example, H^+ and

OH$^-$ ions might be assigned to one plane, Na$^+$ and Cl$^-$ to another. It is then assumed that a mean value for the electrostatic potential in that plane exists. This assumption is common to a range of models. The models differ in the definition and assignment of ions to planes of adsorption and in the electrostatic equations used to relate surface potential to surface charge. An excellent summary and comparison of these models was recently provided by Westall and Hohl (1980). This summary was concerned with the determination of charge and potential in the absence of adsorption of trace metals such as Cu. When Cu is present, the adsorbed Cu species must be allocated to an appropriate plane to complete the model. In models published to date, the allocation has been to the same plane as the surface complexes of electrolyte ions (Fig. 5a). This choice was made for mathematical simplicity (Davis and Leckie 1978a). However, when metal ions are adsorbed onto metal oxides, the adsorbed ions may well approach the surface very closely, becoming almost part of the surface. An alternative model therefore allocates adsorption to a further plane closer to the surface (Fig. 5b). The model represented in Fig. 5b is analogous to that used to describe adsorption of anions by Bowden *et al.* (1980). For convenience we will adopt a terminology similar to that of Westall and Hohl (1980) and refer to the model of Fig. 5a as a three-layer model and Fig. 5b as a four-layer model.

A further assumption in many models of adsorption is that sites on the surface of particles may react with ions in solution in a manner analogous to that for soluble ligands (Schindler *et al.* 1976; Bowden *et al.* 1973, 1974, 1977, 1980; Davis and Leckie 1978a). The models have therefore been termed "surface complexation models" (Westall and Hohl 1980). In the approach used by Schindler *et al.* (1976) and by Davis and Leckie (1978a), the reactions at the surface are depicted by equations such as:

$$\text{SOH} + \text{Cu}^{2+} \rightleftharpoons \text{SO}^-.\text{Cu}^{2+} + \text{H}^+ \qquad (3)$$

and

$$\text{SOH} + \text{Cu}^{2+} \rightleftharpoons \text{SO}^-.\text{CuOH}^+ + 2\text{H}^+ \qquad (4)$$

where SOH indicates a surface hydroxyl site which may
adsorb protons or metal ions.   These equations are
then treated in a manner analogous that of solution
chemistry and equilibrium constants may be estimated.
By appropriate choice of these constants, either equation
(3) or equation (4) may be favoured and so the number
of protons released, per Cu adsorbed, may vary from 1
to 2.   This was broadly the approach used by Schindler *et
al.* (1976) - though their equation differed slightly
from equation (4).   In this approach, the low adsorption
at low pH is seen solely as a consequence of the protons
on the right-hand side of equation (3) and (4).   As
these enter the equilibrium equation for equation (4)
as a squared term, the effect is especially powerful.

Figure 6 shows that, using this approach, adsorption
of Cu and of other ions could be closely described.
Nevertheless this can only be a partial model because

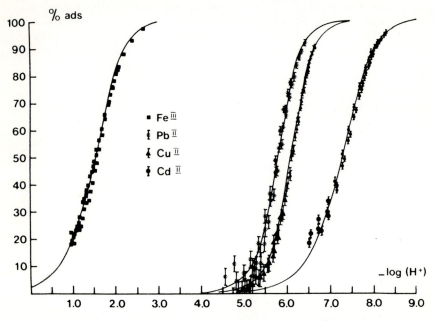

FIGURE 6.   *Adsorption of Cu(II), Fe (III), Pb(II) and
Cd(II) as a function of -log [H$^+$].   The solid lines are
calculated from a simplified surface complex model which
considers that Cu^{2+} is the adsorbing species.   (From Schindler
et al., 1976)*

an important effect has been neglected. This is that
adsorption takes place at a charged surface. The
effects of this are considered by both the three-layer
and the four-layer model and these will be considered
in turn.

A. The three-layer model

The activity of an ion near the surface of a
charged particle will be different from that in the
bulk solution. We can take this into account in adsorption
calculations if we make the assumption that there is
an invariant number which we call the intrinsic equilibrium
constant that describes the tendency of ions to react
with a surface group in the absence of an interfacial
electric field. In the presence of a field, ions in
the solution interphase are distributed according to
the Boltzmann distribution:

$$\left[M^{z+} \right]_j = \left[M^{z+} \right] \exp(-z \ \psi_j / RT)$$

(5)

where M is an ion of valency z, ψ_j is the electrostatic
potential in the plane j and F, R and T have their
usual physicochemical meaning. Then if the intrinsic
equilibrium constant for equation (3) is designated
$^*K^{int}_{Cu^{2+}}$, the concentration of Cu found at the surface
as a result of equation (3) is:

$$\left[SO^-. \ Cu^{2+} \right] = \ ^*K^{int}_{Cu^{2+}} \ \frac{[SOH] \ [Cu^{2+}] \ \exp(-2 \ F\psi_\beta / RT)}{[H^+] \ \exp \ (-F\psi_s / RT)}$$

(6)

Similar equations may be written for equation (4) or
for other reaction mechanisms that may be postulated
(Davis and Leckie 1978a). Evaluating such equations
and so calculating the surface concentration requires
values for the electrostatic potentials. These are
obtained by simultaneously satisfying all of the equations
which relate: the mass and charge balances of all the
components in the system; the charge balance for the
interface; and the potential and charge relations
across the interface. Full descriptions of the equations
involved are given by Davis and Leckie (1978a) and by
James *et al.* (1981).

The exponential terms describe the effects of the
charge carried by the surface. If the surface is
positively charged, cations tend to be repelled from
the surface. Because μ_β is positive, the exponential
term may be thought of as decreasing the effective
concentration of the cation. However exponential
terms appear in both the numerator and the denominator
of equation (6). The terms do not cancel because the
Cu^{2+} ions and the H^+ ions are allocated to different
planes of potential and so different values of μ apply.

A characteristic of this model is that the number of
protons released depends on the relative contribution of
reactions (3) and (4). As these reactions are differentially
affected by pH, their relative importance will determine
the slope of plots of Cu adsorbed against pH - if reaction
(4) were favoured, the plots would tend to be steep.
Some of the differences in slope between the metals in
Figure 1 could be explained this way. The relative position
of metals in plots such as Figure 1 is explained by differences
in the relative magnitude of the intrinsic equilibrium
constants: for haematite, the constants for Pb would be
large relative to those for Cu; for goethite the opposite
would be the case. Finally, if all else were equal, a
positively charged surface would tend to favour reaction
(4). This is because there is less repulsion between the
lower charged ion and the surface. Therefore one would
predict that for adsorption on a positively charged surface,
such as iron oxides at pH 4-6, the number of protons
released per Cu adsorbed would approach 2. This is consistent
with the results of Forbes *et al*. (1976) for goethite.
For adsorption on negatively charged surfaces, such as
manganese oxides at pH 4-6, reaction (3) is favoured
because of the greater attraction for the divalent ion at
the surface. Protons released would therefore approach
1. This is consistent with the observations of Murray
(1975) and McKenzie (1980).

The foregoing has shown that the triple-layer model
appears to have the potential to describe many of the
observations of adsorption of metals. However rigorous
testing of the model requires that all the predictions of
the model be tested quantitatively and this has not yet

been done. It has been shown that if is capable of describing
the adsorption of metals on a range of adsorbents (Davis
and Leckie 1978a) and the fit to data for Cu adsorption
on amorphous iron oxyhydroxide is reproduced in Figure 7.

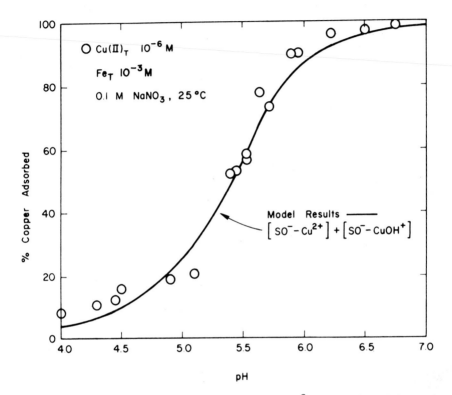

FIGURE 7. *Adsorption of Cu(II) on $10^{-3}M$ amorphous iron oxyhydroxide in 0.1 mol/dm³ NaNO₃. Circles are experimental data and the line is a model calculation considering Cu^{2+} and $CuOH^+$ as adsorbing species. (From Davis and Leckie, 1978)*

This fit is obtained by assuming that both reaction (3)
and reaction (4) are involved. It would be desirable to
test this by comparing predicted values for protons released
against observed values but appropriate data was not
available. However a similar model has been shown to
describe adsorption of cadmium onto titanium dioxide and,
at the same time, data for both the surface charge and
the electrokinetic potential (James *et al.* 1981).

B.   Four-layer Model

The model of Bowden et al. (1973, 1974, 1977 and 1980) differs in the description of the adsorption equation. No specific reaction mechanism is postulated. Instead, the electrochemical potential of an ion in the bulk solution is assumed to be equal to the electrochemical potential of the ion in the adsorbed state. From this the following equation is obtained:

$$a_{is} = K_i \, a_i \, \exp(-z_i \, F \, \psi_s/RT) \tag{7}$$

where a_{is} is the activity of the ion i at the surface, a_i its activity in the bulk solution, and K_i is a binding constant and is characteristic of the ion and the surface. The surface activity is then set equal to the ratio of occupied sites to vacant sites. If we postulate that $CuOH^+$ ions in solution are in equilibrium with the surface, the surface concentration of adsorbed copper (Cu_a) would be given by:

$$Cu_a = \frac{N_T \, K_{CuOH^+} \, a_{CuOH^+} \, \exp\,(-F \, \psi_a/RT)}{1 + K_{CuOH^+} \, a_{CuOH^+} \, \exp\,(-F \, \psi_a/RT)} \tag{8}$$

where N_T is the total concentration of surface sites capable of adsorbing copper. Note that N_T is defined differently from SOH of expression (6), the equilibrium concentration of uncomplexed surface hydroxyls. As for the three-layer model, evaluating this expression involves solving a large set of simultaneous equations as described by Barrow et al. (1980).

Adsorption is seen as depending on three components: the magnitude of the binding constant, the activity of the ion in the bulk solution, and the electrostatic effects represented by the exponential term. Adsorption of the monovalent $CuOH^+$ ion is seen as feasible even though the proportion present be very low, because, on positively charged surfaces, the electrostatic effect is not as unfavourable as for the divalent ion, and because the binding constant might be higher than for the hydrated divalent ion. Variation in the number of protons released per Cu adsorbed is achieved by varying the position of the a plane (Fig. 5). If this plane is close to the S plane, the charge is largely balanced by displacing protons from the S plane and the ratio of protons released to Cu adsorbed is high. As the a plane is moved further from the surface the charge is increasingly balanced by changes in the "outside" layers and fewer protons are released.

With the exception of some preliminary work with zinc (Bowden *et al.* 1973) there has been no detailed description of the application of this model to metal ions. However the model is general and should be well suited to the adsorption of Cu by mineral surfaces.

C. *Comparison of the Three-layer and the Four-layer Models*

Both the three-layer and the four-layer models seem to be capable of describing all the phenomena of adsorption of metals by oxides. Whether this apparent capability can, in fact, be realized has not yet been tested - let alone which model is best. To describe any of the phenomena in isolation is fairly easy. An adequate test of the models is only obtained if all the observable phenomena are simultaneously explained. Studies are therefore needed in which as many as possible of the phenomena of adsorption are measured using a range of adsorbates on a range of adsorbents and in which these observations are compared with models. Ideally the observations would include spectroscopic and electrokinetic observations.

V1. ADSORPTION OF Cu BY SOIL

Because adsorption of Cu is so complicated, understanding has come from the study of simplified systems. Much of this understanding is still new, and, as indicated above, is far from complete. Application to the much more-complex environment of soils has hardly begun. It would be expected that the adsorption processes described here would be most important in soils wth a large content of variable-charge minerals such as iron and aluminium oxides and a relatively low content of organic matter. Adsorption of cations by andisols (which have a high content of variable charge minerals) has been studied by Wada and his colleagues (summarised by Wada 1981). Such reactions are

not confined to the somewhat extreme soil properties of
such soils but should extend through all soils.
Nevertheless, within a limited area, and with a consequently
limited portion of the possible spectrum of soils, there
may not be a strong correlation between copper adsorption
and the aluminium content of soils (Harter 1979).

From the several observations of increased adsorption
of Cu with increasing pH, it would be expected that
increasing the pH of the soil would decrease the
availability of Cu to plants.   However this has not always
been observed (Jarvis 1981).   Apart from the possibility
that there may be effects on the plant *per se*, two comments
may be made.   The first is that, in soil, oxides usually
bear a net negative charge even at pH values at which the
pure oxide is positively charged.   This seems to be due to
adsorption of organic matter and perhaps of silicates.
Judging from the work of McKenzie (1980), adsorption on
negatively charged surfaces may be less sensitive to
increases in pH than adsorption on positively charged
surfaces.   The second comment is that liming a soil
stimulates mineralization of organic matter and this may
release Cu.   The effect of modifying the pH on Cu
availability would therefore be expected to vary from soil
to soil - but in a way that should be predictable from
knowledge of the way that Cu is mainly held in that soil.

A further characteristic of adsorption in soils, as
distinct from adsorption on well-crystalized oxides, is
that a slow reaction following adsorption is more marked.
As a result the adsorbed ligand becomes more tightly bound
and less available to plants.   This is well documented for
anions and especially for phosphate.   There is increasing
evidence that it also occurs for metals such as Zn (Kuo and
Mikkelson 1979) and Cu (Brennen *et al*. 1980).   The
mechanisms involved are not understood.

VII. ACKNOWLEDGMENT

We would like to pay a tribute to the late Professor
Alan Posner who, by example, encouraged us, and many
others, to study adsorption of ions at interfaces.

REFERENCES

Baes, C.F., and Mesmer, R.E. (1976). "The Hydrolysis of
Cations" A. Wiley Interscience, New York.
Bowden, J.W., Bolland, M.D.A., Posner, A.M. and Quirk, J.P.
(1973). *Nature 245*, 81.
Bowden, J.W., Posner, A.M., and Quirk, J.P. (1974). Trans.
10th Int. Congress of Soil Science II, 29.
Bowden, J.W., Posner, A.M., and Quirk, J.P. (1977). *Aust.
J. Soil Res. 15*, 121.
Bowden, J.W., Nagarajah, S., Barrow, N.J., Posner, A.M.,
and Quirk, J.P. (1980). *Aust. J. Soil Res. 18*, 49.
Cavallaro, N., and McBride, M.B. (1978). *Soil Sci. Soc.
Am. J. 42*, 550.
Davis, J.A. (1977). Ph.D. Thesis, Stanford University.
"Adsorption of Trace Metals and Complexing Ligands at
the Oxide/Water Interfaces".
Davis, J.A., and Leckie, J.O., (1978a). *J. Colloid Interface
Sci. 67(1)*, 90.
Davis, J.A., and Leckie, J.O. (1978b). *Environ. Sci. Technol.
12(12)*, 1309.
Farrah, H., and Pickering, W.F. (1976a). *Aust. J. Chem.
29*, 1167.
Farrah, H., and Pickering, W.F. (1976b). *Aust. J. Chem.
29*, 1177.
Farrah, H., and Pickering, W.F. (1977). *Water Air Soil
Pollut.8*, 189.
Farrah, H., and Pickering, W.F. (1979). *Chem. Geol. 25*, 317.
Forbes, E.A. (1973). Ph.D. Thesis, University of Western
Australia. "The specific adsorption of heavy metal
cations on goethite".
Forbes, E.A., Posner, A.M., and Quirk, J.P. (1974).
J. Colloid Interface Sci. 49(3), 403.
Forbes, E.A., Posner, A.M., and Quirk, J.P. (1976). *J.
Soil Sci. 27*, 154.
Grimme, H. (1967). *Zeitschrift für Pflanz. u. Bodenkunde
121(1)*, 59.
Hem, J.D., and Skougstad, M.W. (1960). U.S. Geological
Survey Supply Paper 1459-E. U.S. Printing Office,
Washington, D.C.
Harter, R.D. (1979). *Soil Sci. Amer. J. 43*, 679.
James, R.O. (1971). Ph.D. Thesis, University of Melbourne.
"The Adsorption of Hydrolysable Metal Ions at the
Oxide/Water Interface".
James, R.O., and Healy, T.W. (1972). *J. Colloid Interface
Sci. 40*, 65

James, R.O., Stiglich, P.J., and Healy, T.W. (1980).
A.C.S. Symposium Series "Adsorption from Aqueous Solutions".

Jarvis, S.C. (1981). Chapter 12 (this volume).

Koppelman, M.H., and Dillard, J.G. (1977). *Clays and Clay Miner. 25,* 457.

Kuo, S., and Mikkelsen, D.S. (1980). *Soil Sci. 128,* 274.

Libus, Z., and Tiałowska H. (1975). *J. Soln. Chem. 4,* 1011.

MacNaughton, M.G., and James, R.O. (1974). *J. colloid Interface Sci. 47,* 431.

Mann, A.W., and Deutscher, R.L. (1977). *Chem. Geol. 19,* 253.

McBride, M.B. (1976). *Soil Sci. Soc. Am. J. 40,* 452.

McBride, M.B. (1978a). *Clays and Clay Miner. 26(2),* 101.

McBride, M.B. (1978b). *Soil Sci. Soc. Am. J. 42,* 27.

McBride, M.B., and Blasiak, J.J. (1979). *Soil Sci. Soc. Am. J.43 ,* 866.

McKenzie, R.M. (1979). *Geochim. Cosmochim. Acta 43,* 1855.

McKenzie, R.M. (1980). *Aust. J. Soil Res. 18,* 61.

Murray, J.W. (1975). *Geochim. Cosmochim. Acta 39,* 505.

Nortia, T., and Laitinen, S. (1968). *Suom. Kemistil. B41,* 136.

O'Connor, T.P., and Kester, D.R. (1975). *Geochim. Cosmochim. Acta 39,* 1531.

Richardson, P.W., and Hawkes, H.E. (1958). *Geochim. Cosmochim. Acta 15,* 6.

Schindler, P.W., Fürst, B., Dick, R., and Wolf, P.U. (1976). *J. Colloid Interface Sci.* 55, 469

Steger, J.F. (1973). *Clays and Clay Miner. 21,* 429.

Sylva, R.N., and Davidson, M.R. (1978). *J. Chem. Soc. Dalton Trans. 232.*

Vuceta, J. (1976). Ph.D. Thesis, California Institute of Technology. "Adsorption of Pb(II) and Cu(II) on α-Quartz from Aqueous Solutions: Influence of pH, Ionic Strength and Compexing Ligands".

Vuceta, J., and Morgan, J.J. (1978). *Environ. Sci. Technol. 12(12),* 1302.

Wada, K. (1981). In "Soils with variable charge" Ed. B.K.G. Theng: (N.Z. Soc. Soil Sci.: Lower Hutt).

Westall, J., and Hohl, H. (1980). *Adv. Colloid Interface Sci. 12,* 265.

4

REACTIONS WITH ORGANIC MATTER

F.J. Stevenson
Alanah Fitch

Department of Agronomy
University of Illinois
Urbana, Illinois

I. INTRODUCTION

Many aspects of the chemistry and mobility of Cu in soils are related to the formation of complexes with organic matter (Stevenson and Ardakani, 1972). The ability of the organic fraction to complex Cu is well recognized but little data are available regarding the fraction of the soil Cu which occurs in organic forms or of the factors affecting the availability of the bound Cu to plants.

II. ORGANICALLY BOUND Cu IN SOILS

Organic constituents in soils form both soluble and in-soluble complexes with Cu (McBride, this symposium). A wide variety of compounds are involved, including simple aliphatic acids, amino acids, phenolic acids, amino acids, peptides and proteins, polysaccharides, and humic and fulvic acids (Stevenson and Ardakani, 1972). Thus far, greatest attention has been given to binding by humic and fulvic acids.

A. Fractionation of Soil Cu

One of the more successful schemes for fractionating soil Cu is that of McLaren and Crawford (1973). The organically bound form was estimated on the basis of extraction with 0.1 M sodium pyrophosphate. From about one-fifth to one-half of the Cu in 24 contrasting soil types was accounted for in organically bound forms. Using another sequential extraction procedure, Tessler *et al.* (1979) found that approximately 25% of the Cu in two sediments was accounted for in organically bound forms.

The percentage of the soil Cu in association with organic matter would be expected to be high in soils rich in organic matter (*e.g.*, Histosols and Mollisols), and to be influenced by such soil properties as pH and kind and amount of clay.

Clay and organic colloids (*e.g.*, humic acids) are major soil components involved in Cu retention. However, individual effects are not as easily ascertained as might be supposed, for the reason that, in most mineral soils, organic matter is intimately bound to the clay, probably as a clay-metal-organic complex. Thus, two major types of adsorbing surfaces are normally available for interaction with Cu, namely clay-humus and clay alone. Accordingly, clay and organic matter function more as a unit than as separate entities and the relative contribution of organic and inorganic surfaces to adsorption will depend upon the extent to which the clay is coated with organic substances. As can be seen from the schematic diagram shown in Fig. 1, the interaction of organic matter with clay still provides an organic surface for complexation of Cu.

Data obtained for retention of weakly basic pesticides by soil (reviewed by Stevenson, 1972) suggest that, up to an organic matter content of about 8%, both organic and mineral surfaces are involved in adsorption; at higher organic matter contents, adsorption occurs mostly on organic surfaces. It

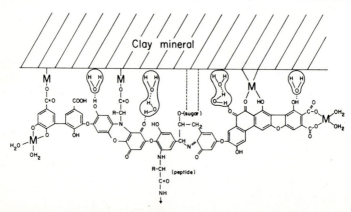

FIGURE 1.   *Schematic diagram of clay-organic matter-metal complex.*

should be noted, however, that the amount of organic matter required to coat the clay will vary from one soil to another and will depend on both kind and amount of clay in the soil. For soils having similar clay and organic matter contents, the contribution of organic matter to the binding of Cu will be highest when the predominant clay mineral is kaolinite and lowest when montmorillonite is the main clay mineral.

Another factor of importance in the solid phase complexation of Cu is the extent to which binding sites on organic surfaces are occupied by trivalent cations, noticeably Fe and Al. These cations generally form stronger complexes with organic molecules than divalent cations. The difficulty of removing Fe and Al from extracted organic matter is well known.

The quantity of Cu in the soil solution at any one time (often in the ppb range) is normally trivial in comparison to the total amount held by clay and humus colloids, or as precipitates. However, from the standpoint of plant nutrition, soluble Cu is of special importance. McLaren and Crawford (1973) concluded that the bulk of the available Cu reserves in soil resided in the organically-bound fraction and that the amount in forms available to plants (exchangeable- and soluble-Cu) was controlled by equilibria involving specifically adsorbed forms. The postulated relationship between the three forms was as follows:

$$\text{Exchangeable- and} \atop \text{soluble-Cu} \rightleftharpoons {\text{Specifically} \atop \text{adsorbed-Cu}} \rightleftharpoons {\text{Organically} \atop \text{bound-Cu}}$$

Stevenson (1979) observed a nonlinear relationship between organically bound Pb and total Pb in a Mollisol artificially enriched with Pb in the field. Lead was selectively retained by inorganic soil components at low application rates but as the fixing sites approached saturation more and more of the Pb was associated with organic matter. The conclusion was reached that contamination of the test soil with Pb (*e.g.*, from automobile exhaust) would have little effect on available forms until soil concentrations exceeded about 100 µg/g, following which availability would be controlled by the organic pool. Copper would be expected to behave in a similar fashion. Although much is known about the accumulation of Cu and other heavy metals in soils near mine smelters, little is known regarding the chemical forms that are present or of the factors affecting availability. For forest soils, most of the Cu (and other heavy metals) accumulates in the organic-rich surface layer (Boggess, 1977).

B. *Soluble Complexes of Cu*

1. *Chemical Speciation of Cu.* The most common method for determining Cu in aqueous solution is by atomic absorption spectrometry. However, this technique determines only total concentration without differentiating whether the Cu is in free or bound forms.

Several indirect approaches have been used to estimate the fraction of the Cu in the soil solution which occurs in complexed forms. One technique involves removal of free Cu with a competing complexing agent (Hodgson *et al.*, 1965). An alternate approach has been to pass the soil extract through a cation exchange resin, which adsorbs cationic forms and allows complexed forms to pass through. In both approaches, the amount of complexed Cu is taken as the difference between the amount removed and total concentration in solution. Results using these approaches suggest that as much as 99% of the Cu in displaced soil solutions is complexed with organic matter (Hodgson *et al.*, 1966).

Direct determination of free Cu^{2+} in soil extracts and natural water can be done using the Cu^{2+} ion selective electrode (ISE) and by anodic stripping voltammetry (ASV). A major limitation of ISE is its rather low sensitivity (concentrations of Cu^{2+} < 10^{-6} M cannot be accurately measured). In both methods, the response is affected by pH and ionic strength (Ernst *et al.*, 1975); furthermore, errors can arise through sorption of organics on the electrode surface (Bresznik *et al.*, 1976; Blutstein and Smith, 1978; Greter *et al.*, 1979).

Advantages of ASV for determining free Cu^{2+} include the following (Blutstein and Smith, 1978):

(1) The method is highly sensitive and is capable of measuring free Cu^{2+} at the low levels found in soil extracts and natural waters.

(2) Only free and labile species of Cu^{2+} are measured. Some information about chemical forms may be obtained from potential shifts and analysis of the voltammetric waveshape obtained.

(3) A wide range of natural aqueous solutions can be analyzed.

(4) Small volumes of solution can be analyzed, using as little as 2 ml of sample.

(5) The method is nondestructive, thereby allowing response to changes in chemical and physical parameters to be monitored on the same sample.

Oxidation of lake and river water by ultraviolet irradiation in the presence of H_2O_2 generally increases the concentration of Cu^{2+} detectable by ASV and the amount released by oxidation has been designated organically bound (Gardiner and Stiff, 1975; Blutstein and Smith, 1978). For many samples, not all of the Cu can be accounted for in this manner and rather elaborate fractionation schemes, in conjunction with ASV, have been employed for determining combined forms of Cu, some of which may have application to soil extracts.

In the scheme of Batley and Florence (1976), Florence (1977), Florence and Batley (1977), and Batley and Gardner (1978), ASV labile and total Cu are determined in:

(1) The original water sample.
(2) Water passed through a chelating resin.
(3) Water subjected to ultraviolet irradiation in the presence of H_2O_2.
(4) Ultraviolet irradiated water passed through a chelating resin.

The following operational defined species were obtained, where the suffixes L and A refer to organic and inorganic species.

ML + ML1 + MA1 ASV labile Cu removed by chelating resin. Includes free Cu^{2+} plus organic and inorganic complexes which readily dissociate under the measurement conditions.

ML2, MA2 ASV labile Cu associated with organic and inorganic species which are not removed by a chelating agent.

ML3, MA3 Nonlabile organic and inorganic species which are removed by a chelating resin.

ML4, MA4 Nonlabile organic and inorganic species which are not removed by a chelating resin.

Application of the fractionation scheme to river and reservoir water has shown that substantial amounts of the Cu are associated with organic compounds of fractions ML2, ML3, and ML4 (Florence, 1977; Batley and Gardner, 1978).

2. *Nature of Soluble Organic Ligands.* Very little is known of the kinds and amounts of soluble organic constituents in soil. Based on results obtained with lake and river waters, summarized by Jackson (1975), Reuter and Perdue (1977), and Mantoura *et al.* (1978), over 80% of the soluble

organic matter would be expected to consist of complex, yellow or brown polymeric substances resembling fulvic acids. The abundance of humic substances in river water showed marked geographical variation; values of the order of from 0.1 to 10 mg per liter are common (Reuter and Perdue, 1977).

Simple biochemical substances having the ability to chelate Cu^{2+} (*e.g.*, simple aliphatic acids, sugar acids, amino acids, and phenols) may be of some importance in soil because of their production by microorganisms. Although these constituents have only a transitory existence, significant amounts may be present in the soil solution during periods of intense biological activity. For example, high concentrations may be attained in localized zones favorable for the growth of microorganisms, such as near decomposing residues and in the rhizosphere (soil zone directly influenced by plant roots). Anaerobic soils and soils amended with organic wastes may also be relatively rich in metal-binding biochemical compounds. A wide array of organic compounds have been identified in soil leachates and in litter decomposition products. Since natural complexing substances have the capability of transforming solid phase forms of micronutrient cations into soluble metal complexes, their production during decay of plant and animal residues may increase the availability of Cu to plants. For information on organic compounds in water extracts of soils and in natural waters the reviews of Stevenson and Ardakani (1972), Jackson (1975), and Reuter and Perdue (1977) are recommended.

Chromatographic studies have shown that, while the concentration of any given compound or group of compounds in the soil solution may be slight, the total concentration of all complexing species may be appreciable. In many agricultural soils, particularly those well supplied with organic matter, the combined total of potential chelate formers in the aqueous phase is probably sufficient to account for the trace quantities of metal ions normally present. The concentration of individual biochemical species in the soil solution (20% moisture level) is approximately as follows (Stevenson and Ardakani, 1972):

$$\text{Simple aliphatic acids} \ldots 1 \times 10^{-3} \text{ to } 4 \times 10^{-3} \text{ M}$$
$$\text{Amino acids} \ldots 8 \times 10^{-5} \text{ to } 6 \times 10^{-4} \text{ M}$$
$$\text{Aromatic acids} \ldots 3 \times 10^{-4} \text{ to } 5 \times 10^{-5} \text{ M}$$

The quantity of Cu that occurs in available forms at any one time will be affected not only by the synthesis and destruction of biochemical chelating substances, but on consumption by microorganisms. Bacteria, actinomycetes, and fungi will compete with higher plants for available Cu and net immobilization may result when levels are suboptimum for plant growth. The relationship is analogous to N immobilization when crop residues with wide C/N ratios undergo decomposition in soil. Thus, microorganisms have two opposing effects on Cu availability, namely, solution through production of ligands and competition.

The possibility that some forms of Cu in the soil solution are less available to plants than others has not been adequately explored. It appears likely that Cu availability will depend upon both the nature of the ligands and the amounts present. For example, Cu complexes with low molecular weight compounds may be more available to plants than those high in molecular weight.

Finally, the decomposition of crop residues and organic wastes by microorganisms may lead to the liberation of significant quantities of Cu. The Cu content of farmyard manure (7-40 µg/g) is significantly higher than for herbage (2-15 µg/g) and the application of 5,000 kg of farmyard manure per ha will lead to the addition of about 0.08 kg of Cu (see Stevenson and Ardakani, 1972). Sewage sludges may contain as much as 4,000 µg/g of Cu on a dry-weight basis (McCalla *et al.*, 1977).

C. *Binding of Cu by Humic and Fulvic Acids*

Abundant evidence exists for the complexing of Cu^{2+} by humic and fulvic acids, including:

(1) Inability of K^+ and other monovalent cations to replace adsorbed Cu^{2+} from mineral and organic soils.
(2) Correlation between humus content and Cu^{2+} retention.
(3) Ability of known chelating agents to extract Cu while solubilizing part of the soil humus.
(4) Selective retention of Cu^{2+} by humic and fulvic acids in the presence of a cation-exchange resin.

Geering and Hodgson (1969) concluded that both fulvic acid-type constituents and individual biochemical compounds were involved in the movement of micronutrients to plant roots, with the fulvic acids being the most efficient in complexing metals. As noted earlier, fulvic acids are considered to be the constituents primarily responsible for the binding of Cu^{2+} in natural waters.

   1. Cu-Binding Capacities. Methods used to determine the maximum binding capacities of humic substances include coagulation (Rashid, 1971), proton release (van Dijk, 1971; Stevenson, 1976, 1977), metal ion retention as determined by competition with a cation-exchange resin (Zunino *et al.*, 1972; Crosser and Allen, 1977), dialysis (Zunino and Martin, 1977), ASV (Guy and Chakrabarti, 1976; O'Shea and Mancy, 1976), and ISE measurements (Bresnahan *et al.*, 1978; Buffle *et al.*, 1977). These studies suggest that, in general, the maximum amount (meq) of Cu^{2+} that can be bound to humic and fulvic acids under optimum conditions is approximately equal to the content of acidic functional groups as measured by base titration. The exchange acidities of humic acids vary greatly but generally fall within the range of 1.5 to 5.0 meq/g. This corresponds to the sorption of from 48 to 160 mg of Cu per g of humic acid. Assuming a C content of 56% for humic acids, one Cu atom would be bound per 20 to 60 C atoms in the fully saturated complex. Lees (1950) arrived at a value of one Cu atom per 60 C atoms for a peat humic acid. Factors influencing the quantity of Cu bound by humic substances include pH, ionic strength, molecular weight, and functional group content (Rashid, 1971; Guy *et al.*, 1975).

   2. Solubility Characteristics. Humic and fulvic acids form both soluble and insoluble complexes with Cu^{2+} depending upon degree of saturation. Because of their high acidities and relatively low molecular weights, Cu complexes of fulvic acids are more soluble than those of humic acids.
   Copper can affect the solubility characteristics of humic substances in several ways. When humic and fulvic acids are dissolved in water, dissociation of COOH occurs and the molecule assumes a stretched configuration due to repulsion of charged groups. Upon addition of Cu^{2+}, the charge is reduced and the molecule collapses, thereby reducing solubility. Being divalent, Cu^{2+} also has the potential for linking individual molecules together to produce chain-like structures. Stevenson (1976, 1977) concluded that metal complexes of humic acid are soluble at low metal-humic acid ratios (few combined molecules in the chain) but precipitation occurs as the chain-like structure grows and the isolated COOH groups become neutralized through salt bridges. The point at which visible precipitation occurs will be influenced by such factors as ionic strength, pH, humic acid concentration, and type of metal cation.

The following diagram shows the metal ion as a 2:1 chelate complex holding two molecules together in a chain, as a 1:1 complex at the end of a chain, and in a salt-type linkage with an isolated COOH group.

Immobilization of Cu^{2+} by interaction with humic substances can occur through either the formation of insoluble complexes as described above or through solid-phase complexation to humate present as a coating on clay surfaces.

3. Mechanisms of Cu^{2+} Binding. The complexing ability of humic and fulvic acids is due largely to their content of oxygen-containing functional groups, such as COOH, phenolic OH, and ketonic C=O (Stevenson and Ardakani, 1972). Amino and imino groups may also be involved.

Structures commonly considered to be present in humic substances, and which have the potential for binding Cu, include the following:

Although generally assumed to be present in humic substances, proof has yet to be obtained for complexation at 1,2-dihydroxy or hydroxyquinone sites.

Schnitzer (1969), Gamble *et al.* (1970), and Cheam and Gamble (1974) concluded that two main types of reactions were involved in Cu^{2+}-fulvic acid interactions, the most important one involving both phenolic OH and COOH groups. A reaction of lesser importance involved less acidic COOH groups only.

The probable presence of multiple complexing sites in humic substances has been indicated by infrared (IR) and electron spin resonance (ESR) spectroscopy, as well as by measurements of stability constants (discussed later).

Infrared spectroscopy has been used in attempts to determine the nature of the reactive sites and the extent to which carboxylate linkages are covalent. The findings have shown that COOH groups, or more precisely carboxylate (COO^-), play a prominent role in the complexing of Cu by humic and fulvic acids but that OH, C=O, and NH groups may also be involved (Banerjee and Mukherjee, 1972; Vinkler *et al.*, 1976; Boyd *et al.*, 1979; Piccolo and Stevenson, 1981). Considerable controversy exists as to the extent to which COO^- linkages are covalent or ionic. The asymmetric stretching vibration of COO^- in ionic bonds occurs in the 1630-1575 cm^{-1} region; when coordinate linkages are formed, the frequency shifts to between 1650-1620 cm^{-1}. Frequency shifts with Cu-humate complexes have been variable and slight, a result which may be due in part to formation of mixed complexes. Interpretations in the 1620 cm^{-1} region are further complicated due to interference from covalent bonding with other groups (Piccolo and Stevenson, 1981).

Characterization of the binding of Cu^{2+} by humic acids using ESR has also given somewhat variable results. Lakatos *et al.* (1977) found that Cu^{2+} was bound by two ligands containing N donors and two carboxylates. In contrast, McBride (1978) concluded that only oxygen donors (COO^-) are involved; furthermore, a single bond was formed between Cu^{2+} and humic acid. Goodman and Cheshire (1973, 1976) and Cheshire *et al.* (1977) obtained evidence which indicated that Cu retained by a peat humic acid after extensive acid washing was coordinated to porphyrin groups, from which they concluded that a small fraction of the Cu in peat was strongly fixed in the form of porphyrin-type complexes and thereby unavailable to plants. Spectra obtained by Bloom and McBride (1979) for acid-washed peat failed to show the participation of groups other than COO^- in the binding of Cu^{2+}. In other work, Bresnahan *et al.* (1978) observed that the nature of the spectra for a Cu-fulvic acid system was strongly influenced by the Cu-fulvic acid ratio. At low ratios, two binding sites were shown whereas at high ratios only one peak for hydrated Cu^{2+} was indicated.

The observation that the strength of binding of Cu by humic acids decreases with an increase in the amount of Cu applied (Davies *et al.*, 1969; Goodman and Cheshire, 1973, 1976; Cheshire *et al.*, 1977) is in agreement with other observations suggesting that Cu, when present in low amounts in highly organic soils, is so tightly complexed that it cannot be taken up by crop plants. Ennis and Brogan (1961) prepared a Cu-humic acid complex by saturation with $CuSO_4$ and found that the Cu became increasingly unavailable to oat plants as more and more of the Cu was removed by chemical extraction (increasing concentrations of HCl were used).

Mention should be made that some investigators have challenged the ability of humic substances to form highly stable complexes with metal ions on the basis of steric hinderance. According to one concept, soil humic substances are spherical, cross-linked polymers which, although capable of retaining metal ions in readily exchangeable forms, are incapable of forming structures characteristic of chelate complexes. Many of the discordant results recorded in the literature may be due to variation in experimental conditions, including metal-humate and metal-fulvate ratio. Greater emphasis in the future needs to be given to complexation at low degrees of Cu saturation (binding at strongest sites).

In concluding this section, it can be said that humic and fulvic acids contain multiple binding sites for Cu. A prominent role is played by COO^-, which probably forms covalent bonds at low degrees of saturation but which may become increasingly ionic in character as binding capacity is approached.

III. STABILITY CONSTANTS OF Cu^{2+} COMPLEXES WITH HUMIC AND FULVIC ACIDS

A principle way of characterizing a metal-organic complex is through the determination of its stability or formation constant. Accurate stability constants of the complexes of Cu^{2+} with humic and fulvic acids would provide a better understanding of the behavior of Cu in soil and water systems.

Stability constants are classified according to the type of reaction they describe. The basic reaction is:

$$jM + iA \rightleftharpoons M_jA_i \qquad\qquad [1]$$

where M is the metal ion and A is the ligand.

The overall formation constant, K, is defined as:

$$K = \frac{(M_j A_i)}{(M)^j (A)^i}$$
[2]

Two general approaches have been utilized in investigations of metal complexes of humic substances. In the first case, the metal ion has been regarded as the central group (formation of MA_i complexes); in the second, the macromolecule is the central group (formation of M_jA complexes). In a modification of the second approach, over-all constants are obtained for binding at one or more classes of identical combining sites. A summary of the assumed reactions is presented in Table I.

All approaches for determining stability constants can be considered equally valid. As Sposito *et al.* (1979) pointed out, the disadvantage of treating the macromolecule as a

TABLE I. Resume of Assumed Complexation Reactions

Central group	Reaction[a]	Constant
Cu^{2+}		
1	$Cu^{2+} + HA \rightleftharpoons CuA^+ + H^+$	$b_1 = (CuA^+)(H^+)/(Cu^{2+})(HA)$
2	$CuA^+ + HA \rightleftharpoons CuA_2 + H^+$	$b_2 = (CuA_2)(H^+)/(CuA^+)(HA)$
3	$Cu^{2+} + 2HA \rightleftharpoons CuA_2 + 2H^+$	$B_2 = (CuA_2)(H^+)^2/(Cu^{2+})(HA)^2$
4	$Cu^{2+} + A^- \rightleftharpoons CuA^+$	$k_1 = (CuA^+)/(Cu^{2+})(A^-)$
5	$CuA^+ + 2A^- \rightleftharpoons CuA_2$	$k_2 = (CuA_2)/(CuA^+)(A^-)$
6	$Cu^{2+} + 2A^- \rightleftharpoons CuA_2$	$K_2 = k_1 k_2 = (CuA_2)/(Cu^{2+})(A^-)^2$
7[b]	$Cu^{2+} + A^* \rightleftharpoons CuA$	$K_{apparent} = (CuA)/(Cu^{2+})(A^*)$
Macromolecule		
8	$jCu + A \rightleftharpoons Cu_jA$	K_o^1 (*j sites, class 1*)
9	$Cu_jA + lCu \rightleftharpoons Cu_{j+l}A$	K_o^2 (*l sites, class 2*)

[a]HA = *undissociated reactive site*; A^- = *dissociated form*.
[b]A^* = *uncomplexed A (see text)*.

monofunctional ligand is that a conditional stability constant is obtained that exhibits the undesirable property of varying strongly with the extent of Cu complexation per mole of the macromolecule. When treated as a polyfunctional ligand, discrete classes of combining sites can be identified. It should be noted, however, that there is probably a continuous variation in site strength for any given class (Saar and Weber, 1979). The possibility that mixed complexes are formed has not been adequately explored.

Numerous problems are encountered in determining stability constants of metal-soil organic matter complexes. Some of the approaches are those developed for the coordination chemistry of simple molecules and they apply in only a superficial way to complex macromolecules, such as humic and fulvic acids. Humic substances from whatever source are heterogeneous with respect to molecular weight, and a pH effect will determine the degree of ionization of acidic groups and thereby the number of sites available for binding. Furthermore, several classes of binding sites may be present, in which case the site forming the most stable complex will be the first to react. The possibility also exists that humic and fulvic acids contain combining sites which are identical but which react in such a way that binding at one site may accompany changes in pH or concentration of neutral salt. Under certain circumstances, the selection of background ionic media can be critical due to formation of inorganic complexes (Sposito *et al.*, 1979).

In all methods, some determination must be made of the concentration of macromolecule or reactive site. As will be evident later, several approaches have been used for this purpose, including:

(1) Direct determination of molecular weight, such as by osmometry.
(2) Estimation of molecular weight by graphical means.
(3) Determination of reactive site concentration by potentiometric titration.
(4) Estimation of total complexing sites from metal ion retention data.

An estimate must also be made of the amount of complex formed, either by the quantity of ligand bound or Cu^{2+} bound. Approaches used for this purpose include ion selective electrode (ISE) measurements (Bresnahan *et al.*, 1978), proton release data (Stevenson, 1976, 1977), spectrophotometric titration (Blaser *et al.*, 1980), gel filtration (Mantoura and Riley, 1975), spectrofluorometry (Saar and Weber, 1980), and anodic stripping voltammetry (ASV).

Direct calculation of K from equation [2] also requires that measurements be made for A and M, the concentrations of free ligand and free metal ion, respectively. This has not always been done and values for these parameters have been taken as the difference between the total amounts present and the amounts tied up in the complex. Stability constants obtained in this way are apparent or conditional stability constants (Hartley *et al.*, 1980).

$$K' = \frac{(MA)}{(M^*)(A^*)}$$

[3]

where M^* and A^* are the uncomplexed amounts of M and A including other species not involving M or A. A stability constant calculated in this way does not yield information about the number or types of sites at which Cu^{2+} may be complexed.

Apparent or conditional stability constants obtained for Cu^{2+} complexes of fulvic acids are recorded in Table II. An increase in log K' is apparent with an increase in pH, ranging from 2.34 at pH 3.0 to 5.20 at pH 7.0.

TABLE II. Apparent or Conditional Stability Constants (log K') for Cu²⁺ Complexes of Fulvic Acids[a]

Source	pH	Salt Concentration	Log K'	Reference
Water	7.0	not given	5.20	Shuman and Woodward (1977)
Ibid.	6.5	Ibid.	5.71	Ibid.
Ibid.	6.5	Ibid.	5.14	Ibid.
Ibid.	6.5	Ibid.	4.80	Ibid.
Ibid.	6.5	Ibid.	4.50	Ibid.
Ibid.	6.0	Ibid.	4.99	Ibid.
Ibid.	5.7	Ibid.	4.87	Ibid.
Soil	5.0	0.1 M NaClO₄	4.35	Cheam and Gamble (1974)
Ibid.	3.0	Ibid.	3.22	Ibid.
Ibid.	3.0	Ibid.	2.34	Cheam (1973)

[a]*For assumed reaction see equation [3]. Most measurements at room temperature.*

When the metal ion functions as the central group, which is usually the case with complexes of small molecules, a series of species of the type MA_i are formed. Thus, the reaction between the dissociated form of a ligand (A^-) and a divalent cation (*e.g.*, Cu^{2+}) can be illustrated as follows:

$$A^- + M^{2+} \xrightleftharpoons{k_1} MA^+ \qquad A^- + MA^+ \xrightleftharpoons{k_2} MA_2 \qquad [4]$$

Two successive constants are required, one for each ligand that adds to the metal ion:

$$k_1 = \frac{(MA^+)}{(A^-)(M^{2+})} \qquad\qquad k_2 = \frac{(MA_2)}{(A^-)(MA^+)} \qquad [5]$$

The overall constant K_2 is given by:

$$K_2 = k_1 k_2 = \frac{(MA_2)}{(A^-)^2(M^{2+})} \qquad\qquad [6]$$

When the potentiometric titration method is used for determining stability constants, complex formation is regarded as a competitive reaction between H^+ and the metal ion for the reactive site (A^-) and a separate set of constants is obtained. The relationship between the constants thus obtained and those given by equations [4]-[5] will be shown later.

Investigations have also been carried out in which the macromolecule is assumed to be the central group. This approach will be described later.

A. *The Ion-Exchange Equilibrium Method*

Until recently, the ion-exchange equilibrium method was the most popular approach for determining stability constants of metal-soil organic matter complexes. Two assumptions are made, namely, the metal ion is the central group (j of equation [1] is unity), and the concentration of the metal ion in solution is negligible in comparison to the ligand. In this approach, free and bound metal ions in the equilibrium solution are estimated from the amounts of metal ion retained by a competing cation-exchange resin. If at equilibrium, A^- (initial ligand concentration) is in large excess, A^- is essentially constant and equal to A_o and log K is calculated from the equation:

$$\log \left[(\lambda_o/\lambda) - 1 \right] = \log K - i\log (A_o^-) \qquad [7]$$

where λ is the coefficient for the distribution of the metal ion between the resin and the solution phase in the presence of the ligand and λ_o is the distribution coefficient in the absence of the ligand.

From an experimental standpoint, λ is measured for several concentrations of ligand. A plot of $\log (\lambda_o/\lambda - 1)$ vs $\log A_o^-$ is subsequently prepared, yielding i as the slope and $\log K$ as the intercept. According to Clark and Turner (1969), equation [7] applies only when i is an integral number (*i.e.*, slope obtained from $\log (\lambda_o/\lambda - 1)$ vs $\log (A_o^-)$ plot is unity or a multiple thereof). This has not always been the case in studies using humic and fulvic acids.

Other errors arise from the assumption that the concentration of metal ion in solution is negligible compared to that of the ligand, and that λ_o is constant for the range of metal ion concentrations used. Modifications have been made in the ion-exchange equilibrium method in attempts to avoid the above-mentioned problems (Ardakani and Stevenson, 1972; Zunino *et al.*, 1972).

Log K values obtained by the ion-exchange equilibrium method (assumed 1:1 complexes) are recorded in Table III. Because of variations in experimental conditions, the observed differences cannot be ascribed entirely to differences in the chemical or physical properties of the humic or fulvic acids. In general, the results indicate an increase in log K with increasing pH. A similar trend was noted for log K' (see Table II).

The higher log K values with increasing pH have been attributed to increased ionization of functional groups, especially COOH, with increasing pH. Due to competition of H^+ and metal ions for binding sites, less metal ion would be expected to be bound at the lower pH.

The results given in Table III also indicate that Cu^{2+} forms stronger complexes with humic acids than with fulvic acids. Further study is required on this point.

B. *Base Titration*

This method is based on the competition of metal ions and protons for complexing sites on the ligand. The technique was modified by Gregor *et al.* (1955) for determining stability constants of metal complexes with synthetic polyelectrolytes, and it is this modified approach that has been used to

TABLE III. *Stability Constants (Log K) for Cu^{2+} Complexes of Humic and Fulvic Acids as Obtained by the Ion-Exchange Equilibrium Method (1:1 Complexes Assumed)*

Ligand	pH	Salt Concentration	Log K	Reference
Fulvic acid				
Water	7.6	0.01 M KNO_3	7.82	van den Berg and Kramer (1979)
$Soil^a$	5.0	0.1 M KCl	4.00	Schnitzer and Hansen (1970)
Ibid.	3.0	Ibid.	3.30	Ibid.
Humic acid				
Soil	5.0	not given	9.10	Rosell et al. (1977)
Ibid.	5.0	Ibid.	9.00	Ibid.
Ibid.	5.0	Ibid.	8.40	Ibid.
Ibid.	5.0	Ibid.	8.10	Ibid.
Ibid.	4.0	Ibid.	2.78	Adhikari et al. (1977)

^{a}*Earlier work of Schnitzer and Skinner (1966) had given a log K of 8.69 at pH 5.0 and 5.78 at pH 3.5.*

determine stability constants of metal-soil organic matter complexes (Stevenson, 1976, 1977).

The reactions between a weak acid polyelectrolyte (such as humic and fulvic acids) and a divalent cation can be illustrated as follows:

$$HA + M^{2+} \underset{\longleftarrow}{\overset{b_1}{\longrightarrow}} MA^+ + H^+; \qquad HA + MA^+ \underset{\longleftarrow}{\overset{b_2}{\longrightarrow}} MA_2 + H^+ \qquad [8]$$

where HA represents an acidic functional group. Two successive formation constants (b_1 and b_2) are required, one for each ligand group that adds to the complex.

The overall constant B_2 is given by:

$$B_2 = b_1 b_2 = \frac{(MA_2)(H^+)^2}{(HA)^2(M^{2+})} \qquad [9]$$

Stability constants obtained by this approach are not identical with those obtained by the ion-exchange method, which describes the interaction of a metal ion with a dissociated functional group. However, a functional relationship exists between constants obtained by the two approaches and the ionization constant, K_a, of the acidic functional group ($b_i = K_a k_i$).

Essentially, the potentiometric method requires calculations for (A^-), the concentration of free chelating species, and $\bar{n}$, the average number of ligand molecules bound per metal ion. The latter, referred to as Bjerrum's (1941) formation function, is as follows:

$$\bar{n} = \frac{(A_t) - (HA) - (A^-)}{(M_t)} = \frac{(MA^+) + 2(MA_2)}{(M^{2+}) + (MA^+) + (MA_2)} \qquad [10]$$

where A_t and M_t are the overall concentrations of ligand and metal ion, respectively. The value of $\bar{n}$ provides an estimate of the number of complexing sites bound per mole of metal ion.

For 2/1 complexes the desired constants can be obtained from the following equation by regression analysis.

$$\frac{\bar{n}}{(\bar{n} - 1)(HA/H^+)} = \frac{(2 - \bar{n})(HA/H^+)}{(\bar{n} - 1)} B_2 - b_1 \qquad [11]$$

A complication in using titration data for calculating stability constants is that titrations carried out in the presence of metal ions result in horizontal displacement of the titration curves, apparently due to release of an otherwise nontitratable H^+ from the humic acid and/or protons from hydration water of the metal bound in 1:1 complexes. To solve this problem, a procedure has been used involving sequential additions of the metal ion at constant pH (pH returned to initial starting point after each addition). The assumption is made that the metal ion reacts with the dissociated form (A^-) of the reactive group and that protons are subsequently released as ionization equilibrium is reestablished ($HA \rightleftharpoons A^- + H^+$). Variations with the total number of sites does occur and suggests the presence of polynuclear complexes (Fleck, 1966).

Overall stability constants obtained by the revised approach for the Cu^{2+} complexes of three humic acids are illustrated in Fig. 2. A pronounced increase in K_2 values is evident with a decrease in the concentration of supporting electrolyte, which is in agreement with results obtained by the ion-exchange equilibrium method. Lower apparent log K values result from competition with the electrolyte for binding sites and from a decrease in available sites due to folding of the macromolecule at high electrolyte concentration.

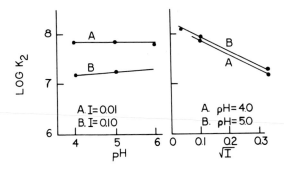

*FIGURE 2.   Variation of log K with pH and ionic strength
for a Cu^{2+}-humic acid complex.   Adapted from
Stevenson (1977).*

Log K values obtained at low salt concentrations are more
likely to represent conditions existing in the soil solution
and in natural waters than those at high salt concentrations.
     Constants obtained by potentiometric titration represent
overall stability constants and cannot be compared directly
with those determined by the ion-exchange equilibrium method.
It is of interest, however, that log k_1 values ($\sim$ log $K_2/2$)
were of the same order of magnitude as those obtained using
the ion-exchange equilibrium method (1:1 complexes assumed).
Unlike the ion-exchange equilibrium method, log K_2 values (as
well as log k_1) were uniform over the pH range 4.0 to 5.0
(see Fig. 2).
     A closely related approach was utilized by Takamatsu and
Yoshida (1978). In this case, free M^{2+} was estimated from
ion-selective electrode measurements and the number of reactive sites was taken as being equivalent to the content of
acidic functional groups as estimated by potentiometric
titration.
     The overall stability constant (log k_1k_2 or log K_2) for
a Cu^{2+}-humic acid complex as a function of pH (I = 0.1) was
described by the equation:

$$\log k_1k_2 = 8.65 + 0.65 \ (pH{-}5) \qquad\qquad [12]$$

     The value at pH 5 (8.65) is slightly higher than was
obtained by potentiometric titration (see Fig. 2). The change
with pH was also greater.
     Explanations given by Takamatsu and Yoshida (1978) for
the variation in log K_2 with pH were:

(1) Complexing at one site affects binding of Cu^{2+} at an adjacent site.

(2) In the higher pH region, Cu^{2+} is bound selectively to functional groups that form the strongest complexes but which are protonated in the lower pH region.

(3) Steric stabilization occurs through the formation of chelate rings.

A comparison of results obtained for successive and overall stability constants with Cu^{2+} as the central group is given in Table IV.

C. Scatchard Plot Method

The macromolecule is assumed to function as the central group (formation of M_jA complexes) and binding is characterized by a formation function, $\bar{v}$, defined as:

TABLE IV.   Successive and Overall Stability Constants for Cu^{2+} Complexes of Humic and Fulvic Acids (Cu^{2+} as Central Group).

Ligand	pH	Salt Concentration	Log K_i		Reference
			k_1	k_1k_2	
Fulvic acid					
Water	6.0	0.1 M NaNO$_3$	5.0	9.5	Buffle et al. (1977)
Ibid.	6.0	Ibid.	4.8	10.1	Ibid.
Humic acid					
Soil[a]	6.8	0.1 M KNO$_3$	7.8	–	Ernst et al. (1975)
Ibid.	5.0	0.1 M NaClO$_4$	–	8.65	Takamatsu and Yoshida (1978)
Ibid.	5.0, 4.0	0.1 M KCl	–	7.22	Stevenson (1977)
Ibid.	5.0	0.1 M NaNO$_3$	4.0	–	Buffle et al. (1977)[b]
Ibid.	4.0	0.1 M KClO$_4$	4.14	8.38	Fitch and Stevenson[b]

[a] A value of 16.8 for log $k_1k_2k_3$ also recorded.
[b] Unpublished observations.

$$\bar{v} = \frac{\text{molar conc. of bound metal ion}}{\text{molar conc. of polymer}} \tag{13}$$

The quantity $\bar{v}$ can be regarded as the average number of metal ions associated with each macromolecule A.

When all combining sites are identical, $\bar{v}$ is related to the overall binding constant, K_o, by the equation:

$$\bar{v} = \frac{jK_o(M^{2+})}{1 + K_o(M^{2+})} \tag{14}$$

Rearrangement gives:

$$\frac{\bar{v}}{(M^{2+})} = jK_o - K_o\bar{v} \tag{15}$$

A plot of $\bar{v}/(M^{2+})$ vs $\bar{v}$ yields K_o as the slope and jK_o as the intercept.

A variation of the formation function is θ, defined as:

$$\theta = \frac{\text{molar conc. of bound metal ion}}{\text{total number of reactive sites}} = \frac{\bar{v}}{j} \tag{16}$$

By substituting into equation [15], the following is obtained:

$$\frac{\theta}{(M^{2+})} = K_o - K_o\theta \tag{17}$$

Equation [17] can be used when the molar concentration of the macromolecule is unknown but where the number of reactive sites can be estimated, such as by potentiometric titration for total acidity or from Cu^{2+} retention data for maximum binding ability.

The Scatchard plot approach (equation [15]) has been used by Bresnahan *et al.* (1978), Saar and Weber (1979), and Sposito *et al.* (1979). A modified approach has been used by Cheam (1974) and Cheam and Gamble (1973). Application of equation [17] was made by Fitch and Stevenson (unpublished). In all studies, the extent of metal binding was ascertained from ion-selective electrode measurements. Scatchard plots give evidence for two classes of combining sites. Thus, the formation function of equation [14] is given by:

$$\bar{v} = \frac{J_1 K_1 (M^{2+})}{1 + K_1 (M^{2+})} + \frac{J_2 K_2 (M^{2+})}{1 + K_2 (M^{2+})} \qquad [18]$$

Saar and Weber (1979) attempted to separate stability constants into discrete classes based on Scatchard analysis of their data. They concluded that there was a continuous variation in site strength due to changes in fulvic acid and metal ion concentration which affected folding of the macromolecule and exposure of reactive sites. Incremental stability constants were calculated which varied with the fulvic acid to metal ion ratio. As Bresnahan *et al.* (1978) suggested, the strongest class of sites will react first, followed by the weaker ones.

Stability constants obtained by a Scatchard-type treatment of the experimental data are presented in Table V. By application of the modified approach (equation [17]), Fitch and Stevenson (unpublished) obtained a log K_1 of 6.08 and a log K_2 of 5.15 for a Cu-humate complex.

D. *Other Approaches*

Buffle *et al.* (1977) considered a system where more than one humic acid molecule binds to the metal (*e.g.*, metal ion acts as the central group). A unique feature of their approach is that a graphical method was used to estimate molecular weight. A degree of complexation function was determined ($\alpha = M/M_b$), from which the following equation was derived relating α, M_t, and A_t to the stability constant for the reaction

$$B_1^* = (MA)(H^+)^x/(M^{2x})(H_x A) \qquad [19]$$

Results obtained by Buffle *et al.* (1977) were reported as log (B_1^*/H^+). Values of 3.30, 4.00, and 8.42 were obtained at pH's of 3, 5, and 8, respectively.

Zunino and Martin (1977) described complexation in terms of a Langmuir-type adsorption equation. Values for bound ligand (A_b) were obtained from computation of the maximum binding ability (MBA). The stability constant was given by the equation:

$$K = \frac{M_b}{(M^{2+})^j (MBA - M_b)} \qquad [20]$$

TABLE V. *Stability Constants of Cu²⁺ Complexes with Humic and Fulvic Acids Using the Scatchard Plot Approach*

Ligand	pH	Supporting Electrolyte	Log K			Reference
			Class 1	Class 2	Overall	
Fulvic acid						
Water	8.0	0.01 M NaCl	8.80	8.05	8.42	Mantoura and Riley (1975)
Peat	8.0	Ibid.	8.51	7.16	7.85	Ibid.
Soil	8.0	Ibid.	—	—	3.30	Ibid.
Water (6)	8.0	Ibid.	—	—	8.42 – 9.83	Mantoura et al. (1978)
Peat	8.0	Ibid.	—	—	7.85	Ibid.
Water	6.0	0.01 M KNO₃	6.11	3.85	—	Bresnahan et al. (1978)
Soil	6.0	Ibid.	6.30	3.07	—	Ibid.
Water	5.0	Ibid.	5.95	3.70	—	Ibid.
Soil	5.0	Ibid.	6.00	4.08	—	Ibid.
Sewage	5.0	0.1 M KClO₄	3.85	2.09	—	Sposito et al. (1979)
Soil	4.0	0.1 M KNO₃	5.60	3.95	—	Bresnahan et al. (1978)
Water	4.0	Ibid.	5.48	4.00	—	Ibid.
Humic Acid						
Sediment	8.0	0.01 M NaCl	—	—	9.91 – 11.37	Mantoura et al. (1978)[a]
Peat	8.0	Ibid.	—	—	8.29	Ibid.

[a]Values represent range for four samples.

An apparent overall log K of -0.254 was obtained for the Cu^{2+} complex of a fungal humic acid-like polymer.

E. *Practical Implications*

The strong binding of Cu^{2+} (expecially when present in low amounts) has been verified. The observation that humic acids bind Cu more tightly than fulvic acids, and that stability constants of Cu-humate complexes increase with an increase in degree of humification, has implications regarding the mechanisms whereby Cu is maintained in soluble forms in soil. A sequence is envisioned in which Cu present in low concentrations is immobilized by complexation to humic acids. However, as the stronger binding sites become saturated, an increasing amount of the Cu will be solubilized through the action of fulvic acids and simple biochemical compounds. Overall stability constants of Cu-humate complexes appear to be of the order of those recorded for biochemical compounds.

IV. SUMMARY

Organic substances of various types play a prominent role in the binding of Cu in soils. A key role is played by humic and fulvic acids, which probably form highly stable complexes when Cu is present in trace amounts (*e.g.*, Cu deficient soils). The formation of Cu-organic matter complexes would have the following effects in soil.

(1) The concentration of Cu in the soil solution can be lowered through complexation to clay-humus or to the formation of insoluble complexes with humic acids. Soluble ligands may be of considerable importance in transforming solid phase forms of Cu into soluble forms.

(2) For conditions under which Cu tends to precipitate (*e.g.*, in calcareous soils), complexation will serve to maintain Cu in soluble forms.

(3) When excess Cu is present, complexation may reduce the concentration of ionic Cu^{2+} to a nontoxic level.

(4) Natural complexing agents may be involved in the migration and transport of Cu. An interesting ecological relationship has been described by Fraser (1961), who attributed the accumulation of toxic amounts of Cu in a forest peat to a series of events that included uptake from the surrounding soil by plant roots, translocation to leaves, incorporation

into humus of the forest floor following leaf fall, and migration to the swamp in seepage waters as soluble organic complexes.

REFERENCES

Adhikari, M., Chakrabarti, G., and Hazra, G. (1977). *Agrochim.* *21*, 134.

Ardakani, M.S., and Stevenson, F. J. (1972). *Soil Sci. Soc. Amer. Proc. 36*, 884.

Banerjee, S. K., and Mukherjee, S. K. (1972). *J. Indian Soc. Soil Sci. 20*, 91.

Batley, G. E., and Florence, T. M. (1976). *Marine Chem. 4*, 347.

Batley, G. E., and Gardner, D. (1978). *Estuarine and Coastal Marine Sci. 7*, 59.

Bjerrum, J. (1941). "Metal Amine Formation in Aqueous Solutions". P. Haase and Sons, Copenhagen.

Blaser, P., Fluhler, H., and Polomski, J. (1980). *Soil Sci. Soc. Amer. J. 44*, 709.

Bloom, P. R., and McBride, M. B. (1979). *Soil Sci. Soc. Amer. J. 43*, 687.

Blutstein, H., and Smith, J. D. (1978). *Water Res. 12*, 119.

Boggess, W. R. (1977). "Lead in the Environment". National Science Foundation Special Publication NSF/RA-770214, Washington, D.C.

Boyd, S. A., Sommers, L. E., and Nelson, D. W. (1979). *Soil Sci. Soc. Amer. J. 43*, 893.

Bresnahan, W. T., Grant, C. L., and Weber, J. H. (1978). *Anal. Chem. 50*, 1675.

Brezonik, P. L., Bauner, P. A., and Stumm, W. (1976). *Water Res. 10*, 605.

Buffle, J., Greter, F. L., and Haerdi, W. (1977). *Anal. Chem. 49*, 216

Cheam, V. (1973). *Can. J. Soil Sci. 53*, 377.

Cheam, V., and Gamble, D. S. (1974). *Can. J. Soil Sci. 54*, 413.

Cheshire, M. V., Berrow, M. L., Goodman, B. A., and Mundie, C. M. (1977). *Geochim. et Cosmochim. Acta 41*, 1131.

Clark, J. S. and Turner, R. C. (1969). *Soil Sci. 107*, 8.

Crosser, M. L., and Allen, H. E. (1977). *Soil Sci. 123*, 176.

Davies, R. I., Cheshire, M. V., and Graham-Bryce, I. J. (1969). *J. Soil Sci. 20*, 65.

Ennis, M. T., and Brogen, J. C. (1961). *Irish J. Agr. Res. 1*, 35.

Ernst, R., Allen, H. E., and Mancy, K. H. (1975). *Water Res. 9*, 969.

Fleck, G. M. (1966). "Equilibrium in Solution". Holt, Reinhart and Winston, New York.

Florence, T. M. (1977). *Water Res. 11*, 681.

Florence, T. M., and Batley, G. E. (1977). *Talanta 24*, 151.

Fraser, D. C. (1961). *Econ. Geol. 56*, 1163.

Gamble, D. S., Schnitzer, M., and Hoffman, I. (1970). *Can. J. Chem. 48*, 3197.

Gardiner, J., and Stiff, M. J. (1975). *Water Res. 9*, 517.

Geering, H. R., and Hodgson, J. F. (1969). *Soil Sci. Soc. Amer. Proc. 33*, 54.

Goodman, V. A., and Cheshire, M. V. (1973). *Nature, New Biology 244*, 158.

Goodman, V. A., and Cheshire, M. V. (1976). *J. Soil Sci. 27*, 337.

Gregor, H. P., Luttinger, L. B., and Loebl, E. M. (1955). *J. Phys. Chem. 59*, 34.

Greter, F. L., Buffle, J., and Haerdi, W. (1979). *J. Electroanal. Chem. 101*, 211.

Guy, R. D., and Chakrabarti, C. L. (1976). *Can. J. Chem. 54*, 2600.

Guy, R. D., Chakrabarti, C. L., and Schramm, L. L. (1975). *Can. J. Chem. 53*, 661.

Hartley, F. R., Burgess, C., and Alcock, R. M. (1980). "Solution Equilibria". Ellis Harwood, Ltd., London.

Hodgson, J. F., Geering, H. R., and Norvell, W. A. (1965). *Soil Sci. Soc. Amer. Proc. 29*, 665.

Hodgson, J. F., Lindsay, W. F., and Trierweiler, J. F. (1966). *Soil Sci. Soc. Amer. Proc. 30*, 723.

Jackson, J. A. (1975). *Soil Sci. 119*, 56.

Jackson, K. S., Jonasson, I. R., and Skippen, G. B. (1978). *Earth-Sci. Rev. 14*, 97.

Lakatos, B., Tibai, T., and Meisel, J. (1977). *Geoderma 19*, 319.

Lees, H. (1950). *Biochem. J. 46*, 450.

Mantoura, R. F. C., and Riley, J. P. (1975). *Anal. Chim. Acta. 78*, 193.

Mantoura, R. F. C., Dickson, A., and Riley, J. P. (1978). *Estuarine Coastal Mar. Sci. 6*, 387.

Matsuda, K., and Ito, S. (1970). *Soil Sci. Plant Nutr. (Japan) 16*, 1.

McBride, M. B. (1978). *Soil Sci. 126*, 200.

McCalla, T. M., Peterson, J. R., and Lue-Hing, C. (1977). In "Soils for Management of Organic Wastes and Waste Waters" (L. F. Elliott and F. J. Stevenson, eds.), p. 11. American Society of Agronomy, Madison, Wis.

McClaren, R. G., and Crawford, D. V. (1973). *J. Soil Sci. 24*, 172.

O'Shea, T. A., and Mancy, K. H. (1976). *Anal. Chem. 48*, 1603.

Piccolo, A., and Stevenson, F. J. (1981). *Geoderma* (in press).

Rashid, M. A. (1971). *Soil Sci. 111*, 298.

Reuter, J. H., and Perdue, E. M. (1977). *Geochim. et Cosmochim. Acta 41*, 325.

Rosell, R. A., Miglierina, A. M., and DeNovilla, L. Q. (1977). In "Soil Organic Matter Studies", p. 15. FAO/IAEA, Vienna.

Saar, R. A., and Weber, J. H. (1979). *Can. J. Chem. 57*, 1263.

Saar, R. A., and Weber, J. H. (1980). *Anal. Chem. 52*, 2095.

Schnitzer, M. (1969). *Trans. 9th. Intern. Congr. Soil Sci. 1*, 635.

Schnitzer, M., and Hansen, E. H. (1970). *Soil Sci. 109*, 333.

Schnitzer, M., and Skinner, S. I. M. (1966). *Soil Sci. 102*, 361.

Shuman, M.S., and Woodward, G. P. (1977). *Environ. Sci. Tech. 11*, 809.

Sposito, G., Holtzclaw, K. M., and Levesque-Madore, C. S. (1979). *Soil Sci. Soc. Amer. J. 43*, 1148.

Stevenson, F. J. (1972). *J. Environ. Qual. 1*, 333.

Stevenson, F. J. (1976). *Soil Sci. Soc. Amer. J. 40*, 665.

Stevenson, F. J. (1977). *Soil Sci. 123*, 10.

Stevenson, F. J. (1979). *Soil Biol. Biochem. 11*, 493.

Stevenson, F. J., and Ardakani, M. S. (1972). In "Micro-nutrients in Agriculture" (J. J. Mortvedt *et al.*, eds.), p. 79, American Society of Agronomy, Madison, Wis.

Takamatsu, T., and Yoshida, T. (1978). *Soil Sci. 125*, 377.

Tessler, A., Campbell, P. G. C., and Bissom, M. (1979). *Anal. Chem. 51*, 844.

van den Berg, C. M. G., and Kramer, J. R. (1979). *Analytica Chemica Acta 106*, 113.

van Dijk, H. (1971). *Geoderma 5*, 63.

van Holde, K. E. (1971). "Physical Biochemistry". Prentice-Hall, Englewood Cliffs, New Jersey.

Vinkler, V., Lakatos, B., and Meisel, J. (1976). *Geoderma 15*, 231.

Zunino, H., Pierano, M., Aguilera, M., and Iscobar, I. (1972). *Soil Sci. 114*, 414.

Zunino, H., and Martin, J. P. (1977). *Soil Sci. 123*, 188.

5

BEHAVIOUR OF Cu ADDITIVES — FERTILISERS

R.J. Gilkes

Department of Soil Science and Plant Nutrition
University of Western Australia
Nedlands, Western Australia

I. INTRODUCTION

A wide range of copper sources have been evaluated for
their effectiveness as fertilisers when applied to soils or
foliage. Early research by workers in Western Australia
showed that bluestone ($CuSO_4 \cdot 5H_2O$, 25% Cu), oxidised copper
ore (18% Cu) and pyrite roaster residues (2.2% Cu) were equally effective sources for oats (Teakle *et al.* 1941). Table I
lists some of the many Cu compounds that have been investigated since this time and found to be effective fertilisers.
The most commonly used compound appears to be $CuSO_4 \cdot 5H_2O$
which Teakle (1939) reports farmers having used for oats in
Western Australia as long ago as 1879.
Little is known of the chemical behaviour of these compounds after their application to the soil. The water-
soluble compounds presumably quickly enter soil solution and
the subsequent availability of Cu will depend mainly on soil
and plant factors. The behaviour of poorly soluble compounds
will in addition depend on intrinsic properties of the fertiliser such as the solubility of constituents, particle size
and crystallinity. Soil properties (pH, texture, etc.) also
influence dissolution rate. For example, malachite (a basic
copper carbonate) might be expected to be a more effective
source in acid soils where it dissolves rapidly, than in
calcareous soils where dissolution would be slow. Thus a
knowledge of the chemical and physical properties of Cu
compounds will aid their evaluation as fertilisers. This
review presents a summary of published data on the properties
of Cu fertilisers and their evaluation using chemical and

biological tests.

TABLE I. *Cu Compounds Used as Fertilisers*

Source	Formula	%Cu	H_2O solubility
Cu metal	Cu	100	Insoluble
Cuprite	Cu_2O	89	Insoluble
Tenorite	CuO	75	Insoluble
Covellite	CuS	66	Insoluble
Chalcocite	Cu_2S	80	Insoluble
Chalcopyrite	$CuFeS_2$	35	Insoluble
Malachite	$CuCO_3 \cdot Cu(OH)_2$	57	Insoluble
Azurite	$2CuCO_3 \cdot Cu(OH)_2$	55	Insoluble
Chalcanthite	$CuSO_4 \cdot 5H_2O$	25	Soluble
Copper sulfate monohydrate	$CuSO_4 \cdot H_2O$	35	Soluble
Basic copper sulfates	$CuSO_4 \cdot 3Cu(OH)_2$ (general formula)	13-53	Insoluble
Copper nitrate	$Cu(NO_3)_2 \cdot 3H_2O$		Soluble
Copper acetate	$Cu(C_2H_3O_2)_2 \cdot H_2O$	32	Slightly
Copper oxalate	$CuC_2O_4 \cdot 0.5H_2O$	40	Insoluble
Copper oxychloride	$CuCl_2 \cdot 2CuO \cdot 4H_2O$	52	Insoluble
Copper ammonium phosphate	$Cu(NH_4)PO_4 \cdot H_2O$	32	Insoluble
Copper chelates	Na_2Cu EDTA	13	Soluble
Copper chelates	$NaCu$ HEDTA	9	Soluble
Copper polyflavanoids	--	5-7	Soluble
Copper-sulfur frits	--	varies	Varies
Copper-glass fusions	--	varies	Varies

II. THE PROPERTIES OF Cu FERTILISERS

In many countries the properties of macronutrient fertilisers are subject to stringent regulations governing both
nutrient contents and the analytical methods to be used for
their measurement. For example, regulations may require the
phosphate content of fertilisers to be separately reported as
water-soluble, citrate-soluble and citrate-insoluble fractions
(AOAC 1970). Strict limits are specified for the variations
permitted in the contents of these fractions and penalties
specified for non-compliance with registered compositions. In
marked contrast the Cu content of fertilisers is normally
defined solely in terms of the total Cu concentration with no
specifications relating to solubility. The nominal total Cu

concentration of the fertiliser is registered with the appropriate statutory authority which usually enforces regulations by a system of spot checks and analyses of samples. Labelling and packaging of fertilisers is also frequently subject to regulation. The absence of detailed specifications relating to the solubility of Cu fertilisers results from the lack of information as to whether complete, medium or minimal solubility of micronutrients is necessary for best agronomic performance. Silverberg *et al.* (1972) consider that micronutrients in granular compound fertilisers should be at least partly water-soluble as are the common macronutrient carriers, but this view has yet to be substantiated by plant trials.

Although Cu fertilisers are used in many countries, statistical data on the types and amounts of Cu fertilisers produced are not generally available. Cunningham (1972) has provided figures on the use of Cu fertilisers in various regions of the USA between 1967 and 1970. Most Cu was used on the sandy and organic soils of the South Atlantic region with the total national consumption decreasing from 2169t to 781t Cu p.a. during this period. Much of this Cu was probably incorporated into granular, fluid and suspension compound fertilisers. In Australia most Cu is applied to podzolised and calcareous soils in South Australia and Western Australia (Gartrell 1980) although significant amounts are also used in Victoria, Queensland and Tasmania. Between 1973 and 1978 annual Australian consumption of fertiliser Cu varied widely between extremes of 440 and 1500t Cu p.a., the great variation being in response to economic factors including the amount of new land cleared for agriculture. Approximately equal amounts were supplied to fertiliser manufacturers as $CuSO_4 \cdot 5H_2O$ and CuO during this period with most Cu being granulated with OSP (ordinary or single superphosphate) together with other micronutrients to produce a range of fertilisers containing 0.2-2.0% Cu depending on local requirements. The practice of incorporating Cu into granular macronutrient fertilisers as well as into fluid and suspension fertilisers provides an opportunity for undesirable chemical reactions which are discussed in the following sections.

A. *Dry Blended and Coated Fertilisers*

Most copper compounds may be dry-blended with various mixtures of finished granular fertilisers to provide an almost infinite range of formulations for specific agronomic requirements. Segregation of constituents may occur during handling leading to non-uniform application in the field

(Silverberg *et al.* 1972). Although segregation can be avoided by careful matching of the particle sizes of all constituents, these fertilisers will still give rise to a sparse distribution of Cu in the field due to the large particle size of the copper source.

Coated fertilisers have been developed to eliminate segregation and promote more uniform fertilisation with micronutrients. The finely-ground micronutrient is coated onto macronutrient granules using oils, waxes, water or fertiliser solutions as binders. There is little published information on the properties of Cu fertilisers prepared in this way.

B. Granulated Compound Fertilisers

Solid macronutrient fertilisers are generally produced as uniform granular materials to aid handling and application to soils. Micronutrients may be conveniently added to solids (e.g. phosphate rock) or solutions (e.g. sulfuric acid) prior to reaction or they may be added after reaction to recycled material entering the granulator. Segregation of micronutrients is avoided by these procedures. Representative analytical data are shown in Figures 1 and 2 for a West Australian Cu-OSP (copper-ordinary superphosphate) containing approximately 0.7% Cu (Gilkes and Sadleir 1981). Copper was added as coarsely powdered $CuSO_4 \cdot 5H_2O$ to recycle fines entering a rotary drum granulator.

The quite uniform distribution of Cu between the various granule size fractions is similar to that for P and demonstrates the efficacy of this method of micronutrient addition. Clearly there is no marked accumulation of Cu in a single size fraction apart from a relative concentration in the minor very fine (<0.25mm) size fraction. The low water solubility of Cu in this fertiliser is due to the crystallisation of insoluble copper phosphates. In marked contrast, insoluble ZnO added to OSP alters to a soluble Zn phosphate (Gilkes 1975).

The Cu concentration in individual fertiliser granules is not constant and exhibits a much greater range than is observed for P and the other major constituents. However, about 90% of all granules contain more than half the average Cu concentration so that non-uniform application to soils is not likely to result from this degree of variation. A similar degree of homogeneity should occur for most Cu-macronutrient fertilisers manufactured in this way.

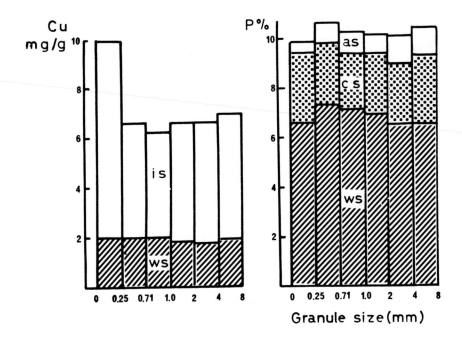

FIGURE 1. *Concentrations of copper and phosphorus in different size granules of a sample of commercially-produced CuOSP. For Cu; ws = water soluble Cu, is = Cu insoluble in water. For P; ws = water soluble P, cs = P soluble in neutral ammonium citrate, as = acid soluble P. Most granules were 1-4 mm in diameter (unpublished data of author).*

C. Fluid and Suspension Fertilisers

Segregation and non-uniform application of Cu can be eliminated by use of clear fluid and suspension fertilisers which may be applied to either soils or foliage. Sufficient Cu for normal rates of application may be dissolved in solutions of NH_3-NH_4NO_3, NH_4NO_3-H_2O, NH_4NO_3-urea, ammonium orthophosphate and ammonium polyphosphate (Silverberg et al. 1972). Copper compounds can also be applied as either solid or solution additives to NPK suspension fertilisers which may contain small amounts of clay (attapulgite or smectite) to prevent settling-out of solids. These fertilisers require the use of finely divided powders and providing that good agitation and mixing are maintained, their application to soils should be quite uniform.

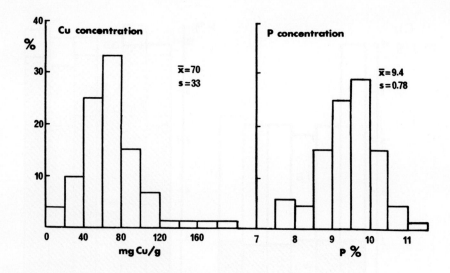

FIGURE 2. *The distributions of Cu and P concentrations for a sample of 72 individual 2.8-4 mm granules of CuOSP. The mean concentrations and standard deviations are shown, they demonstrate the relatively much greater variability of Cu concentration than of P concentration (unpublished data of author).*

III. CHEMICAL REACTIONS BETWEEN FERTILISER CONSTITUENTS

Most of the Cu compounds listed in Table I have been investigated as possible additives to solid, suspension and liquid macronutrient fertilisers. It would appear from the published literature that $CuSO_4 \cdot 5H_2O$ is the most commonly used additive in commercial fertilisers.

Andersen (1969) systematically evaluated the agronomic effectiveness of mixtures of compound fertiliser (N:P:K, 15:4:12) and 0.3% Cu as $CuSO_4.5H_2O$, $CuNO_3$, Cu_2O, CuO, Cu_2O/CuO and fine Cu metal. Using Cu uptake as a criterion he found all the mixtures to be about half as effective for oats and clover as $CuSO_4 \cdot 5H_2O$ plus NPK fertiliser applied separately (Figure 3). This result might be explained by the low water-solubility of Cu in all the mixtures (Table II) and indicates that undesirable chemical reactions had occurred. It is important to note, however, that other workers using similar experimental procedures have been unable to demonstrate a decrease in agronomic effectiveness due to chemical reactions between Cu and macronutrients in fertilisers

TABLE II. *Solubility of Cu Fertilisers and Reaction Products*

Fertiliser	Copper Content	Solubility %
[1]		
Copper sulfate, $CuSO_4 \cdot 5H_2O$	27	100
Copper oxide, CuO	75	0.01
Cu chelate, $Na_2CuEDTA$	14	100
$Cu_{5.8}H_{3.5}(PO_4)_5 \cdot 4.2H_2O(H)$ [a]	39	0.06
$Cu_{1.0}Ca_{0.1}H_{0.8}PO_4 \cdot 1.6H_2O(D)$ [a]	33	0.02
$Cu_{1.2}Ca_{0.2}H_{0.2}PO_4 \cdot H_2O(F)$ [a]	39	0.6
[2] $CuOSP$	0.74	42
[3]		
$NPK-CuSO_4$	0.29	3.4
$NPK-Cu(NO_3)_2$	0.31	1.0
$NPK-CuO-Cu_2O$	0.29	3.0
$NPK-Cu_2O$	0.29	2.0
$NPK-CuO$	0.31	2.3
$NPK-Cu$	0.30	2.0

[a] *Compounds H,D and F are referred to in Figure 7.*

[1] *Gilkes and Lim-Nunez (1979)*
[2] *Gilkes and Sadleir (1978)*
[3] *Andersen (1969)*

(Mortvedt 1965, Gilkes and Lim-Nunez 1979). Thus the nature of Cu reaction products in compound fertilisers and their agronomic evaluation remains unresolved and deserves further attention.

Depending on the composition of the macronutrient fertiliser, a variety of chemical reactions may occur on addition of Cu and these reactions may result in the crystallisation of complex and often insoluble Cu compounds. The reactions cannot be predicted from a knowledge of solubility products of constituents since even the simplest fertilisers contain several major phases and a multitude of minor constituents. Compound Cu-NPK fertilisers provide the potential for crystallisation of a large range of Cu compounds. The types of chemical reaction mechanisms involving micronutrients that may operate during manufacture of compound fertilisers have been reviewed by Lehr (1972). Their application to Cu fertilisers is discussed in the following sections with reference to observed reactions and reference is made to the properties of reaction products.

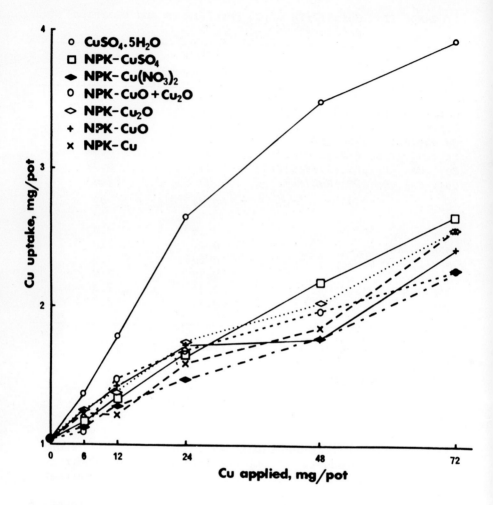

FIGURE 3.   *Total Cu uptake by one crop of oats and 6 crops of clover from soil fertilised with $CuSO_4 \cdot 5H_2O$ plus a separate NPK application and with various Cu compounds that had been first mixed with NPK fertiliser. (Adapted from Andersen 1969).*

1. *Metathetical Reactions.*   In these reactions the added copper compound and a fertiliser salt recombine to form stable salts, e.g.

$$CuSO_4 + (NH_4)_4P_2O_7 + H_2O \rightarrow Cu(NH_4)_2P_2O_7 \cdot H_2O + (NH_4)_2SO_4$$

(soluble)   (insoluble)

This particular reaction may occur on dilution of neutral to acidic APP (ammonium polyphosphate) fluids containing Cu. A similar reaction between copper salts and ammonium ortho-phosphate produces the insoluble compound $CuNH_4PO_4 \cdot H_2O$ (Bridger *et al.* 1962). Compounds of approximate composition $CuNH_4PO_4 \cdot nH_2O$ and $CuHPO_4 \cdot nH_2O$ have been reported to form on addition of Cu to Russian ammonium phosphate fertilisers (Mukhamedzharov *et al.* 1970).

When Cu salts are added to freshly made OSP they rapidly dissolve due to the high residual acidity and moisture content of the OSP. Subsequently during curing a variety of Cu phosphates may crystallize within granules as void-fillings or coatings. Some of these compounds have crystal structures resembling that of the rare mineral $NaCaCu_5(PO_4)_4Cl \cdot 5H_2O$, sampleite (Guillemin 1956), but exhibit a range of chemical compositions that may include acid hydrogen ions (for example $NaCu_6H_3(PO_4)_5Cl \cdot 6H_2O$). Another acid copper phosphate compound $(Cu_{1.0}Ca_{0.1}H_{0.8}PO_4 \cdot 1.6H_2O)$ has been observed in many specimens of CuOSP; it has a quite different and as yet undetermined crystal structure (Gilkes and Lim-Nunez 1979).

Clear evidence for the crystallization of Cu phosphates in voids during curing of CuOSP as well as confirmation of the chemical composition of these compounds is provided by electronmicroprobe analysis (Figure 4 taken from Gilkes 1977b). Such direct evidence on whole untreated granules is particularly valuable where exhaustive washing or chemical extraction is necessary to isolate compounds for conventional chemical analysis. These treatments can alter the composition of Cu compounds through absorption or incongruent dissolution.

2. Crystallization of Addition Products. These reactions occur between a copper additive and an ionic component of the aqueous phase of the compound fertiliser. Under neutral to alkaline conditions the following reactions occur:

$$CuSO_4 + 6(OH) \rightarrow Cu_4SO_4(OH)_6 \text{ or } Cu_3SO_4(OH)_4$$

(soluble)   (insoluble)

The insoluble compounds $Cu_2NO_3(OH)_3$ and $Cu_2PO_4(OH) \cdot H_2O$ are among other basic copper salts that may form in non-acidic fertilisers. The agronomic effectiveness of these compounds has not been determined.

Double salts may crystallize in fertilisers if they are less soluble than the constituent simple salts. However, no examples of Cu-containing double salts have been isolated from fertilisers.

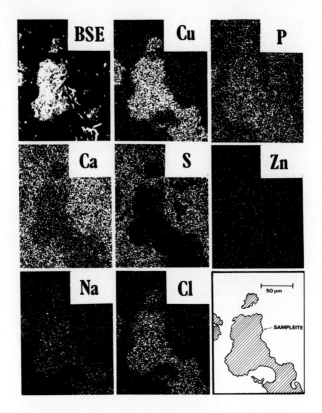

FIGURE 4. *Electron microprobe analyzer x-ray images of
an inclusion of a copper phosphate in CuOSP, the inclusion
being surrounded by the MCP (monocalcium phosphate) and
anhydrite matrix. The compound has an approximate composition
$NaCu_6H_3(PO_4)_5Cl \cdot 6H_2O$ and resembles the mineral sampleite.
BSE denotes back-scattered electron image. (Adapted from Gilkes
(1977b)).*

3. *Formation of Adducts with Urea.* The precipitation of
a soluble Cu-containing adduct of urea occurs when Cu_2O is
added to fertilisers containing APP (ammonium polyphosphate),
NH_4NO_3, $CO(NH_2)_2$ and KCl. Although it is soluble in water this
compound $|Cu(NH_4)_6(P_2O_7)_2 \cdot 2CO(NH_4)_2 \cdot 2H_2O|$ is an undesirable
constituent of fluid fertilisers since it causes segration
of Cu and handling difficulties.

4. *Formation of Ionic and Molecular Complexes.* Micronutrient metals may form a variety of ionic complexes in compound fertilisers including orthophosphates, pyrophosphates,

tripolyphosphates, hydrated metal complexes and ammine complexes. Furthermore, Cu may form soluble molecular chelates with various organic acids including EDTA (Ethylenediaminetetraacetic acid) and EDDHA (Ethylenediaminedi (O-hydroxyphenylacetic acid)) which are sometimes incorporated into fluid fertilisers.

Soluble ionic complexes in fluid fertilisers may only be in quasi-equilibrium, so crystal nucleation of equilibrium solid phases may be induced by mechanical or thermal disturbance during storage and handling. These solids may include the insolubles e.g. $(NH_4)_6Cu(P_2O_7)_2 \cdot 6H_2O$, $Cu_5(P_3O_{10})_2 \cdot 13H_2O$ and $Na_6Cu(P_2O_7)_2 \cdot 16H_2O$ (Van Wazer 1958) which give rise to undesirable segregation of fertiliser Cu.

Where several divalent micronutrients (e.g. Cu, Fe, Mn, Zn) are simultaneously added to fluid fertilisers, the Cu may isomorphously substitute in the various reaction products of the other metals. Lehr (1972) considers that such substitution is most likely to occur in the metal ammonium pyrophosphates and metal ammonium orthophosphates that crystallize from fluid fertilisers. Most phosphate ores contribute Fe, Al, F, Na and other impurity elements to OSP, CSP (concentrated superphosphate) and to the phosphoric acid used to manufacture compound fertilisers. These elements may promote the crystallization of insoluble phosphates (Lehr *et al.* 1967, White 1976, Gilkes and Lim-Nunez 1980) which could incorporate Cu as an isomorphous substituent.

Where other macronutrient cations (K, Mg) or lime are added to NP fertilisers the range of possible reaction products is increased. In NPK fertilisers the insoluble compounds $Cu_3K_4(P_2O_7)_2(OH)_2 \cdot 4H_2O$ and $CuK_2P_2O_7 \cdot 2H_2O$ may precipitate.

No firm evidence is available for the agronomic value of the various compounds described above, but they are believed to be effective sources of Cu (Mortvedt 1965).

5. Dehydration and Oxidation. Other possible reactions in fertilisers that may involve Cu include hydration-dehydration and oxidation-reduction of Cu compounds. Addition of more than 2% Cu as copper sulfate (i.e. 10% $CuSO_4 \cdot 5H_2O$) to OSP may adversely affect granulation due to the high water content of this salt. Oxidation of metallic Cu and Cu_2O to soluble divalent Cu may occur by reaction with H^+ in acidic fertilisers or by reaction with NO_3^-.

IV. IDENTIFICATION OF REACTION PRODUCTS

A wide range of modern analytical techniques has been
employed to identify Cu reaction products in fertilisers.
Most techniques require the compounds to be isolated from the
major fertiliser constituents before analysis. Separation of
insoluble Cu compounds can often be achieved by dissolution of
soluble fertiliser constituents in water followed by heavy
liquid separation, magnetic separation and hand-picking under
a binocular microscope. Once sufficient pure material has
been isolated (a few milligrams), it can be identified by
optical, chemical, IR, DTA, TGA, XRD and electron-optical
measurements. The strongest x-ray diffraction lines for most
of the Cu compounds known to occur in fertilisers and others
that may occur are listed in the JCPDS (1979) Alphabetical
Index to the Powder Diffraction File. Further diagnostic data
for some of these compounds may be found in the references
cited in the text.

V. BEHAVIOUR OF Cu FERTILISERS IN SOILS

The general behaviour of Cu in soils is examined elsewhere
in this monograph (McBride, Stevenson & Fitch, this symposium),
but the release of Cu from granular fertilisers to soil solution depends partly on intrinsic properties of the fertiliser
and is therefore discussed here. Fertiliser granules applied
to soils experience changed environmental conditions due to
soil pH, water content and reactive soil constituents. This
change in chemical environment may induce further chemical
reactions within granules.

A. *Simple Dissolution*

Water soluble Cu compounds ($CuSO_4 \cdot 5H_2O$, CuEDTA, etc.)
quickly dissolve after application to wet soils. Dissolution
of poorly soluble forms (CuO, Cu-frits, etc.) is slower and
probably depends on such factors as soil pH, rainfall and
soil texture as well as the particle size of the copper compound. Finely powdered Cu fertilisers will generally be most
effective since they present a large surface area for dissolution as well as enabling a more uniform application to soils.
Dissolution of Cu compounds which are constituents of
macronutrient fertilisers may be controlled by the rate of

dissolution of the macronutrient matrix. The dissolution of
Cu in CuOSP granules in moist soils has been studied in laboratory, glasshouse and field experiments (Figure 5) (Gilkes
1977a, Gilkes and Sadleir 1979). Superphosphate granules
rapidly (within one day) become wet as a consequence of hygro-
scopic and capillary uptake of water from the soil by MCP
(monocalcium phosphate). An acid solution (pH 1.8) develops
within the granule due to incongruent dissolution of MCP and
this solution flows into the soil under capillary forces
leaving a residual or relic granule consisting mainly of
anhydrite (CaSO), with lesser DCP (dicalcium phosphate) and
impurities (Williams 1971, Gilkes 1975).

Although most Zn is lost from CuZnOSP granules during the
rapid initial reaction only about one half of the Cu dissolves
and flows from the granule with the acid phosphate solution.
The remaining Cu in the relic granule is mainly present as
insoluble Cu phosphates and its dissolution will depend to a
large extent on the dissolution rate of the poorly soluble
anhydrite matrix. Thus in low rainfall regions relic granules
containing copper phosphates may be recovered from sandy soils
one year or more after application (Gilkes 1975, 1977b).
When rainfall promotes leaching more rapid dissolution of the
relic granule occurs and a higher proportion of Cu is released
although this may only be in the form of particles of insoluble copper phosphates freed by dissolution of surrounding
matrix. Some NPK fertilisers also leave relic granules which
might be expected to restrict dissolution of included copper
compounds (Pocharoff *et al.* 1967). In contrast to OSP there
is little insoluble residue left at the granule site after
dissolution of MAP and DAP fertilisers. However, the higher
pH of the saturated solutions of these compounds (pH4 and 8
respectively) may be less effective in dissolving included
Cu compounds than the very acid solution (pH 1.8) that develops in OSP granules.

The Cu that dissolves and flows out of a fertiliser
granule into the soil is probably retained in the closely
adjacent soil since several workers have demonstrated the
immobility of Cu in soils. Radioactively-labelled Cu applied
to the soil surface both with and without OSP was leached to
a maximum depth of 5 cm in a light textured soil by 46 cm of
water (Jones and Belling 1967). The data in Figure 6 show
that despite being leached with 90 cm water, the Cu in a
surface application of finely ground Cu OSP was retained within 5 mm of the surface of a sandy (8% clay) soil (Gilkes *et
al.* 1975). Clearly soil constituents exhibit a strong affinity for Cu including that Cu supplied by fertilisers.

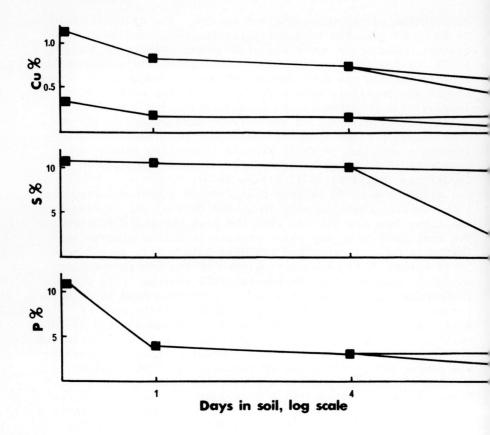

FIGURE 5. *Losses of Cu, S and P by diffusion from CuOSP fertiliser granules in Boyanup soil at field capacity and during leaching with 1 cm H_2O/day. Quantities are expressed as percentages of the initial granule weight.*
■ *coincident data points,* ☐ *diffusion,* ● *leaching.*
The two curves for Cu are for high (1.2% Cu) and low (0.35%Cu) analysis Cu fertilisers. (Adapted from Gilkes 1977a).

B. Incongruent Dissolution

Copper compounds in fertilisers may dissolve incongruently as the fertiliser is wetted by soil solution. The crystalline residues of incongruent dissolution may be poorly soluble and potentially unavailable to plants. For example the compound $Cu(NH_4)_6(P_2O_7)_2 \cdot 2CO(NH_4)_2 \cdot 2H_2O$ dissolves incongruently to leave an insoluble residue of $Cu(NH_4)_2P_2O_7 \cdot H_2O$.

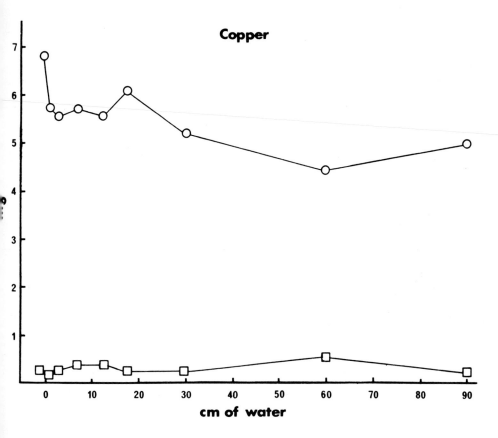

FIGURE 6. Amounts of Cu(mg) remaining in the surface
0-5mm○ and 5-10mm□ depth levels of soil columns after
various amounts of leaching. The Cu was applied as a uniform
layer of finely ground CuOSP. (Adapted from Gilkes et al.1975)

C. Reactions with Other Constituents in Banded Fertilisers

In a laboratory experiment Fiskell *et al.* (1967) observed
that Cu was less mobile when mixed with DAP, AOP (ammoniated
orthophosphate) and OSP. They also showed in field experiments that the agronomic effectiveness of Cu was reduced when
applied in a band with these macronutrient fertilisers. Insoluble copper compounds were believed to have formed in the
band and may have been responsible for this reduction.
Evaluation of CuO and $CuSO_4 \cdot 5H_2O$ mixed with 0:4:20 (NPK)
fertiliser using wheat in a glasshouse experiment showed that

mixing did not reduce the agronomic effectiveness of these Cu
sources, thus indicating that undesirable chemical reactions
did not occur in this case (Kretschmer and Forsee 1954).
Many of the reactions between Cu and fertiliser constituents
that occur during manufacture could also occur when fertilisers are applied in bands.

VI. AGRONOMIC EFFECTIVENESS OF Cu FERTILISERS AND
    REACTION PRODUCTS

The fertiliser effectiveness of CuOSP and three copper
phosphate reaction products occurring in CuOSP has been investigated in glasshouse trials using wheat as the indicator
crop (Gilkes and Lim-Nunez 1979). The finely ground copper
phosphates $|Cu_{5.8}H_{3.5}(PO_4)_5 \cdot 4.2H_2O$, $(Cu_{1.0}Ca_{0.1})H_{0.8}PO_4 \cdot 1.6H_2O$,
$(Cu_{1.2}Ca_{0.2})H_{0.2}PO_4 \cdot H_2O$, compounds H,D and F respectively$|$
were only slightly soluble (Table II). Copper compounds were
applied at four rates (0, 0.01, 0.033, 0.67 µgCu/g soil) and
thoroughly mixed through a Cu-deficient, acid, sandy soil.
Identical rates of the common copper fertilisers CuO, CuEDTA
and $CuSO_4 \cdot 5H_2O$ were also used to permit a comparison of the
fertiliser effectiveness of the reaction products with these
compounds.
     Both yield and Cu uptake were similar for all six comp-
ounds indicating that crystallization of insoluble Cu reaction products during manufacture does not reduce fertiliser
effectiveness (Figure 7). An identical conclusion was
reached by Gilkes and Sadleir (1968) who used the same glasshouse procedure to compare the effectiveness of Cu in CuEDTA,
$CuSO_4 \cdot 5H_2O$, CuO, CuOSP and $NaCu_6H_3(PO_4)_5Cl \cdot 6H_2O$. It must be
emphasized that under practical agricultural conditions
insoluble Cu phosphates in CuOSP are not applied as fine
powders and are not thoroughly mixed through the soil. They
occur within millimetre-size fertiliser granules as relatively large inclusions and coatings, and might be much less
effective sources in this form. Gartrell (1980) has shown in
field trials that the effectiveness of CuOSP is greater in
seasons subsequent to the season of application. This increase may be partly due to the disaggregation and dispersion
through the soil of insoluble copper phosphates by cultivation
and pedoturbation.
     As mentioned in the introduction to this chapter, a wide
variety of Cu sources are effective fertilisers when applied
as fine powders. Despite a 10,000 fold range in water-
solubility (Table II), the Cu compounds investigated by

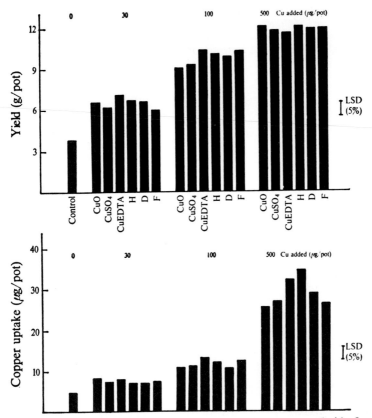

FIGURE 7. Yield and copper uptake by tops of 40 day old wheat plants. The various fertilisers were applied at rates of 0, 0.01, 0.033 and 0.167 µgCu/g soil. Compounds of H, D and F are described in Table II. (Adapted from Gilkes and Lim-Nunez 1979).

Gilkes and Lim-Nunez (1979) were almost equally effective in promoting plant growth and Cu uptake (Figure 7). Anderson (1969) also observed that a variety of Cu compounds were equally effective when first granulated with NPK fertiliser although in his work the separate application of NPK fertiliser and $CuSO_4 \cdot 5H_2O$ was superior (Figure 3). Bridger *et al.* (1962) found the insoluble reaction product $CuNH_4PO_4 \cdot H_2O$ to be an effective Cu source. This type of evidence led Mortvedt (1965) and Lehr (1972) to conclude that although the chemical reactions occurring in compound fertiliser may produce insoluble Cu compounds, these do not reduce the agronomic effectiveness of the Cu.

It seems probable therefore that soil reactions, fertiliser particle size and root interception play major roles in
determining the effectiveness of Cu fertilisers, a conclusion

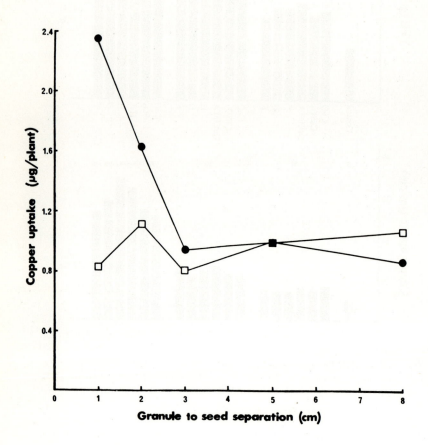

FIGURE 8. *Uptake of Cu by tops of 45 day old wheat*
plants growing at various distances from single granules of
Cu-OSP(●) and OSP(□).   Adapted from data of Gilkes and
Sadleir (1979).

that has been reached by other workers (Gartrell, McBride this symposium). This situation is analogous to that described for Zn fertilisers by Giordano and Mortvedt (1966) who convincingly demonstrated the importance of these factors in glasshouse trials. Similarly Sharpee *et al.* (1969) found that the availability of Cu from Cu-S fusions increased with decreasing particle size.

The Cu concentration of granular fertilisers should not be so high as to result in a low density of application in the field. The importance of root interception with Cu fertiliser granules for the optimum utilisation of Cu has been demonstrated in a glasshouse trial with wheat (Gilkes and Sadleir 1979). Only those plants growing within two centimeters of a CuOSP granule were able to obtain Cu from this source (Figure 8). Copper uptake and weights of both tops and roots were greater for these plants than for plants growing at greater distances from the granule. The authors point out that marked root proliferation occurs adjacent to macronutrient fertiliser granules (Drew 1975) so that the probability of root contact with Cu in a compound fertiliser is enhanced relative to Cu that is applied separately.

VII. CONCLUDING REMARKS

Although some workers have demonstrated that Cu reaction products reduced the efficiency of Cu fertilisers relative to $CuSO \cdot 5H_2O$, such effects are probably of minor significance and are certainly not in proportion to the reduced water-solubility of the Cu in these products (i.e. about 10,000-fold). Soil reactions with Cu and the extent of root interception with Cu enriched zones adjacent to fertiliser granules are probably the dominant factors controlling Cu uptake. Thus most efficient use of fertiliser Cu requires the establishment of a large number of Cu fertilised sites in the soil. Manufacturers should therefore aim to produce granules of Cu-macronutrient fertiliser mixtures of the smallest size consistent with easy handling and large enough to avoid excessive losses of the macronutrients due to rapid leaching (e.g. S,N,K) and fixation (e.g. P) (Williams 1971). Best utilisation will also be assisted by drilling the Cu fertiliser with the seed rather than by broadcasting. Cultivation after top dressing Cu fertilisers to promote thorough mixing of fertiliser Cu through the soil should increase its effectiveness.

REFERENCES

Anderson, A.J. (1969). *Tidsskr. F. Planteavl., Bd 47*, 278.
A.O.A.C. (1970). "Official Methods of Analysis," 11th Edn.
   (W. Horwitz, ed.), Association of Official Analytical
   Chemists, Washington.
Bridger, G.L., Salutsky, M.L., and Starastka, R.W. (1962).
   J. Agr. Fd. Chem. 10, 181.
Cunningham, H.G. (1972). *In* "Micronutrients in Agriculture"
   (J. Mortvedt *et al.* eds.), p. 419. Soil Science Society
   of America Inc., Madison.
Drew, M.C. (1975). "Letcombe Laboratories Annual Report,
   Wantage, U.K. pp 63-73.
Fiskell, J.G.A., Breland, H.L., Lacascio, S.J., and
   Everett, P.H. (1967). *Proc. Soil Crop Sci. Soc. Fla.* 27,
   35.
Gartrell, J.W. (1980). *Aust. J. Exp. Agric. Anim. Husb.* 20,
   370.
Gilkes, R.J. (1975). *Aust. J. Soil Res.* 13, 203.
Gilkes, R.J. (1977a). *J. Agric. Fd. Chem.* 25, 667.
Gilkes, R.J. (1977b). *J. Soil Sci.* 28, 103.
Gilkes, R.J., and Lim-Nunez, R. (1979). *Aust. J. Soil Res.* 17,
   325.
Gilkes, R.J., and Lim-Nunez, R. (1980). *Aust. J. Soil Res.*
   18, 85.
Gilkes, R.J., and Sadleir, S.B. (1978). *Aust. J. Soil Res.*
   16, 113.
Gilkes, R.J., and Sadleir, S.B. (1979). *Aust. J. Soil Res.*
   17, 121.
Gilkes, R.J., and Sadleir, S.B. (1981). *J. Soil Sci. Soc.
   Amer.* (in press).
Gilkes, R.J., Young, R.C., and Quirk, J.P. (1975). *Aust. J.
   Soil Res.* 13, 33.
Giordano, P.M., and Mortvedt, J.J. (1972). *In* "Micronutrients
   in Agriculture" (J. Mortvedt *et al.* eds.), p. 505.
   Soil Science Society of America Inc., Madison.
Guillemin, C. (1956). *Bull. Soc. Fr. Mineral. Cristallagr.*
   79, 219.
J.C.P.D.S. (1979). Search Manual and Powder Diffraction Data,
   Joint Committee on Powder Diffraction Standards,
   Swarthmore, Pennsylvania.
Jones, G.B., and Belling, G.B. (1967). *Aust. J. Agric. Res.*
   18, 733.
Kretschmer, A.E., and Forsee, W.T. (1954). *Proc. Soil Sci.
   Am.* 18, 471.

Lehr, R.L. (1972). *In* "Micronutrients in Agriculture"
    (J. Mortvedt *et al.* eds.), p. 459. Soil Science Society
    of America Inc., Madison.
Lehr, J.R., Brown, E.H., Frazier, A.W., Smith, J.P., and
    Thrasher, R.D. (1967). "Crystallographic Properties of
    Fertilizer Compounds." National Fertilizer Development
    Centre, Muscle Shoals, Alabama.
Mortvedt, J.J. (1965). *Agric. Chem.* 20, 40.
Mukhamedzharov, M., Khakimova, V.K., and Vishnyakova, A.A.
    (1970). *Uzb. Khim. Zh.* 14, 8.
Pocharkoff, E.D., Kuczynski, S., and Macchia, S. (1967).
    J. Agric. Fd. Chem. 15, 123.
Sharpee, K.W., Ludwick, A.E., and Attol, O.J. (1969).
    Agron. J. 61, 746.
Silverberg, J., Young, R.D., and Hoffmeister, G. (1972).
    In "Micronutrients in Agriculture" (J. Mortvedt *et al.*,
    eds.), p. 666. Soil Science Society of America Inc.,
    Madison.
Teakle, L.J.H. (1939). *J. Agric. West. Aust.* 16, 116.
Teakle, L.J.H., Thomas, I., and Turton, A.G. (1941).
    J. Agric. West Aust. 18, 70.
Van Wazer, J.R. (1958). "Phosphorus and its Compounds,"
    Vol. 1. Interscience Publishers, Inc., New York.
White, M.S. (1976). *N.Z. J. Sci.* 19, 421.
Williams, C.H. (1971). *Aust. J. Soil Res.* 9, 83.

6

COPPER POLLUTION OF AGRICULTURAL SOILS

Kevin G. Tiller and Richard H. Merry

Division of Soils,
CSIRO,
Glen Osmond, South Australia

I. INTRODUCTION

Copper has been used by man since about 8000 B.C., initially as native Cu, but smelting and the production of bronze implements dates from about 4000 B.C. (Butts, 1980). In modern life, Cu has many applications. It is alloyed with various metals, *e.g.* Sn, Pb, Zn, Ni, Al and Mn, and used in the production of wire, electrical apparatus, boilers and pipes, as well as in fertilizers and fungicides. The total world production of Cu from ancient times to 1980, estimated to be about 170×10^6 tonnes, is expected to double during the next decade or so (Nriagu, 1979a). This provides ample scope for the pollution of the environment from many sources, *viz.*, the burning of coal, oil, wood and urban wastes, mining and smelting operations, manufacturing industry including plating and metal processing, the direct disposal of wastes such as sewage sludge onto land, and agricultural activities.

II. EXTENT OF COPPER POLLUTION AND ITS IMPACT

Contamination of soils by Cu-containing material will be discussed in relation to that resulting from local management decisions concerning fertilizers, sprays, and utilization of wastes as well as to sources which are incidental to modern civilization. The latter are to a degree beyond local control yet may contribute significantly to Cu levels in the environment. Estimates of Cu which has been emitted globally

COPPER IN SOILS AND PLANTS

into the atmosphere from natural sources range from 10 to 18×10^3 tonnes annually and provide a general reference for the assessment of pollution. The largest components are wind-blown dust but materials from bush-fires, volcanoes and plant exudates also contribute (Nriagu, 1979b). Man-influenced sources will tend to have greater local impact and are quite significant globally (56×10^3 t yr^{-1}; Nriagu, 1979b).

A. *Inadvertant Urban/Industrial Pollution*

Much of the Cu pollution associated with industrial/urban activities is released from point sources from which dispersal is strongly dependant on local meteorological and topographical factors in relation to stack height and aerosol size. Some processing plants will have additionally some very localized pollution.

The "background" concentrations of Cu in soils near urban and/or industrial areas may have an appreciable anthropogenic component. Within Europe, regions quite distant from points of emission may be affected (Rühling and Tyler, 1973b) *e.g.* southern Sweden and Norway in relation to the industrial areas in the United Kingdom and Germany. However, no evidence exists in the polar ice caps of accessions of anthropogenic Cu (Weiss *et al.*, 1975; Boutron and Lorius, 1979). If the fallout from the total (all-time) emission of pollutant Cu (estimated as 2.2×10^6 t; Nriagu, 1979c) were uniformly smeared over the earth's land surface, the accumulated application would be about 100g Cu ha^{-1}.

Major point sources of pollution, e.g. smelters, usually exhibit haloes in which concentrations of pollutants decrease exponentially with distance. Studies of the Cu-Ni complex at Sudbury, Ontario, for example, provide clear evidence of pollution of the countryside with Cu and other metals. Most of the Cu is deposited within 32km but soils in areas closer than 7.5km frequently contained well over 1000 μg Cu g^{-1} (Hutchinson and Whitby, 1974). Elevated concentrations of Cu were found up to 19km from a smelter in Pennsylvania (Buchauer, 1973). The extent of Cu pollution around smelters depends on the kind of ore and the process used; a major Pb-Zn smelter in South Australia released negligible amounts of copper into the surrounding countryside (Cartwright *et al.*, 1977).

Metals arising from mining areas tend to contaminate only the immediate surroundings unless effluents enter streams *etc.* which transport contaminants to alluvial soils downstream, as discussed later.

The hundreds of millions of tonnes of fly ash captured each year around the world when coal is burnt for electricity generation are one potential source of pollution. In relation to soil concentrations, fly ashes are not generally enriched in Cu to an appreciable extent although a range of values from 14 to 2800 μg g^{-1} are quoted (Page *et al.*, 1979). There is some circumstantial evidence (Wangen and Turner, 1980) that fly ash emitted directly into the atmosphere may be transmitted over distances of about 100km downwind but Page *et al.* (1979) consider enrichment of soils and plants by stack-derived fly ash to be negligible. Disposal on agricultural land is not yet a common practice.

The combustion of wood products, fossil fuels and waste incineration within urban areas may result in the atmospheric emission of Cu together with other elements. The impact of such activities, together with those already discussed, will be mainly within urban areas. Purves (1977) and Beavington (1973) have both reported that the concentrations of Cu extracted by EDTA or dilute acetic acid in soils from urban areas are 5 to 10 times those of adjacent rural areas. Hamilton (1976) stated that the Cu concentration of street dusts and soils of urban parks frequently exceed 1000 μg g^{-1} while the Cu concentration of household dusts has been found to be 700-900 μg g^{-1}. Recent work of Gulson *et al.* (1981) near Adelaide, South Australia suggests that pollution of large urban areas may spread into the rural hinterland for up to 40 to 50km; the total atmospheric fallout of Cu within the study area is less than about 50 g ha^{-1} each year (Tiller *et al.* unpublished data).

The incidental introduction of Cu onto soils and crops arises also from other sources such as the corrosion of sheeting, wire, pipes and other equipment. Effluents from plating works, wood preservative treatments and the production of dyes and textiles may contain considerable concentrations of Cu (*e.g.* Netzer and Beszedits, 1979). High-tension electricity cables can contaminate a strip of land up to 20m wide (Hemkes and Hartmans, 1973; Kraal and Ernst, 1976). Pollution of soils along highways by Cu-containing dusts has been reported (McLean and Langille, 1973; David and Williams, 1975; Cannon, 1975) but the effects are relatively minor.

B. *Deliberate Pollution of Agricultural Soils*

Since the application of sewage sludge, municipal composts, pig slurries, *etc.* to agricultural land is user-initiated, their extent and impact can be more readily assessed. However, the activity which produces the highest concentrations of Cu in agricultural soils is the use of fungicidal sprays which reach the soil surface directly or indirectly as leaf litter.

As an impurity of chemical fertilizers, Cu rarely exceeds concentrations of 100 μg g^{-1} (Swaine, 1962) and even with long-term use, fertilizers do not contribute substantially to soil contamination. Sewage sludge and other wastes which are considered to have fertilizer value may have considerable concentrations of Cu and repeated application of these materials to land also contributes to soil contamination.

Sewage sludges from many parts of the world have similar ranges of Cu concentrations as shown in Table 1. A detailed summary of the composition of North American sludges has been prepared by Page (1974). The total Cu contents of 20 Australian sludges were found to range from 230 to 2400 μg g^{-1} (de Vries, unpublished data). In reviewing this subject, Page (1974) has pointed out that moderate rates of application of many sewage sludges may easily exceed normal amounts of Cu and other metals already present in soils. The mean concentration of Cu in soils (20 μg g^{-1}; Swaine, 1955) would be doubled by an application of by only 20 t of sludge containing 2000 μg g^{-1} when applied to one hectare and restricted to about 15 cm depth.

Sewage sludge is added to agricultural soils primarily as a source of organic matter and for its fertilizer value in terms of N and P. The accompanying metals, mainly of industrial origin, compromise its value as a soil amendment because contamination is almost irreversible (Purves, 1977). The significance of the Cu content of sewage sludge is nearly impossible to evaluate because of the accompanying concentrations of Ni, Zn and Cd which are relatively more toxic than Cu (Patterson, 1971). Many studies report that relatively high application rates of Cu-containing sewage sludge result in only small increases in Cu concentrations in

TABLE 1. *Concentration of Copper in Sewage Sludges[a]*

Country	Range	Median	Reference
United Kingdom	200 – 8000	800	Patterson (1971)
Sweden	52 – 3300	560	Berggren and Oden (1972)
United States	84 – 10400	850	Sommers (1977)

[a] *Concentrations in μg g^{-1} (dry basis).*

above-ground parts of crops, but values usually remain in the normal range (Page, 1974; Purves, 1977).

Municipal wastes, either composted or uncomposted, have been used on agricultural land in several countries and may contain variable concentrations of Cu. Purves (1977) reports concentrations of Cu extractable by 0.04M EDTA from 44 to 332 μg g^{-1} and thus appreciable contributions could be made to soil Cu levels. King *et al.* (1974) carried out experiments in which mixtures of sewage sludge and municipal composts were used to balance out undesirable effects of nitrogen immobilization or nitrogen leaching losses which result from each component, respectively, when used alone. Their mixtures resulted in increases in the Cu concentration of crops which were within the normal range, but the possibility exists for appreciable inputs into soils by these means.

Copper is frequently added to the diet of pigs and poultry at rates of 125 to 250 μg g^{-1} to achieve small improvements in rates of food conversion and growth (Baker, 1974; Dalgarno and Mills, 1975). Robinson *et al.* (1971) reported that slurries of pig wastes could contain as much as 750 μg g^{-1} of Cu and Purves (1977) stated that concentrations up to 800 μg g^{-1} (dry basis) were not unusual and thus are similar to the concentration in sewage sludge from industrial areas. Such Cu levels could inhibit biodegradation of the organic matter (see later). Spreading of such materials on land could increase Cu concentrations in soil appreciably and could soon lead to pollution problems in some soil situations, depending on soil type and topography. Batey *et al.* (1972) found in one study that applications of Cu-enriched pig slurry supplied 12kg Cu ha^{-1} yr^{-1}. Higher and lower inputs of Cu from this source have been reported (Baker, 1974; Purves, 1977) which presumably relate to Cu dosage rate and land available for disposal of pig wastes. Sheep on such properties are at some risk because of the adherence of Cu-rich organic matter to pasture foliage and their particular sensitivity to concentrations (see later).

Many formulations of Cu-containing fungicides, such as Bordeaux mixture and copper oxychloride, are used on pome, stone, and citrus fruit as well as grapevines, hops and vegetables. Widespread use of these mixtures began in France in 1885 following observations by Millardet (van Alphen, 1957) that a mixture of lime and $CuSO_4$, which had been used for over 100 years to discourage the stealing of grapes, was also effective in countering downy mildew. Bowen (1966) reports that 7×10^7 kg Cu as Bordeaux mixture is sprayed per year on vines and other crops. The sprays are quite effective and their widespread use is likely to continue.

Copper from these sprays becomes incorporated into the leaf litter and eventually into the soil as a result of biological cycling and cultivation. Some examples of high concentrations of Cu which have accumulated in the surface of contaminated soils are shown in Table 2. These accumulations tend to be largely retained in the surface horizons (Delas, 1963; Rieder and Schwertmann, 1972; Merry, 1980). These values should be viewed in relation to uncontaminated soils which are usually quoted as being about 20 μg Cu g^{-1} (Swaine, 1955).

There have been few extensive surveys of soils affected by Cu-containing sprays but the following means (and ranges) of 39 (11-110) μg g^{-1} for 39 Ontario orchard soils (Frank *et al.*, 1976) and 99 (18-320) μg g^{-1} for 95 Australian orchard soils (Merry, 1980) have been reported. In the above orchards, copper has accumulated at an average rate of 1 to 4 μg g^{-1} yr^{-1}, much less than the 12 μg g^{-1} yr^{-1} reported by Rieder and Schwertmann (1972) for a German hop field.

In Australia, 30,000 hectares of horticultural soils which were formerly used for apples and pears, and which presumably had elevated concentrations of Cu (and Pb and As), have been converted to other uses (Merry, 1980). The possible effects on sensitive crops and grazing animals has not been fully investigated but the hazards to grazing sheep in relation to induced molybdenum deficiency, and possible reduction of yields of sensitive crops must be acknowledged.

There appear to be few other sources of Cu contamination which are of widespread significance in agriculture. Copper sulphate is used as an aquatic herbicide in irrigation ditches (Brown and Rattigan, 1979) and could reach cropland. A more serious matter is the contamination of some rivers and canals with effluents from industrial and mining activities. Use of such water for irrigation as reported by Morishima and Oka (1977) in Japan has led to Cu pollution of up to 450 μg g^{-1} for large areas of rice soils in several districts, and resulted in decreased rice yields. The contamination of valley bottom soils with Cu by flooding of polluted rivers is well documented *e.g.* the lower Rhine Valley (de Groot and Zschuppe, 1972). The use on agricultural land of dredged material from rivers, canals and harbours can also increase appreciably the concentrations of Cu, and other metals, in soils and thus affect the quality and yield of crops. Copper may also be introduced into farm soils in minor ways such as in timber posts treated with heavy metal preservatives, including Cu (de Groot *et al.* 1979), but this source is not considered environmentally significant.

TABLE 2. *Some high concentrations of copper in soils resulting from the accumulation of sprays (µg g^{-1})*

Concentration	Land Use	Location	Reference[a]
110	Apple orchard	Ontario	A
320	Apple orchard	Tasmania	B
700	Apple orchard	Japan	C
1500	Apple orchard	U.K.	D
250	Citrus orchard	Florida	E
460	Citrus orchard	S. Africa	F
845	Vineyard	France	G
1280	Vineyard	Germany	H
520	Hopfield	Germany	I
210	Peach orchard	S. Australia	B

[a] A, Frank et al., 1976; B, Merry, 1980; C, Mochizuki et al., 1975; D, Hirst et al., 1961; E, Ruether and Smith, 1954; F, du Plessis and Burger, 1972; G, Delas et al., 1960; H, Gärtel, 1957; I, Rieder and Schwertmann, 1972.

III. BEHAVIOUR OF COPPER IN POLLUTED SOILS

A. Mobility in Soils and Landscapes

In the time scale of man's activities, movement of metals within landscapes will be mainly associated with the solid phases of earth materials, under the agency of wind, water and gravity. Movement within the soil mass, given the same time scale, will be mainly in the solution phase. Hence, chemical factors which control the distribution of metals between solid and solution phases, and the chemical forms in each phase will control the transport of Cu *etc.* whether by diffusion, mass flow in solution or by mass transport in the solid phase. In cationic form, Cu has a high affinity for soil colloids and is quite immobile. More generally, the strong interactions with solid and soluble organic substances of soils dominate its reactions so that Cu in soil solution is largely non-cationic, occurring mainly as stable complexes with organic compounds often of high molecular weight. The chemistry of Cu, especially in relation to soils, is being reviewed elsewhere in this Symposium.

The overall effect is, however, that when Cu has been added as a pollutant to soils, movement within the soil profile is usually restricted because of its strong affinity for soil colloids. Jones and Belling (1967) using ^{64}Cu, studied the leaching of Cu through various agricultural soils in columns using simulated rainfall. Small movements of Cu were observed only in the more sandy soils with very low clay and organic matter contents. The rates of leaching were somewhat accelerated when CO_2 saturated water and lucerne extracts were used as eluants. Recent experiments by Korkman and Virta (1979), using several soil types in columns, and various forms of added Cu, confirmed the difficulty with which Cu is leached through soils. Using quite a different approach, Tyler (1978) investigated the leaching of Cu (and other elements) in organic soils from a spruce forest. Contaminated soil (2640 µg Cu g^{-1}) from near a brass foundry was compared with a similar, but uncontaminated soil (3.8 µg Cu g^{-1}). The normal pH of rainfall reaching the ground in this environment was 4.2 units and, when sufficient 'percolation' water at this pH equivalent to 30 year's supply was added, only about 1 to 2% of the total Cu was released from the contaminated soil, though losses were greater when more acidic water was used. Tyler (1978) estimated that, if the pH of rainfall remained at 4.2, then 80 to 120 years would be required to lower the Cu concentration of the polluted soil by 10%. This simulation of long-term leaching may minimize the role of soluble complexing substances which may arise under field conditions from the decomposition of organic matter, and which could increase the mobility of Cu. On the other hand, decomposition processes may be greatly retarded in highly contaminated soils due to the inhibition of biological processes (see discussion later).

Delas (1963) and Rieder and Schwertmann (1972), among others, have noted the tendency for applied Cu to be retained in the surface horizon of soil profiles. Merry (1980) has also shown that acidic, sandy soils of southern Australia, some with histories of use of Cu-containing sprays of more than 80 years, have strongly retained added Cu in the upper horizons. No accumulations of Cu were found deeper in the profiles although both As and P were translocated and retained in some B horizons enriched in humus and Fe oxides. Some As appeared to have been completely lost from some soil profiles. Wander (1954) investigated the movement of Cu in very sandy soils in Florida. Where the pH of the surface soil was maintained at about 5.7, 90% of applied Cu was accounted for in the top 30cm of the profile, but when the soil remained more acidic (pH 4.3 or less), there was evidence of leaching to at least 90cm and possibly to 250cm.

Conflicting reports exist in relation to the mobility of sludge-bound metals in soils. Some reports show that Cu associated with sewage sludge remains in the ploughed layer (Andersson and Nilsson, 1972; de Vries and Merry, 1980) but calculations by Page (1974) based on the data of Hinesly *et al.* (1972), indicate appreciable losses of Cu from the layer of sludge application. Simulated trenching of sewage sludge in soil profiles carried out under glasshouse conditions (Taylor *et al.* 1978) did not reveal any movement of Cu into the surrounding soil. The solubility of organic Cu complexes derived from sewage sludge may depend on sludge type, and soil properties, especially pH.

Even though Cu may be relatively immobile in individual soil profiles through high chemical affinity, erosion by wind and water may still result in considerable movement of Cu within the landscape. This situation will be accentuated where contamination arising from mining and smelting operations has led to decreased vegetative cover because of metal toxicity or acidity. Roberts and Goodman (1973) investigated the persistence of heavy metals in soils following the closure of a smelter in south Wales. The time required for 50% depletion of Cu in surface soils varied from 7.7 to 23 years. Since leaching down the soil profile did not explain the results, wind erosion and surface runoff were thought to be largely responsible for the losses observed. Revegetation would retard losses of metals from the polluted area. Rutherford and Bray (1979) presented data for Cu in some soil profiles from Coniston, Ontario, in which distributions could be partly explained in terms of erosion and depositional processes. Buchauer (1973) noted very variable metal concentrations because of erosion of contaminated soils.

Soil and tailings from contaminated areas may contribute to the sediment load of local streams and rivers and this may be deposited well away from the source of pollution. For example, Alloway and Davies (1971) and Davies and Lewin (1974) have shown that erosion of an historically polluted area in Wales may lead to secondary pollution for a long period after cessation of the polluting activities, through continuing fluvial dispersal of old polluted sediments. Leakage of polluted drainage waters from mining wastes deposited near streams can also contribute to secondary dispersal of pollutants. From another point of view, the dispersal of pollutant metals such as Cu by wind and water erosion may prevent its accumulation in toxic concentrations in the surface horizons of soils through dilution or burial with uncontaminated sediments.

B. *Biological Activity*

The subject of heavy metal pollution in terrestrial ecosystems has been extensively reviewed by Hughes *et al.* (1980). With respect to Cu they state 'There is surprisingly little information on this important element in terrestrial food chains'. However, there is evidence that the microflora and fauna of soils may be greatly affected by Cu pollution. Baroux (1972) showed that accumulations of Cu and increasing soil acidity retarded the rate of ammonification in vineyard soils in France. Tyler (1975) reported that nitrogen mineralization in acid forest soils in Sweden which had been contaminated by Cu began to decrease logarithmically at Cu concentrations in the soil of about 50 μg g^{-1}. Similar decreases were noted in contaminated soils from the field by Rühling and Tyler (1973a) and Tyler (1974) for soil enzyme activity and respiration rates as well as in laboratory studies (*e.g.* Quraishi and Cornfield, 1973).

No effects on the mineralization of organic carbon were observed by Cornfield *et al.* (1976) when contaminated sewage sludge was applied to an agricultural soil (pH 6.7) in pots. This apparent difference with the results mentioned above may have been due to the presence of adequate mineral nitrogen and the comparatively low final concentration of Cu (up to 123 μg g^{-1}) and other metals in the soil-sludge mixtures.

Hirst *et al.* (1961) found that Cu had accumulated to concentrations up to 2500 μg g^{-1} in the organic mats at the surface of orchard soils in England. They noted that the earthworm populations were adversely affected and hence the structure of the surface of the soil. Similar effects on the populations of earthworms, insect larvae and isopods were observed by Mochizuki *et al.* (1975) in Japanese orchard soils although Pb and As as well as low soil pH values, may also have contributed.

High concentrations of Cu in soils appear to have profound effects on the microbial, earthworm and insect populations in soils resulting in retardation of normal turnover of N and breakdown of organic matter. These effects are not always caused by Cu pollution alone but can also result from pollution by other metals such as Pb, Zn and Ni.

C. *Effects on Animal Nutrition*

Apart from domestic animals, there appears to be little if any, information on whether pollution of soils by Cu leads to increased body burdens of the element in larger animals.

Elfving *et al.* (1979) found a correlation between soil arsenic and body burdens in rodents living in orchards. Similar uptake could be expected for Cu.

Copper toxicity problems in animals depend on dietary intake not only of Cu, but also Zn, Fe, Ca, Mo, S and Cd (Allaway, 1968; Underwood, 1971; Mills and Dalgarno, 1972). Nor does toxicity depend solely on the concentrations of these elements in fodder. Considerable quantities (sheep up to 100g per day, cattle up to 1-2000 g per day) of soil material may be ingested by grazing animals (Healy, 1973) and contributions of Cu from this source may be substantial, especially during periods of feed shortage when soil intake may increase. Cattle are much less sensitive than sheep to excess Cu in the diet (Todd and Thompson, 1965). Sheep are probably the domestic animals which are most at risk from the pollution of soils by Cu because of the likelihood of Cu-induced molybdenum deficiency (Gartrell, this Symposium).

IV. ASSESSMENT OF POLLUTED SOILS IN RELATION TO TOXICITY IN PLANTS

The prediction of Cu requirements for crop growth, even when Cu concentrations are in the normal range for soils, creates some interpretative problems whether attempted by soil or plant analysis. These problems are magnified in soils polluted by Cu because few comprehensive studies have been undertaken and because polluting elements rarely occur in isolation. The pollutant under review may be accompanied by appreciable concentrations of even more toxic elements.

Page (1974) has stated that no reliable soil tests are available for the diagnosis of heavy metal toxicity to plants. Problems are certainly created for diagnosis by either soil or plant analysis because of the additive, synergistic and antagonistic effects that may occur in contaminated soils (Forster, 1954; Cunningham *et al.*, 1975). In our discussions of contaminated soils we have not included reports of toxicity in the sense of yield reductions that may be brought about by the addition of a single nutrient to crops suffering from multiple deficiencies. An example could be the toxicity ascribed to Teakle's (1942) observation of yield reductions on very sandy soils following applications of "even 5lb. of bluestone per acre".

Most early reports of Cu toxicity were associated with Cu accumulations in the surface horizons of acid soils (Delas, 1963). Reuther and Smith (1953) reported Cu toxicity to citrus seedlings growing in sandy soils with pH less than 5 and total Cu concentrations greater than 150 μg g^{-1}. Similarly Drouineau and Mazoyer (1956) reported Cu toxicity in the sensitive crops, spinach and gladioli, on soils of similar acidity and Cu concentrations higher than 100 μg g^{-1} (extracted by neutral ammonium acetate). With investigations of acidic soils, the results may need qualification by other pH related factors, such as Al toxicity, which affects root development, and Fe and Mn concentrations, which can affect the appearance of symptoms of chlorosis.

The studies of Walsh *et al.* (1972) using a Cu-sensitive crop (snapbeans), give some indication of lower limits of soil levels likely to depress plant growth when high concentrations of Cu salts are applied under field conditions. Linear regressions limit derivation of critical values, but yields decreased by about one third when Cu extracted by 0.1M HCl or DTPA (Lindsay and Norvell, 1978) was greater than 40 μg g^{-1} or by 0.01 M EDTA was greater than 30 μg g^{-1}. Appreciably higher values could be expected for less susceptible crops, other soil types, where Cu has been added in different chemical forms and where Cu has equilibrated with the soil over some decades.

The results of Purves (1977) obtained in pot experiments indicate that the toxicity of Cu to oats may be expected at Cu concentrations between 100 and 200 μg g^{-1} (0.04M EDTA), depending on soil and moisture conditions. For clover, a corresponding value of about 100 μg g^{-1} seemed appropriate although a more conservative value of 30 μg g^{-1} was quoted. On the other hand, Patterson (1971), quoting data of N.A.A.S. Laboratories at Wolverhampton, reports only slight toxic symptoms in maize and mustard at total concentrations of Cu in soil of 360 μg g^{-1}, equivalent to 245 μg g^{-1} extracted by 0.05 M EDTA. Patterson (1971) also described work of Webber at Leeds in which soils adjusted to pH values ranging from 5 to 8 were treated with increasing rates of Cu. Stunting of barley occurred at total Cu concentrations of 200 μg g^{-1} (pH 5) and 400 μg g^{-1} (pH 7); values extracted by EDTA would be correspondingly lower.

Korcak and Fanning (1978) found good correlations between Cu extracted by either dilute acids or DTPA and concentrations in plants grown in soils treated with metal salts, but not for soils treated with Cu-enriched sewage sludge, although they improved when lime treatments were taken into account. McLean and Dekker (1978) showed that DTPA-Cu was a poor indicator of likely yield depression of Cu-treated soils, whether the Cu was added as a salt or after incubation with sewage sludge. Such data do not encourage optimism for the use of DTPA or similar extractions of contaminated soils. Generally speaking, it is not these situations where pure metal salts are applied for which tests for potential toxicity are required. The added complication of predicting toxicity of an element in the presence of several other potentially toxic elements is somewhat akin to the prediction of a deficiency of a particular nutrient by soil test when other nutrients are also severely limiting.

In many contaminated soils, the total Cu content, which can be closely approximated by strong acid digests, and that extracted by complexing agents (*e.g.* DTPA and EDTA), may serve equally well in predicting the uptake of Cu by plants in pot experiments (Davis, 1979; Merry, 1980). The latter author found that Cu extracted by both DTPA (Lindsay and Norvell, 1978) and 0.1M EDTA (Clayton and Tiller, 1979) were highly correlated with total soil Cu (aqua regia digests) in orchard soils. However total Cu had no predictive value for the concentration of Cu in perennial ryegrass or white clover in the field. EDTA solutions of sufficient concentration may extract most of the total contaminant Cu (Patterson, 1971; Rieder and Schwertmann, 1971; Clayton and Tiller, 1979; Merry, 1980).

Some extraction procedures are unsuited for universal application. Soils or soil mixtures which are alkaline, and especially if calcareous, are unsuited to extraction by dilute acids and highly polluted soils, especially if acidic, may give unsatisfactory results with dilute complexing agents because of reagent saturation. The well-accepted effect of pH on uptake of trace metals by plants must be distinguished from pH-related artefacts connected with the soil test itself. Miles and Parker (1979) suggested that DTPA tests may be unsuited for assessing availability in soils of different pH. This concern was shared by Merry (1980) whose results showed that although the Cu concentration in radish and silver beet plants grown on contaminated orchard soils decreased as soil pH increased, the amount of Cu extracted by DTPA increased with increasing pH on the most contaminated soils (Figure 1). Although DTPA was originally specified for near-neutral and alkaline soils (Lindsay and Norvell, 1978), it has not been restricted in practice to these soil types.

The many soil tests used in different countries and organizations make it difficult to compare critical values for toxicity obtained in different investigations. In order to evaluate some studies using different reference procedures, we extracted a range of Cu-contaminated agricultural soils with several commonly used chemical reagents. The results are given in Table 3. The comparison was restricted to acidic soils in order to obtain meaningful data for the dilute acidic reagents. Under these conditions 0.1 M HCl, 0.05M EDTA and 0.005M DTPA extracted similar proportions of total Cu; the other reagents extracted smaller and more variable proportions. The different bases of studies involving different plant species and soil types in the glasshouse or field, with Cu recently applied or in long-term equilibrium, also creates additional problems of interpretation. Other factors such as the effectiveness of mycorrhizal fungi may further modify the ability of plants to absorb Cu (Menge *et al.*, 1977). Finally we must recognize that all diagnostic tests gain acceptance through successful calibration in the field.

V. CONCLUDING REMARKS

Pollution of agricultural soils has been quite extensive and is likely to continue. Even in areas where the sources of contamination have been removed, *e.g.* by cessation of either industrial operations or of the use of Cu-containing sprays in agriculture, the permanency of earlier contamination, as stressed by Purves (1977) and Davies (1977), must be recognized. Copper additions to soils are characterized by the relative immobility of the reaction products arising from the high chemical affinity of Cu species for soil surfaces. Soil organic matter has an acknowledged key role in controlling Cu solubility, but the question of 'reversion' of pollutant Cu in relation to its reaction with organic matter and its time - scale, as raised by Leeper (1978), has still to be resolved. This matter remains of importance because of the potential of high Cu concentrations in soils to decrease crop yields and to interfere with biochemical reactions in soils leading to decreased turnover of organic matter.

The varied reports of 'critical' values arising from soil testing for the prediction of Cu toxicity are not encouraging. Results are often so crop- and soil-specific that the development of a universal soil test seems unlikely. Foliar analysis in the toxicity range has received less attention than has deficiency (Jones, 1971). The value of about 20 µg g^{-1} of Cu in dry plant tissue which is often quoted as the limit for healthy plants does not seem justified in view of the review of Gupta (1979). Values are dependent on complex species/part/maturity factors and are affected by interaction with other elements.

Procedures for the amelioration of the undesirable effects of additions of Cu to soils are largely related to the soil factors which control Cu toxicity to plants (Chaney, 1973). Practical measures (Delas, 1963; Chaney, 1973) include decreasing the Cu concentration by mixing or covering the

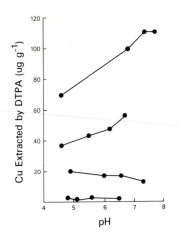

FIGURE 1.   *Cu extracted by DPTA from 4 orchard soils as a function of pH.*

TABLE 3.   *Percentage of total copper extracted by various reagents from contaminated soils[a].*

	0.1M ammonium acetate	2.5% acetic acid	0.005M DTPA	0.1M HCl	0.05M EDTA
Range	<.2-3.6	5.1-34	23-41	37-58	43-66
Mean ±s.d.	1.6±1.0	19±9	30±5	47±8	54±7

[a] *n=12; range of soil pH was 4.7-6.7; range of total copper was 27-320 μg g^{-1}.*

Two soils were contaminated with sewage sludge and the remainder with orchard sprays.

contaminated layers with uncontaminated soil or by deep cultivation of Cu concentrated at the soil surface. Soil amendments which increase pH, cation exchange capacity, organic matter content, or phosphate may help in overcoming toxicity. Indirect procedures may also be effective, such as applying iron chelates to overcome Cu-induced Fe chlorosis (Stewart and Leonard, 1952) or by growing Cu-tolerant plant species (Antonovics *et al.*, 1971).

Pollutant Cu is frequently accompanied by other elements in toxic concentrations such as Zn, Pb, As and Ni. These associated elements can influence the biological effectiveness of the added Cu (Chaney, 1973; Page, 1974; Cunningham *et al.*, 1975) and complicate the assessment of hazard by both soil and plant analysis.

REFERENCES

Allaway, W.H. (1968). *Adv. Agron. 20,* 235.

Alloway, B.J., and Davies, B.E. (1971). *Geoderma 5,* 197.

Andersson, A., and Nilsson, K.O. (1972). *Ambio 1,* 176.

Antonovics, J., Bradshaw, A.D., and Turner, R.G. (1971). *Adv. Ecol. Res. 7,* 1.

Baker, D.E. (1974). *Fed. Proc. Fed. Am. Soc. Exp. Biol. 33,* 1188.

Bavoux, J. (1972). *C.R. Acad. Sci., D275,* 499.

Batey, T., Berryman, C., and Line, C. (1972). *J. Brit. Grassl. Soc. 27,* 139.

Beavington, F. (1973). *Aust. J. Soil Res. 11,* 27.

Berggren, B., and Oden, S. (1972). 1. Analys resultat Rorande Fung metaller Och Klorerade Kolväten I Rötslam Fran Svenska Reningsverk 1968-1971. Institutionen für Markvetenskap Lantbrukshogskolan, Uppsala, Sweden.

Boutron, C. and Lorius, C. (1979). *Nature (London), 277,* 551.

Bowen, H.J.M. (1966). "Trace Elements in Biochemistry," Academic Press, New York.

Brown, B.T., and Rattigan, B.M. (1979). *Environ. Pollut. 20,* 303.

Buchauer, M.J. (1973). *Environ. Sci. Technol. 7,* 131.

Butts, A. (1980). *In* "Encyclopedia Britannica" Vol. 5, p. 148 (H.H. Bentch : Chicago).

Cannon, H.L. (1975). International Conference on Heavy metals in the Environment, Toronto 1975, Vol. II Pt. 1, p.1.

Cartwright, B., Merry, R.H., and Tiller, K.G. (1977). *Aust. J. Soil Res. 15,* 69.

Chaney, R.L. (1973). *In* "Recycling municipal sludges and effluents on land" pp 129-141. Nat. Assoc. of State Universities and Land-Grant Colleges, Washington, D.C.

Clayton, P.M., and Tiller, K.G. (1979). CSIRO Aust. Div. Soils Techn. Paper No. 41.

Cornfield, A.H., Beckett, P.H.T., and Davis, R.D. (1976). *Nature (London) 260,* 518.

Cunningham, J.D., Ryan, J.A., and Keeney, D.R. (1975). *J. Environ. Qual. 4,* 455.

Dalgarno, A.C., and Mills, C.F. (1975). *J. Agric. Sci. 85,* 11.

David, D.J., and Williams, C.H. (1975). *Aust. J. Exp. Agric. Anim. Husb. 15,* 414.

Davies, B.E. (1977). *In* "Proceedings, International Seminar on Soil Environment and Fertility Management in Intensive Agriculture. (SEFMIA) Japan.

Davies, B.E., and Lewin, J. (1974). *Environ. Pollut. 6,* 49.

Davis, R.D. (1979). *J. Sci. Food Agric. 30,* 937.

de Groot, R.C., Popham, T.W., Gjovik, L.R., and Forehand, T. (1979). *J. Environ. Qual. 8,* 39.

de Groot, A.J. and Zschuppe, K.H. (1972). *T.N.O. Views, 27,* 437.

de Vries, M.P.C., and Merry, R.H. (1980). *Aust. J. Exp. Agric. Anim. Husb. 20,* 470.

Delas, J. (1963). *Agrochimica 7,* 258.

Delas, J., Delmas, J., and Demias, C. (1960). *C.R. Sceances Acad. Agric. Fr. 45,* 651.

Drouineau, G., and Mazoyer, R. (1956). *Trans. 6th Int. Congr. of Soil Sci., Paris, Di419.*

du Plessis, S.F., and Burger, R.D. (1972). *Agrochemophysica 4,* 47.

Elfving, D.C., Stehn, R.A., Pakkala, I.S., and Lisk, D.J. (1979). *Bull. Environ. Contam. Toxicol. 21,* 62.

Forster, W.A. (1954). *Ann. Appl. Biol. 41,* 637.

Frank, R., Braun, H.E., Ishida, K., and Suda, P. (1976). *Can. J. Soil Sec. 56,* 436.

Gärtel, W. (1957). *Weinb. und Keller 4,* 221.

Gulson, B.L., Tiller, K.G., Mizon, K.J., and Merry, R.H. (1981). *Environ. Sci. Technol. 15, in press.*

Gupta, U.C. (1979). *In* "Copper in the Environment" (J.O. Nriagu, ed.) p. 255. John Wiley and Sons Inc., New York.

Hamilton, E.I. (1976). *Sci. Total Environ. 5,* 1.

Healy, W.B. (1973). *In* "Chemistry and Biochemistry of Herbage" (G.W. Butler and R.W. Bailey, eds.) Vol. 1, p. 567. Academic Press, London.

Hemkes, O.J., and Hartman, J. (1973). *Tijdschr. Diergeneesk 98,* 446.

Hinesley, T.D., Jones, R.L., and Ziegler, E.L. (1972). *Compost Sci., 13,* 26.

Hirst, J.M., Le Riche, H.H., and Bascomb, C.L. (1961). *Plant Path. 10,* 105.

Hughes, M.K., Lepp, N.W., and Phipps, D.A. (1980). *Adv. Ecol. Res. 11,* 218.

Hutchinson, T.C., and Whitby, L.M. (1974). *Environ. Conserv. 1,* 123.

Jones, G.B., and Belling, G.B. (1967). *Aust. J. Agric. Res. 18,* 733.

Jones, J.B. (1972). *In* "Micronutrients in Agriculture" p. 319. Soil Sci. Soc. Am. Inc., Madison.

King, L.D., Rudgers, L.A., and Webber, L.R. (1974). *J. Environ. Qual. 3*, 361.

Korcak, R.F., and Fanning, D.S. (1978). *J. Environ. Qual. 7*, 506.

Korkman, J., and Virta, P. (1979). *J. Sci. Agric. Soc. Finl. 51*, 51.

Kraal, H., and Ernst, W. (1976). *Environ. Pollut. 11*, 131.

Leeper, G.W. (1978). "Managing the Heavy Metals on Land." Marcel Dekker, Inc., New York.

Lindsay, W.L., and Norvell, W.A. (1978). *Soil Sci. Soc. Am. J. 42*, 421.

McLean, K.S., and Langille, W.M., (1973). *Commun. Soil Sci. Plant Anal. 4*, 495.

McLean, A.J., and Dekker, A.J., (1978). *Can. J. Soil Sci., 58*, 381.

Menge, J.A., Lembright, H., and Johnson, E.L.V. (1977). *Proc. Int. Soc. Citriculture, 1*, 129.

Merry, R.H. (1980). "The Accumulation of Copper, Lead and Arsenic in Orchard Soils and its Effects on Plants", M. Ag. Sc. Thesis, University of Adelaide.

Miles, L.J., and Parker, G.R. (1979). *J. Environ. Qual. 8*, 443.

Mills, C.F., and Dalgarno, A.C. (1972). *Nature (London) 239*, 171.

Mochizuki, T., Chiba, S., Hanada, S., and Saitoh, H. (1975). *Nippon Dojo-Hiryogaku Zasshi 46*, 45.

Morishima, H., and Oka, H-I. (1977). *Agro-Ecosystems 3*, 131.

Netzer, A., and Beszedits, S. (1979) *In* "Copper in the Environment. Part 1 : Ecological Cycling" (J.O. Nriagu, ed.), p. 123. John Wiley and Sons, Inc., New York.

Nriagu, J.O. (1979a). *In* "Copper in the Environment" (J.O. Nriagu, ed.), p.13. John Wiley and Sons Inc., New York.

Nriagu, J.O. (1979b). *In* "Copper in the Environment" (J.O. Nriagu, ed.), p.44. John Wiley and Sons Inc., New York.

Nriagu, J.O. (1979c). *Nature (London) 279*, 409.

Page, A.L. (1974). U.S. Environmental Protection Agency, EPA-670/2-774-005, Cincinnati, Ohio.

Page, A.L., Elseewi, A.A., and Straughan, I.R. (1979). *Residue Rev. 71*, 83.

Patterson, (1971). *In* "Trace Elements in Soils and Crops" p. 193. Techn. Bull. No. 21. Ministry of Agriculture, Fisheries and Food, London.

Purves, D. (1977). "Trace-Element Contamination of the Environment." Elsevier, Amsterdam.

Quraishi, M.S.I., and Cornfield, A.H. (1973). *Environ. Pollut. 4*, 159.

Reuther, W., and Smith, P.F. (1953). *Soil Sci. 75*, 219.

Reuther, W., and Smith, P.F. (1954). *Soil Sci. Soc. Fla., Proc.* *14*, 17.

Rieder, W., and Schwertmann, U. (1972). *Landwirtsch. Forsch.* *25*, 170.

Roberts, R.M., and Goodman, G.T. (1973). *Proc. Univ. Mo. Annu. Conf. Trace Subst. Environ. Health VII*, 117.

Robinson, K., Draper, S.R., and Gelman, A.C. (1971). *Environ. Pollut. 2*, 49.

Rühling, A., and Tyler, G. (1973a). *Oikos 24*, 402.

Rühling, A., and Tyler, G. (1973b). *Water, Air, Soil Pollut.* *2*, 445.

Rutherford, G.K., and Bray, C.R. (1979). *J. Environ. Qual. 8*, 219.

Sommers, L.E. (1977). *J. Environ. Qual. 6*, 225.

Stewart, I., and Leonard, C.D. (1952). *Science 116*, 564.

Swaine, D.J. (1955). *Commonw. Bur. Soil Sci., Tech. Commun. No. 48.*

Swaine, D.J. (1962). *Commonw. Bur. Soils, Tech. Commun. No. 52.*

Taylor, J.M., Epstein, E., Burge, W.D., Chaney, R.L., Menzies, J.D., and Sikora, L.J. (1978). *J. Environ. Qual., 7*, 477.

Teakle, L.J.H. (1942). *J. Aust. Inst. Agric. Sci. 8*, 70.

Todd, J.R., and Thompson, R.H. (1965). *Br. Vet. J. 121*, 90.

Tyler, G. (1974). *Plant Soil 41*, 303.

Tyler, G. (1975). *Nature (London) 255*, 701.

Tyler, G. (1978). *Water, Air, Soil Pollut. 9*, 137.

Underwood, E.J. (1971). 'Trace Elements in Human and Animal Health', 3rd Edition, Academic Press, New York.

van Alphen, T.G. (1957). Literatuuroverzicht NR. 20, Centrum voor Landbouwdocumentatie/Wageningen.

Walsh, L.M., Erhardt, W.H., and Siebel, H.D. (1972). *J. Environ. Qual. 1*, 197.

Wander, T.W. (1954). *Soil Sci. Soc. Fla., Proc. 14*, 27.

Wangen, L.E., and Turner, F.B. (1980). *Water, Air, Soil Pollut. 13*, 99.

Weiss, H., Bertine, K., Doide, M., and Goldberg, E.D. (1975). *Geochim. Cosmochim. Acta 39*, 1.

COPPER IN PLANTS

7

ABSORPTION OF COPPER BY PLANT ROOTS

Robin D. Graham

Department of Agronomy
Waite Agricultural Research Institute
The University of Adelaide
Glen Osmond, South Australia

INTRODUCTION

In the fifty years since Cu was established as an
essential element for higher plants, numerous studies both
by pure and applied biologists have greatly increased our
knowledge of this fascinating metal in nature; yet there is
much more to discover both in the realms of agronomy, where
deficiency and toxicity of Cu are significant problems, and
in the biochemical arena. In particular the subject of
this paper is one that is poorly researched to date and my
aim is to define the limits of present knowledge and to set
guidelines for future research.
The return to the land of animal wastes containing Cu
in high concentrations has led to renewed interest in the
absorption of Cu by plants. Since the amounts required by
the plant are so small, it has always appeared possible that
Cu, and other trace elements, may enter without the plant
having exerted any control. A gentle debate has been
sustained over the years, centred on whether absorption of
minor nutrients is 'passive' or 'active', the latter implying
the intervention of metabolism in the absorption process.
As we shall see, the evidence for active absorption of Cu is
increasing but the case is not yet complete.

COPPER IN SOILS AND PLANTS

I.　THEORETICAL CONSIDERATIONS

The chemistry of Cu has already been described in the
early chapters of this book.　The two oxidation states are
together important in biology but it is generally the Cu^{2+}
state which is dominant in the soil environment of plant
roots.　The ready reduction of Cu^{2+} by even weak reducing
agents means, however, that we ought not to exclude Cu^+ from
our thinking about the absorption process.

It is of primary importance in studying absorption of Cu
to note that most chelates bind Cu^{2+} strongly, due to its
high ligand field stabilization energy (Parker, 1981;　Walker
and Webb, 1981) which, with the strong preference of Cu^{2+} for
tetragonal co-ordination geometry, and its affinity for
peptide nitrogen provides a potential basis for selectivity in
absorption,firstly in distinguishing $Cu^{2+}$from $Cu^+$and secondly
in selecting Cu^{2+} from Zn^{2+} (which prefers tetrahedral
geometry - Frausto da Silva and Williams, 1976) or from
Mn^{2+} which has a low affinity for such chelates.　It is
also important to note the affinity of Cu^{2+} for carbonyl,
hydroxyl, thiol, carboxylic and phenolic groups since these
are important in cell walls.　Cell walls contain sites for
the strong and specific adsorption of Cu which is not easily
desorbed.

For a root cell which is absorbing Cu from a dilute
external solution, we can formalize the resultant (or nett)
force tending to move Cu into the cell using equilibrium
thermodynamics, ignoring the fluxes between the cell and
its environment.　The resultant driving force on Cu ions
is known as the electrochemical potential difference for the
ion in the two different phases:

$$\Delta \bar{\mu}_j = z\ FE\ +\ RT\ \ln \frac{a_j{}^i}{a_j{}^o}$$

where in this case $a_j{}^i$ and $a_j{}^o$ are the chemical activities
of the free ion inside and outside the root cell respectively,
and E the electrical potential difference between the
two phases.　R is the gas constant, T the absolute
temperature, z the valency (including its sign) and F the
faraday.　The first term in this equation is the electrical,
and the second term the chemical potential component of the
total electrochemical potential difference.　When $\Delta \bar{\mu}_j$ is
negative, there is a driving force tending to move ions into
the cell, and out of the cell when $\Delta \bar{\mu}_j$ is positive.　The

second term (chemical potential) is usually positive, the concentration being greater inside, but if the absorbed species is Cu^{2+} and bearing in mind the high negative membrane potential (PD) which exists in root cells (Graham and Bowling, 1977), the first term can be negative and dominant (Graham, 1979). When $\Delta\bar{\mu}_j$ is negative the ion may move into the cell down an energy gradient *provided a diffusion pathway exists* and no further energy beyond that required to maintain the membrane PD should be needed for absorption. Where $\Delta\bar{\mu}_j$ is positive, as for absorption of anions, it is clear that energy must be expended to 'pump' the ions up an energy gradient into the cell. An ion pump located in the membrane and coupled to the cell's metabolism is usually envisaged as the means of absorption of ions against a gradient of $\bar{\mu}_j$.

Where we are dealing with the Cu^{2+} ion, we are fortunate to have available a specific ion electrode which permits measurement of the activity (concentration) of this ion. The cupric electrode gives a close to theoretical change in PD of 29 mV per 10-fold change in $a_{Cu^{2+}}$.

The question of whether active transport is involved in the absorption of trace elements has been discussed more frequently over the last decade or so. Seen simply, active transport of ions into plant cells involves the expenditure of metabolic energy from the cell to effect the absorption. However, scientists argue about the details of how to define active transport (Bowling, 1976). A useful definition is that of Ussing (1949) who defined active transport as transport against an electrochemical potential gradient. Such transport would involve expenditure of energy. However, transport *down* a gradient in $\Delta\bar{\mu}_j$ may also involve expenditure of some energy to maintain the full transmembrane PD (Graham and Bowling, 1977). Therefore, evidence that absorption of Cu is inhibited by metabolic inhibitors or uncouplers such as CN^- or DNP does not rule out the possibility that Cu may be entering the cell by diffusion of Cu^{2+} down its electrochemical potential gradient. For the hypothesis that absorption of Cu occurs by diffusion under an electrochemical potential gradient, it is virtually essential that Cu^{2+} be the absorbed species for which there is considerable indirect evidence (see section 5).

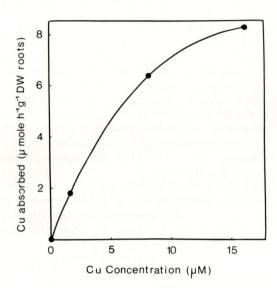

FIGURE 1.   *Absorption of Cu by intact citrus*
seedlings as a function of external
Cu concentration.
(After Hiatt, Amos and Massey, 1963).

Mineral elements do not readily leak out through the
plasmalemma of root cells once they are absorbed, yet for
absorption they must penetrate this fairly impermeable
membrane.   This paradox has led to the Carrier concept
which involves the binding of an ion to a carrier molecule
on the outer surface of the membrane, the subsequent
movement of this ion-carrier complex through the membrane
and finally the re-configuration of the complex to cause
release of the element/ion into the cytoplasm.   In the case
of Cu this carrier could easily be a Cu-specific polypeptide
to which binding would be very strong.
The binding of an ion to an ionophore in the membrane is
analogous to the binding of substrate and enzyme.
Consequently enzyme kinetics are frequently used to describe
the kinetics of absorption of ions (Epstein, 1972).   The
rate of absorption is given by

$$v = \frac{V_m [Cu]}{K_m + [Cu]}$$

where V_m is the maximum rate of absorption and $[Cu]$ is the concentration of Cu in the external medium. K_m, the rate constant, is equal to $[Cu]$ when $v = \frac{1}{2}V_m$. It is convenient to describe this type of uptake kinetics in terms of K_m and V_m. Figure 1 shows the type of uptake data which is consistent with this absorption model.

Epstein (1972) has described a dual carrier model of absorption, commonly found for K^+ in which two carriers are envisaged, the second operating at a higher external concentration than the first, and giving higher rates of absorption. This idea has been extended to the multiphasic concept of absorption and reinterpreted by Nissen (1973) as a single carrier operating at a different rate for each concentration range under study. Dual and multiphasic absorption characteristics for Cu have been suggested (see section 5).

II. METHODS OF STUDYING ABSORPTION OF COPPER

The rate of absorption of Cu from external concentrations in the physiological range is small indeed and accordingly difficult to measure, especially against the background level in the plants and its variability from plant to plant. A simple way out of this dilemma is to allow the absorption time to be long, thus magnifying the total amount of Cu taken up. Data of this type are commonly handled by the mathematical procedure described by Williams (1948), which involves the integration over a time interval of the differential rate equation. The instantaneous rate of intake (I_{Cu}) of Cu per unit weight of root W_R can be written as

$$I_{Cu} = \frac{1}{W_R} \frac{dM}{dT}$$

where M is the amount of Cu in the plant at time t. To integrate this equation, it is necessary to make an important assumption: that over the interval $t_1 \rightarrow t_2$ there exists a linear relationship between W_R and M. Then

$$I_{Cu} = \frac{1}{t_2 - t_1} \frac{dM}{dW_R} \int_{t_1}^{t_2} \frac{1}{W_R} dW_R = \frac{\ln W_{R_2} - \ln W_{R_1}}{t_2 - t_1} \cdot \frac{M_2 - M_1}{W_{R_2} - W_{R_1}}$$

It is clear that this assumption is not always met but the equation is a useful one.   It is also assumed that the change in the amount of Cu adsorbed in root cell walls, which can be considerable, is nevertheless small in comparison with the total absorption over the interval. This method is particularly suitable for studies which encompass all or a large part of the life cycle of a plant, where root systems become rather large (Jarvis and Robson, 1981).   Data may be expressed on a fresh or dry root weight basis.   Although less common, fresh weight basis is to be preferred since it is highly correlated with root volume and reasonably correlated with root surface area (Graham, Anderson and Ascher, 1981).   The integrated form of the above equation should generally be used over time intervals not exceeding two weeks (see Williams, 1946;   1948).

An interesting variation on this approach is that of Graves (1978) who measured uptake of Cu by *Chrysanthemum* by sequential harvests over a period of about 80 days.   Total content of Cu per plant was plotted against time and a second order regression equation fitted to the curves (one for high and one for low concentration of external Cu). Differentiating these equations gave the rate of Cu absorption for any time t and by dividing by root weight at the appropriate time, an absorption rate of μg Cu h^{-1} g^{-1} (root D.W.) could be calculated.   His results show the absorption rate decreasing with increasing plant age for both high and low concentration systems.

A number of other workers have taken the simpler approach of assuming linearity over the time interval both of absorption rate and root growth and calculated:

$$I_{Cu} = \frac{\Delta M}{\bar{W}_R \, \Delta t}$$

Nielsen (1976) used $\Delta M / \Delta t$ as the mean rate of absorption of Cu on a per plant basis and from data obtained over a range of Cu concentrations, expressed the results in terms of Michaelis–Menten enzyme kinetic constants (Table I).

Excised roots have been used to study the absorption of Cu and the data have usually been analysed using Michaelis-Menten kinetics (Hiatt *et al.*, 1963; Veltrup, 1976, 1977, 1979; Harrison, Lepp and Phipps, 1978, 1979; Cathala and Salsac, 1975). With excised roots, uptake periods are short but controlled and reproducible conditions permit the detection of the amount of Cu absorbed by standard analytical techniques, provided external concentrations are reasonably high. With these studies a desorption treatment must follow the uptake period in order to remove Cu adsorbed to specific binding sites in the root cell walls. These sites are numerous and highly specific for Cu. Half or more of the total Cu in roots may be bound to cell wall sites (Hiatt *et al.*, 1963; Brams and Fiskell, 1971; Cathala and Salsac, 1975; Jarvis, 1978; Jarvis and Robson, 1981). While Cu may be adsorbed onto negative cation exchange sites along with other ions, it is clear that much of this Cu in the apoplast is strongly attached to multidentate organic ligands involving N, S and O atoms; Cu in coordinate complexes of this kind cannot be readily desorbed in the normal way with cold Ca^{2+}, or indeed by Mg^{2+}, Ba^{2+}, Cd^{2+}, Co^{2+}, Zn^{2+} or Ni^{2+}; but it can be desorbed by Cu itself (^{63}Cu-^{64}Cu exchange) and by Pb^{2+} which has similar coordination chemistry to Cu^{2+} (Larsen, 1966; Harrison *et al.*, 1979). Because of the magnitude, intensity and specificity of this adsorbed apoplasmic Cu, it is recommended that Cu^{2+} or Pb^{2+} be used to desorb the apoplast in all short-term studies of absorption of Cu into the symplast of roots. Results using excised roots with Pb^{2+} desorption have recently been reported by Harrison *et al.* (1979a). The conclusions from many other studies must therefore remain under a cloud, since most have used Ca^{2+} for desorption or used no desorption at all. Hiatt *et al.* (1963) used 0.1M HCl for desorption. H^+ appears to be reasonably effective for desorbing Cu^{2+} (Fiskell and Brams, 1965; Cavallaro and McBride, 1980) but there is a risk of damage to cell membranes (Fiskell and Brams, 1965; Epstein, 1972).

A few students of this subject have resorted to the only useful radioisotope of Cu for this purpose, ^{64}Cu, which has a very short half life and is a moderate β and γ emitter. The energy of these particles make ^{64}Cu suitable for counting in intact tissues without prior digestion and this saving in time is important with a short-lived isotope. Graham, Anderson and Ascher (1981) made use of the advantages of ^{64}Cu to study absorption from concentrations as low as 1 nM Cu. This is approximately the lower limit for this

method, being set by the specific activity of the ^{64}Cu
available, its half life and the amounts of ^{64}Cu absorbed by
the roots in a reasonable time. Desorption with Pb^{2+} or
$^{63}Cu^{2+}$ (instead of Ca^{2+} as used by them) would undoubtedly
decrease ^{64}Cu counts in the roots even more. Nielsen
(1976), Wallace, Sufi and Romney (1971), Brams and Fiskell
(1971), Dokiya *et al.* (1968) and Larsen (1966) have used
^{64}Cu for absorption studies on intact plants of a range of
genera. With ^{64}Cu studies, Δt is short and rates of
absorption, after the initial rapid uptake into the apoplast,
are usually linear with time, or assumed to be; I_{Cu} becomes

$$I_{Cu} = \frac{\Delta M}{W_R \Delta t}$$

where ΔM is determined from the ^{64}Cu content of the tissues
at the end of the time interval Δt. Often, M_1 and t_1 may
be taken as zero if the roots are effectively desorbed.
Graham *et al.* (1981) found insignificant amounts of ^{64}Cu
in the shoot of cereal seedlings after absorption periods
of 1 hour. Unpublished data suggest further that the shoot
could be ignored for absorption periods up to 4 hours or
more in such studies using ^{64}Cu.

Absorption of Cu has been studied in solution culture
systems by measuring the depletion of the Cu in the medium
over time (Larsen, 1966; Nielsen, 1976). Both workers used
^{64}Cu. A theoretical treatment of this technique has been
given by Claassen and Barber (1974).

Jarvis (1981) and Jarvis and Robson (1981) have reported
rates of absorption of Cu from flowing culture solutions.
In fact, rates of absorption of Cu are so low that serious
depletion is not often a problem in still cultures so that
flowing cultures are in that sense unnecessary. However,
data from flowing cultures are valuable in that the macronutrients are kept at more realistic, low concentrations as
well, and very low concentrations of Cu in the medium can be
maintained for long periods of time (e.g. Jarvis and
Robson, 1981).

There is a remarkable dearth of data on absorption of Cu
from soil systems where all parameters needed for
calculations of absorption rates have been measured, that
is, solution phase Cu concentration, root size, root Cu
content, each at sequential harvests to give the time factor
for the calculations. Commonly, no account is taken of the
effect of time on the growth rate of the absorbing surface.
Again it is difficult to relate absorption to a supply
factor such as Cu concentration of the solution phase.

Nielson (1976) measured the latter by a displacement
technique and reported the corresponding uptake of Cu by the
aerial parts of barley over various intervals of time from 9
to 90 days. Studies of toxic levels of Cu are often
conducted in soil systems but responses are usually interpreted in terms of the concentrations of Cu in plant parts,
usually tops only, after a growth period of one to several
weeks under standard conditions (e.g. Wu *et al.*, 1975;
Morrison *et al.*, 1979).

III. RATES OF ABSORPTION OF COPPER

Table I presents some of the available data on rates of
absorption or uptake of Cu into roots. The techniques of
measurement have varied greatly and it is difficult to decide
whether there is any general agreement among the results.
Rough comparisons may be made among the figures by using
1 cm^3 roots $\equiv$ 1 g F.W. roots $\equiv$ 0.05-0.1 g D.W. roots. Of
the data at high concentration (1 μM or more), the absorption
rates of excised roots (Veltrup, 1976; Harrison *et al.*,
1979) are much higher than in the intact plants of Kausar
et al. (1976). The latter are, however, within a factor or
two of the results of Graham *et al.* (1981) at the top end of
their range (10 μM), also with intact plants. This
difference may be associated with the fact that the excised
roots were Cu-starved before the absorption period, or due to
omission of competing ions such as Zn.

At the low concentration end of the range, there is
tolerable agreement among the results of Jarvis (1981), Jarvis
and Robson (1981) and Graham *et al.* (1981), all with intact
plants, although the latter used seedlings in short-term
studies, quite the opposite of the former two. Studies in the
concentration range approaching Cu deficiency in plants are
rare. From flowing culture work, Graves and Sutcliffe (1974)
have indicated that *Chrysanthemum* will not become Cu deficient
until the external concentration is below 10 nM Cu; for wheat
the figure is about 1 nM Cu (Snowball and Loneragan, see
Loneragan, 1975). Using chelation to lower the free Cu^{2+} ion
activity (and assuming Cu^{2+} is the absorbed species) Dragun
et al. (1976) and Graham (1979) in different ways considered
that deficiency would occur at 10^{-13} to 10^{-14} M Cu^{2+} (free
ion) for maize and sunflowers respectively.

The only data for a soil system in Table I is that of
Nielsen (1976) mentioned earlier. Comparison with the data
for solution culture under otherwise similar conditions shows

TABLE I.   Rates of Absorption of Copper by Higher Plant Roots

Conc. of Cu external	Rate of Cu absorption	Species	Comments	Reference
5 μM	10 μ mole $h^{-1}g^{-1}$ D.W. roots	Wheat	excised roots in Cu $(NO_3)_2$ solution; H^+ desorption (Δt 0.5h)	Hiatt et al. 1963
10 μM	40 n mole $h^{-1}g^{-1}$ F.W. roots	Citrus	4 mo. old seedlings (Δt 3h)	Brams and Fiskell 1971
10 μM	6.3, 3.3 n mole $h^{-1}g^{-1}$ F.W. roots	Rice,wheat	excised roots; solution; no EDTA (Δt 2h); Ca^{2+} desorption	Kausar et al. 1976
75 nM	2.6 n mole h^{-1} plant^{-1}	Barley	intact plants; K_m and V_m (Δt 15 days); extracted soil solution	Nielsen 1976
150 nM	0.05 n mole h^{-1} plant^{-1}	Barley	plants in soil; K_m and V_m (Δt 15 days); uptake into shoot	Nielsen 1976
9 μM	1.5 μ mole $h^{-1}g^{-1}$ D.W. roots	Barley	excised roots; K_m and V_m (Δt 2h); no desorption; no EDTA	Veltrup 1976

Concentration	Plant	Rate	Conditions	Reference
5-10 nM ~800 nM	Chrysanthemum	0-1 1.3-4 n mole h^{-1} lg^{-1} D.W. roots	rooted cuttings (Δt 2-10 days)	Graves 1978
19 μM	Barley	4 $μ$ mole h^{-1} lg^{-1} D.W. roots	excised roots; K_m and V_m (Δt 0.5h) Pb^{2+} desorption; no EDTA	Harrison et al. 1979
10 nM	Rye, wheat	0.4, 0.4 p mole h^{-1} cm^{-1} root	))	
10 nM	Rye, wheat	4.2, 3.3 p mole h^{-1} cm^{-2} root	)) intact plants in solution; (Δt 1h) Ca^{2+} desorption; EDTA at 10 μM	Graham et al. 1981
10 nM	Rye, wheat	510, 310 p mole h^{-1} cm^{-3} root	)))	
X nM	Wheat-rye hybrids	100 $X^{½}$ p mole h^{-1} lg^{-1} F.W. roots	as above; range of X = $1-10^4$ nM;	Graham et al. 1981
80 nM	Ryegrass	3.2 n mole h^{-1} lg^{-1} D.W. roots	intact plants in flowing cultures (Δt 7 days)	Jarvis 1981
<1 nM ~100-300 nM	(Wheat, red clover)	0.7, 1 3.0, 2.6 n mole h^{-1} lg^{-1} D.W. roots	intact plants (mixed) in flowing cultures (Δt 2-13 days)	Jarvis and Robson 1981

the absorption rate from stirred solutions is much higher
than from soil for comparable concentrations of Cu in the
solution phase.   As Nielsen (1976) points out, however, the
K_m for the two systems has a similar range.

Absorption into leaf cells can be mentioned here. Bowen
(1969) studied absorption of Cu by leaf discs of sugar cane
and found by Michaelis-Menten kinetic analysis, absorption
characteristics as follows: K_m 15 µM Cu, V_m 5.4 µ mole h^{-1}
g^{-1} D.W. leaf.   These constants are of the same order as
those of Veltrup (1976) and Harrison et al. (1979) for
excised roots, the external concentrations of Cu for which
are all rather high.   However, since Cu is accumulated in
the xylem to concentrations higher than those normal in the
soil solution (Pate et al., 1975; Graham, 1979) the
concentrations used by Bowen seem to be in the normal to high
physiological range.

The available reports divide into those showing a linear
relationship between absorption rate and external Cu
concentration (Brams and Fiskell, 1971; Cathala and Salsac,
1975; Graham et al., 1981) and those showing a curvilinear
relationship such as in Figure 1 (Hiatt et al., 1963;
Cathala and Salsac, 1975; Veltrup, 1976; Harrison et al.,
1978, 1979; Bowen, 1979), all of whom studied a relatively
high concentration range.

IV. FACTORS AFFECTING ABSORPTION OF COPPER BY ROOTS

While aeration (oxygen) is required for absorption of
many macronutrient ions, it is not clear that this is so for
absorption of micro-nutrients, especially for Cu.   The
evidence of Cathala and Salsac (1975) that 'aeration' with
nitrogen had no effect on uptake of Cu by excised roots of
sunflower or maize is not relevant, since there was no
desorption of Cu from the free space, and the effect of
nitrogen on uptake could be dominated by this physical
component of the uptake.   Likewise, their studies of the
effects of temperature and DNP on uptake, like those of
Goren and Wanner (1971) are not relevant to absorption into
the symplasm because of adsorption effects.   In contrast,
Bowen (1979) found that absorption of Cu by excised roots of
barley and sugarcane was largely inhibited (80-85%) by
metabolic poisons.   Absorption of Cu by leaf discs of
sugarcane (after desorption with Ca) was also largely
eliminated by metabolic inhibitors or uncoupling agents.
Further the effect of temperature on Cu absorption by leaf

discs was considerably greater than those described above, but Q_{10} was still obviously less than 2.0.

The evidence for metabolic control of absorption of Cu is still far from being adequate or convincing and it is probably unwise to extrapolate to root systems from the data of Bowen (1969) on sugarcane leaf discs which is so far the most comprehensive study. Data are needed using intact plants, meaningful concentration ranges, effective desorption techniques, and longer absorption times. Use of ^{64}Cu for absorption with ^{63}Cu for desorption is possibly the best approach but the short half life works against longer absorption times. Several reports have shown changing kinetics with increasing time of absorption (Cathala and Salsac, 1975; Veltrup, 1977; Harrison *et al.*, 1978) but none of these has used satisfactory desorption techniques.

The effect of pH on absorption of Cu is also not clear. For one thing, if Cu is largely chelated even in culture solutions, pH will affect the degree of chelation (Cavallaro and McBride, 1980) especially in soil systems. Secondly, both high and low pH extremes can damage tissues and impair their absorption capacity (Bowen, 1969; Epstein, 1972). Thirdly, H^+ ions may compete with Cu^{2+} for adsorption and/or absorption sites. This effect appears to be lessened in the presence of Ca^{2+} as would normally be the case (Bowen, 1969).

There are many effects of other ions on rate of Cu absorption, some of which tell us a little about the nature of the absorption mechanism. Surprisingly, although Cu and N interact most strongly in the nutrition of plants (Loneragan, 1981) there appears to be no dramatic effect of N ions or of phosphate on absorption (DeKock *et al.*, 1971). K decreased absorption of Cu by sunflower (Graham, 1979) in a manner consistent with its depolarizing effect on the membrane PD, thereby decreasing the electrochemical gradient on the Cu^{2+} ion. Ca^{2+} has been reported to inhibit Cu uptake where there was no desorption (Cathala and Salsac, 1975), where desorption with 0.1 M HCl was used (Hiatt *et al.*, 1963) or where Ca desorption was used (leaf discs, Bowen, 1969).

Among the micronutrient interactions, Zn figures prominently, suggesting interactions at the adsorption or absorption site. Bowen (1979) found Zn and Cu each competitively inhibited absorption of the other by roots of two species and concluded (as with leaf discs) that Cu and Zn are absorbed by the same mechanism (see also Schmid *et al.*, 1965; Giordano *et al.*, 1974; Kausar *et al.*, 1976; Hawf and Schmid, 1967; Chaudhry and Loneragan, 1972). The competitive effects observed presumably occur at absorption sites in the plasmalemma but effects at adsorption sites in the cell walls cannot be overlooked because of inadequate desorption of this phase in most studies to date.

Hiatt *et al.* (1963) reported that Al decreased uptake of Cu by wheat roots, affecting both adsorption and absorption, the effect on the former being competitive and on the latter non-competitive, but note that in this study, cell walls were desorbed with HCl.

Wallace and Romney (1977) reported that Cu stimulated strongly the uptake of Mn into roots and shoots of faba beans from solutions.   Concentrations were high.   Synergistic effects between these two ions have been reported a number of times, not all under well defined conditions (e.g. Younts, 1964), but they are numerous enough to be interesting.   On the other hand, there are reports of no effect (Bowen, 1969, leaf discs) and of inhibition (Dokiya *et al.*, 1968).

V.   MECHANISMS OF ABSORPTION OF COPPER

The mechanisms of absorption of copper are far from clear at present, and some definitive experiments are called for to clarify the issue.

To begin with, in both soil and solution culture systems, Cu is almost entirely complexed either by root exudates or organic ligands of soil origin (Geering and Hodgson, 1969; Graham, 1979; and unpublished data).   On the other hand, there is considerable evidence that Cu is not absorbed as a chelate whether the ligand be EDTA or soil humic materials. Cu is absorbed more rapidly from Cu^{2+} solutions than from Cu EDTA or Cu DTPA (DeKock and Mitchell, 1957; Wallace *et al.*, 1971; Dragun *et al.*, 1976; Harrison *et al.*, 1979b), and high molecular weight humic materials from soil also decreased the rate of absorption of Cu (Petruzzelli and Guidi, 1976).   Complementary evidence using a different approach is that of Goodman and Linehan (1979) who studied the absorption of Cu by a root in an electron paramagnetic resonance (EPR) spectrometer.   The spectra were similar for the absorption of Cu^{2+} or Cu EDTA, except that, as expected, the spectrum of Cu^{2+} itself was weak where Cu EDTA was used.   The sequence of spectra during uptake suggested that Cu was bound to a macro-molecule or solid surface prior to or during absorption, that is, the Cu EDTA dissociated.   Thus although Cu exists in the root environment dominantly as a chelate, it appears that normally the ligand is lost prior to absorption.

The question now arises, is the specific adsorption of Cu on the cell walls an important preliminary step in the absorption process? The data of Hiatt *et al.* (1963) are consistent with the view that it is: they showed that the isotherms for adsorption of Cu by the free space (desorbed by 0.1 M HCl) and absorption of Cu into the symplasm of excised wheat roots had basic similarities, and further, that although Al inhibited absorption of Cu, this inhibition arose from competition for adsorption sites in the cell walls and not from competition for absorption sites on the cell membrane. Cathala and Salsac (1975) showed that the much higher rate of uptake of Cu by sunflower compared to maize was associated with a much higher adsorption of Cu in cell walls. It would be interesting to separate their uptake rate into its components to see if similar differences existed in rates of absorption of Cu by the symplasm. Goodman and Linehan (1979) favour the interpretation of their evidence of a bound form of Cu during the uptake process as being this cell wall-bound Cu, rather than membrane-bound Cu.

Of the absorption studies reported to date, many show saturation kinetics of the type shown in Figure 1, that is, the rate of absorption of Cu is a hyperbolic function of the external Cu concentration. Saturation uptake curves for Cu have been reported by Hiatt *et al.* (1963), Goren and Wanner (1971), Cathala and Salsac (1975), Nielsen (1976), Bowen (1979), Veltrup (1976, 1977), and Harrison *et al.* (1978, 1979a). In half of these studies, there was no desorption step but in the studies of Harrison *et al.* the curvilinear isotherm was still obtained when the desorbing ion was changed from Ca to the more satisfactory Pb. Most data of this kind are amenable to Michaelis-Menten enzyme kinetics analysis and interpreted as consistent with the carrier hypothesis of ion transport (Epstein, 1972). In the case of Cu, with its great affinity for binding to polypeptides, a membrane-bound protein is envisaged which is capable of specific, stable binding of Cu^{2+} in competition with cell wall binding sites. Saturation of the system is seen to occur when the availability of Cu^{2+} is so high that all available proteins are working at the fastest possible rate (V_m).

Harrison *et al.* (1978, 1979a) and Veltrup (1976, 1977) reported response curves with inflexions indicative of two or more (dual) carriers operating at different rates and in different concentration ranges. Data of this kind are also interpreted as indicating a single multi-phasic mechanism (Nissen, 1973). However, in the data available to date,

such inflexions, while they may be reproducible, are small
and may not be important.   One such discontinuity reported
by Harrison *et al.* (1978) was eliminated subsequently by
their Pb^{2+} desorption technique (Harrison *et al.*, 1979a).
Further, biphasic absorption characteristics with a 2h
absorption time became mono-phasic when absorption time was
24h (Veltrup, 1976, 1977), again possibly an effect of
adsorption.   Graham *et al.* (1981) interpreted possible
discontinuities in their data as due to environmental
variation in radiation, temperature, aeration rate and
concentration.

The carrier hypothesis envisages the coupling of ion
transport to cellular metabolism, possibly to effect the
movement of the carrier molecule through the membrane and/or
the desorption of Cu^{2+} on the inside where, according to
Goodman and Linehan (1979) Cu becomes bound to amino acids
in complexes such as Cu (glutamate)$_2$ or Cu (glycine)$_2$.

An alternative hypothesis is that Cu^{2+} is the absorbed
species and it is driven into root cortical cells by a
large gradient in electrochemical potential (Graham, 1979).
This mechanism requires that there be a diffusion pathway
for Cu^{2+} through the membrane, which appears to be a
weakness in the hypothesis because of the great tendency of
Cu^{2+} to bind to organic ligands.   A hybrid hypothesis is
possible where Cu^{2+} binds to membrane proteins but has a
greater probability of release on the inside of the membrane
than on the outer, owing to the lower electrochemical
potential of Cu in the cytoplasm.   In any case the
electrochemical gradient hypothesis is consistent with most
available data, since energy is required to maintain the
membrane potential (Graham and Bowling, 1977) which
contributes a major part of the $\Delta\bar{\mu}_j$ across the membrane.
Thus absorption of Cu should be sensitive to metabolic
inhibitors as reported by Bowen (1969, 1979).   Bowen (1969,
1979) considered Cu and Zn to be absorbed by the same
mechanism which was different from that of Mn, as well as
from Al (Hiatt *et al.*, 1963).   The report of Harrison
et al., 1979b) that rate of absorption of Cu increased with
increasing positivity of charge among several Cu chelates is
also consistent with the electrochemical potential hypothesis.
It is probably also consistent with a carrier model in which
specific adsorption in the free space is a preliminary
step.

There are some reports in which there is no suggestion of saturation kinetics, absorption rate being a linear function of concentration (see section 3). Such data do not seem consistent with a carrier hypothesis but better desorption techniques and greater control of variability may change the significance of these results.

VI. GENETIC DIFFERENCES IN ABSORPTION OF COPPER

There are marked genotypic differences in the Cu nutrition of plants. Differences in the rate of absorption of Cu by roots is only one of many possible mechanisms by which genotypic differences may be expressed. Other strategies which are recognised in the evolution of plants include better exploration of soil through greater root length per plant or per hectare, better contact with soil through longer root hairs, modification of availability of Cu in soil adjacent to roots by root exudation, acidification or change in redox potential, more efficient transport of Cu from roots to shoots, more efficient retranslocation of Cu from old tissues to the growing points, and/or lower tissue requirement for Cu. All these strategies, including high rates of absorption of Cu, could be involved in making one genotype better adapted to growth on a soil of low Cu availability. Most of them could be viewed in reverse for adaptation to soils with excessive contents of Cu.

So far there is little evidence for any of these mechanisms operating as the basis for genetic differences between plants adapted to soils of low Cu status. Certainly, the differences in rates of absorption shown in Table I cannot be ascribed to genetic differences in most cases, except where comparisons were made under the same experimental conditions; important differences in techniques among the workers represented are likely to be a major contributor.

One of the classic cases of a genetic difference in Cu nutrition is that between rye and the other temperate small grains, wheat, oats and barley. Rye is markedly more efficient in growth and grain production on Cu deficient soils and this genetic advantage over wheat is inherited by the wheat-rye hybrid, triticale (Graham and Pearce, 1979). Rye extracts up to twice as much Cu as wheat under the same conditions. Further, the so-called Cu efficiency factor of rye is carried on one chromosome (5R) and indeed is now known to be located on its long arm (5RL) (Graham, 1978).

On analogy with other reports of single-gene nutritional factors (Epstein, 1972), we have argued that the probability is high that the Cu efficiency factor of rye is under single gene control. The mechanism of the efficiency factor in rye is, however, not yet clear. Firstly it is well known that rye has a more extensive root system than wheat or other cereals (Nuttonson, 1958) but it is not clear whether this is cause or effect in its better adaptation to poor environments. An attempt to look at the basis of Cu efficiency of rye has so far proved inconclusive (Graham *et al.*, 1981). The rate of absorption of Cu per unit of root depended on the model of ion absorption which is accepted. From solutions in short term uptake studies with intact plants (Graham *et al.*, 1981 and Table I), rye had greater absorption rates than wheat on a per plant or per unit weight or volume of roots basis but not on a per unit root length or surface area basis. We are currently considering the possibility that rye may be more efficient at loading Cu from root cortex to xylem for transport to the shoots. Two reports suggest that Cu is dammed up in the roots around the stele in the endodermis-pericycle region (Chino and Baba, 1979; Brams and Fiskell, 1971). Further, the electrochemical gradient for Cu^{2+} is uphill for the movement of this ion from cortex to xylem (Graham, 1979), thus requiring energy or chelation with an anion to make the transition. This step is therefore a likely one for expression of genetic differences.

If differences in the Cu nutrition of rye and wheat cannot be satisfactorily explained by differences in the absorption and/or translocation of Cu from solutions, there remains still the possibility of a genetically controlled difference at the root-soil interface. Nielsen (1976a) has shown that barley plants can, after a few weeks, promote the availability of Cu in the soil apparently by an increase in organic ligands in the soil solution. Rye produces a great deal of mucilage which binds soil particles to the roots, but soluble ligands have not been measured.

The possibility that Cu efficiency in rye is a single gene effect invokes the concept of a single macro-molecular carrier, a protein probably, located in the plasmalemma of the cortical cells controlling uptake, or in the xylem parenchyma controlling xylem loading. Alternatively, a single gene might code for specific Cu binding sites in the cell walls which may enhance absorption, or code for release of Cu chelates to the soil solution.

Absorption of Cu has been even less extensively studied at the toxic end of the spectrum. There have been a number of studies of the mechanisms of protection from excess Cu in various species tolerant of soils high in Cu such as mine tailings. These include specific adsorption or chelation (binding) of Cu in cell walls, binding of Cu within root cells and within leaves and stems, exclusion of Cu from xylem and thereby from the shoot, alteration of enzyme systems (Antonovics *et al.*, 1971; Wainwright and Woolhouse, 1975). However, there is no evidence that alteration of absorption characteristics or mechanisms is one of the means of control; on the contrary the evidence is that the roots of Cu tolerant genotypes absorb as much Cu as do non-tolerant types or more (Bradshaw *et al.*, 1965). Data on absorption of Cu by Cu-tolerant genotypes leave much to be desired, since most reports give information insufficient to calculate rates of absorption such as those of Table I.

Generally speaking, both Cu-tolerant and non-tolerant plants accumulate Cu into roots to levels which are some function of those in the environment, but concentrations in shoots are less responsive until the roots are damaged, when tolerance breaks down and high levels of Cu may enter the shoot to cause damage. Exclusion from the shoot occurs in tolerant types at a higher critical external Cu concentration than in the non-tolerant plants. Whether Cu is bound in the root cell walls (Turner, 1970) or bound to proteins within root cells as suggested by Reilly (1969), it seems that the major mechanism of protection is exclusion from the shoot. Since Cu is a known inhibitor of photosynthesis (Wainwright and Woolhouse, 1975) such exclusion is essential. The root may control the supply of Cu to the shoot in *all* species over the whole range from deficiency to toxicity. Thus in Cu deficiency situations, 'tolerant' species may be those in which this control is least strongly expressed, and in Cu toxic conditions, the tolerant genotypes are those in which exclusion from the xylem is most strongly expressed.

Morrison *et al.* (1979) discussing the Cu and Co accumulating abilities of three metallophytes from Zaire, made the point that although their requirements for Cu and Co were low (in the normal range), there are two ecological advantages of tolerance of high concentrations of these metals in their roots: a lack of competitor species of higher plants and inhibition of the fungal pathogens which inflicted heavy damage in normal soils. The idea that plants may accumulate Cu in their roots to discourage fungi is an interesting one which may be extended to non-metallophytes. However, although roots may accumulate 10 to

100 times as much Cu as shoots in Cu-adequate soils, the
concentrations may fall into the same range when soil
available Cu is low (Chaudhry and Loneragan, 1970).   In
these soils, the possibility of protection seems unlikely.

VII. CONCLUSIONS

Rates of absorption of Cu are among the lowest of the
essential elements, being of the order of 1 n mole h^{-1} g^{-1}
D.W. of roots in the physiological concentration range.
However, the techniques used in the past for measurement of
these low rates have been inadequate, largely because Cu
in the cell walls is strongly and specifically bound, and
not effectively desorbed by Ca^{2+} as used in classical
studies of ion uptake.   It seems clear that studies of
absorption of Cu by roots should use an effective desorbing
ion such as Pb^{2+} or where ^{64}Cu is used as a tracer, $^{63}Cu^{2+}$
itself.   All aspects of technique need to be established,
for example, length of the absorption period, the
acceptability of excised, low-salt roots, the concentration
and balance of the absorbing medium, as well as the
desorption technique.
The mechanisms of absorption of Cu are far from clear,
but it seems likely that although Cu is almost entirely
complexed in the root environment, it dissociates from the
chelate prior to absorption.   Although it cannot yet be
taken as established, absorption of Cu appears to depend
on the plants' metabolism and is markedly affected by the
presence of many other ions.   A large electrochemical
gradient driving Cu^{2+} ions into root cells appears to exist
so that an alternative hypothesis to a carrier mechanism,
dependent on cell metabolism, cannot yet be excluded.
It seems certain that there are large genetic differences
in rates of absorption of Cu, but at this stage our
knowledge is fragmentary.   Rates of absorption in the toxic
range are as poorly researched as in the normal range and
although genotypic differences clearly exist, again much
greater attention to technique and data collection is needed
for real progress to be made in understanding the
mechanisms involved.

REFERENCES

Antonovics, J., Bradshaw, A.D. and Turner, R.G. (1971).
Adv. Ecol. Res. 7, 1-85.
Bowen, J.E. (1969). *Plant Physiol.* 44, 255-61.
Bowen, J.E. (1979). Int. Symp. "Trace Element Stress in
Plants", p. 24. Los Angeles Nov. 6-9, 1979.
Bowling, D.J.F. (1976). "Uptake of Ions by Plant Roots"
Chapman and Hall, London.
Bradshaw, A.D., McNeilly, T.S. and Gregory, R.P.G. (1965).
In "Ecology and the Industrial Society".
Brit. Ecol. Soc. Symp. 5, 327-43.
Brams, E.A. and Fiskell, J.G.A. (1971). *Soil Sci. Soc.*
Amer. Proc. 35, 772-5.
Cavallaro, N. and McBride, M.B. (1980). *Soil Sci. Soc. Am.*
J. 44, 729-32.
Cathala, N. and Salsac, L. (1975). *Plant and Soil* 42, 65-83.
Chandhry, F.M. and Loneragan, J.F. (1970). *Aust. J. Agric.*
Res. 21, 865-79.
Chandhry, F.M. and Loneragan, J.F. (1972). *Soil Sci. Soc.*
Amer. Proc. 36, 327-31.
Chino, M. and Baba, A. (1979). Int. Symp. "Trace Element
Stress in Plants", p. 23. Los Angeles, Nov. 6-9, 1979.
Claassen, N. and Barber, S.A. (1974). *Plant Physiol.* 54,
564-8.
DeKock, P.C. and Mitchell, R.L. (1957). *Soil Sci.* 84, 52-62.
DeKock, P.C. and Cheshire, M.V. (1968). *Rep. Welsh Soils*
Disc. Grp. 9, 98-108.
DeKock, P.C., Cheshire, M.V. and Hall, A. (1971). *J. Sci.*
Fd. Agric. 22, 437-40.
Dokiya, Y., Owa, N. and Mitsui, S. (1968). *Soil Sci. Pl.*
Nutr. 14, 169-74.
Dragun, J., Baker, D.E. and Risius, M.L. (1976). *Agron. J.*
68, 466-70.
Epstein, E. (1972). "Mineral Nutrition of Plants:
Principles and Perspectives". John Wiley and Sons
Inc. New York.
Fiskell, J.G.A. and Brams, E.A. (1965). *Soil Crop Sci.*
Soc. Fla. Proc. 25, 128-42.
Frausto da Silva, J.J.R. and Williams, R.J.P. (1976).
Structure and Bonding 29, 67-121.
Geering, H.R. and Hodgson, J.F. (1969). *Soil Sci. Soc.*
Amer. Proc. 83, 54-9.
Giordano, P.M., Noggle, J.C. and Mortvedt, J.J. (1974).
Plant and Soil 41, 637-46.

Goodman, B.A. and Linehan, D.J. (1979).   *In* "The Root-Soil
    Interface" (J.L. Harley and R.S. Russell, Eds.) pp. 67-82.
    Academic Press, London.
Goren, A. and Wanner, H. (1971).   *Schweiz Bot. Ges.*
    Berichte 80, 334-40.
Graham, R.D. (1978).   *Proc. 8th Int. Coll. Pl. Anal. Fert.
    Prob.* pp. 165-70.   Auckland, N.Z.
Graham, R.D. (1979).   *Plant, Cell and Env.* 2, 139-43.
Graham, R.D. and Bowling, D.J.F. (1977).   *J. Exp. Bot.* 28,
    886-93.
Graham, R.D. and Pearce, D.T. (1979).   *Aust. J. Agric. Res.*
    30, 791-9.
Graham, R.D., Anderson, G.D. and Ascher, J.S. (1981).
    J. Pl. Nutr. 3, (in press).
Graves, C.J. (1978).   *Ann. Bot.* 42, 117-25.
Graves, C.J. and Sutcliffe, J.F. (1974).   *Ann. Bot.* 38,
    729-38.
Harrison, S.J., Lepp, N.W. and Phipps, D.A. (1978).
    Z. Pflanzenphysiol. 90, 443-50.
Harrison, S.J., Lepp, N.W. and Phipps, D.A. (1979a).
    Z. Pflanzenphysiol. 94, 27-34.
Harrison, S.J., Lepp, N.W. and Phipps, D.A. (1979b).   Int.
    Symp. "Trace Element Stress in Plants".   p. 21,
    Los Angeles.   Nov. 6-9, 1979.
Hawf, L.R. and Schmid, W.E. (1967).   *Plant and Soil* 27,
    249-60.
Hiatt, A.J., Amos, D.F. and Massey, H.F. (1963).   *Agron. J.*
    55, 284-7.
Jarvis, S.J. (1978).   *J. Sci. Fd. Agric.* 29, 12-8.
Jarvis, S.J. (1981).   *Ann. Bot.* (in press).
Jarvis, S.J. and Robson, A.D. (1981).   Unpublished
    manuscript.
Kausar, M.A., Chaudhry, F.M., Rashid, A., Latif, A. and
    Alam, S.M. (1976).   *Plant and Soil* 45, 397-410.
Larsen, S. (1966).   *Agrochimica* 10, 190-6.
Loneragan, J.F. (1975).   *In* "Trace Elements in Soil-
    Plant-Animal Systems" (D.J.D. Nicholas and A.R. Egan,
    Eds.).   Academic Press, New York.
Loneragan, J.F. (1981).   Chapter 8, this volume.
Morrison, R.S., Brooks, R.R., Reeves, R.D. and Malaisse, F.
    (1979).   *Plant and Soil* 53, 535-9.
Nielsen, N.E. (1976).   *Plant and Soil* 45, 659-77.
Nielsen, N.E. (1976a).   *Plant and Soil* 45, 679-87.
Nissen, P. (1973).   *Physiologia Plant.* 29, 298-354.
Nuttonson, M.Y. (1958).   "Rye-Climate Relationships".
    Amer. Inst. Crop Ecol., Washington, D.C.

Parker, J. (1981). Chapter 1, this volume.

Pate, J.S., Sharkey, P.J. and Lewis, O.A.M. (1975). *Planta* 122, 11-26.

Petruzzelli, G. and Guidi, G. (1976). *Z. Pflanzenernahr. Bodenk.* 6, 679-84.

Reilly, C. (1969). *New Phytol.* 68, 1081-7.

Schmid, W.E., Haag, H.P. and Epstein, E. (1965). *Physiologia Plant.* 18, 860-9.

Turner, R.G. (1970). *New Phytol.* 69, 725-31.

Ussing, H.H. (1949). *Physiol. Rev.* 29, 127-55.

Veltrup, W. (1976). *Physiologia Plant.* 36, 217-20.

Veltrup, W. (1977). *Z. Pflanzenphysiol.* 83, 201-5.

Veltrup, W. (1979). *Z. Pflanzenphysiol.* 93, 1-9.

Wainwright, S.J. and Woolhouse, H.W. (1975). *In* "The Ecology of Resource Degradation and Renewal". *Brit. Ecol. Soc. Symp.* 15, 231-57.

Walker, C.D. and Webb, J. (1981). Chapter 9, this volume.

Wallace, A. and Romney, E.M. (1977). *Commun. Soil Sci. Pl. Anal.* 8, 699-707.

Wallace, A., Sufi, S.M. and Romney, E.M. (1971). Proc. 6th Int. Coll. Pl. Anal. Fert. Prob., pp. 545-58. Tel Aviv, 1970.

Williams, R.F. (1946). *Ann. Bot.* N.S. 10, 41-72.

Williams, R.F. (1948). *Aust. J. Sci. Res.* Bl, 333-61.

Wu, L., Thurman, D.A. and Bradshaw, A.D. (1975). *New Phytol.* 75, 225-9.

Younts, S.E. (1964). *Agron. J.* 56, 266-9.

8

DISTRIBUTION AND MOVEMENT OF COPPER IN PLANTS

J.F. Loneragan

School of Environmental and Life Sciences
Murdoch University
Perth, Western Australia

I. INTRODUCTION

The ease and extent to which Cu moves among its various parts are important factors in the plant's utilisation of Cu. They impact on the development and expression of Cu deficiency as Bussler discusses later in this Symposium. They impact, too, on the relationship between yield and Cu levels within the plant as Robson and Reuter will discuss.

This chapter reviews available information on the movement of Cu in xylem and phloem saps and on the distribution, deposition, and remobilisation of Cu among organs, tissues, and cells of plants.

II. MOVEMENT OF Cu IN PLANTS

A. *In Xylem Sap*

In the preceding chapter, Graham (1981) speculated that the process of excretion of Cu from root cells into the xylem sap may be a key process in the movement of Cu to plant shoots accounting for differences among species and cultivars in their efficiency both in obtaining Cu under conditions of low supply and in excluding it at high supply. However, little is known of the processes governing excretion of Cu into the xylem sap or of the factors controlling it.

Within the xylem sap, Cu concentrations of up to 14 µM have been reported (Table I). The values vary widely and

COPPER IN SOILS AND PLANTS

TABLE I.   CONCENTRATIONS OF Cu AND AMINO ACIDS IN
              XYLEM AND PHLOEM SAPS FROM SIX PLANT SPECIES

Species	Xylem Sap		Phloem Sap		Ref.
	µM Cu	mM AA[a]	µM Cu	mM AA[a]	
Arenga saccharifera	–	–	30	–	b
Lupinus albus	trace-9	5-19	6-60	200-230	d,e
L. angusti-folius	trace-4	20	3-12	100	f,g
Lycopersicon esculentum	1.5-2	0.7	–	–	h
Spartium junceum	14	4-8	140	130-300	d,e
Yucca flaccida	–	–	6	–	c

[a] AA amino acids: where data are reported as mg AA/ml they have been converted to mM in this table by assuming that aspartic acid is the only AA present.

[b]   Tammes, 1958   [e]   Pate et al., 1975

[c]   Tammes and van Die, 1964   [f]   Pate, 1975

[d]   Pate et al., 1974   [g]   Hocking et al., 1978

           [h]   Tiffin, 1977a

large fluctuations can be expected with changing levels of Cu supply and other environmental conditions. Indeed, in one study with *Lupinus angustifolius* (Hocking *et al.*, 1978), Cu concentrations in the xylem sap varied from 4 µM at dawn to 0.4 µM at dusk on the same day.

At these very low concentrations and in the absence of suitable Cu radioisotopes, it is difficult to study the nature of Cu compounds in the xylem sap. In 4 dicotyledonous plants, all of the Cu in sap exudates was present in one or more negatively charged complexes possibly involving amino acids (Tiffin, 1972). Xylem sap is acidic (pH 5.2-6.5) and generally contains organic N solutes as major components. One solute, which may be an amide, ureide, alkaloid, or amino acid, generally dominates but a wide range of amino acids is usually also present (Pate, 1976). Amino acids, which are generally present in relatively much higher concentrations than Cu^{2+} in xylem sap (Table I) form very strong complexes with Cu^{2+} as discussed in other papers of this Symposium (Parker, 1981; Stevenson and Fitch, 1981; Walker and Webb, 1981). These complexes are much stronger than those of other metal ions which occur in xylem sap (Sillen and Martell, 1971; Österberg, 1974). Histidine and threonine are also known to form a strong mixed chelate complex with Cu^{2+} and in animals this complex may play a role in transporting Cu from ceruloplasmin in the blood to tissues (Sarkar and Kruck, 1966; Freeman and Martin, 1969).

When an unknown complexing compound extracted from peanut roots was added to solution culture, it depressed the amount of ^{64}Cu held by soybean roots but increased the amount moving to shoots. In the absence of complexing agent, the roots and lower stem retained very large amounts of ^{64}Cu (Tiffin, 1977b). While the concentration of Cu^{2+} used (50 µM in a Hoagland-type nutrient solution) was unphysiologically high, the results serve to demonstrate the effectiveness of complexing agent in increasing the mobility of Cu within plants. Certainly, if it is to move freely to the organs of the shoot in the xylem sap and not be precipitated or be strongly adsorbed on to reactive components in the walls of the xylem vessels, Cu would need to complex strongly with some other component of the xylem sap. Shoot cells requiring Cu would then need some means of removing Cu from the complex to absorb it from the xylem sap in the way that Graham (1981) has discussed for absorption of Cu by root cells.

B.   In Phloem Sap

Three approaches have been used to obtain evidence on the translocation of Cu in phloem:

(1)   applying Cu to plant leaves and measuring its movement or physiological effects,

(2)   analysing phloem sap,

(3)   observing changes in the Cu contents of leaves with time.

None has proved entirely satisfactory.

Bukovac and Wittwer (1957) painted $^{64}Cu^{2+}$ on a primary leaf of bean seedlings and examined its redistribution within the plant. The experiments were short term as the isotope has a very short half life. Moreover, Cu^{2+} reacts strongly with cell walls so that when applied to leaves it might behave differently to Cu entering the leaf in complexed forms in the xylem sap. From their studies, Bukovac and Wittwer (1957) classified Cu as "partially mobile" in plant phloem and suggested that it was intermediate in its behaviour between freely mobile P and immobile Ca. This conclusion has received wide acceptance even though published radio-autograms indicate little if any phloem mobility, showing no radioactivity in any part of the plant other than the distal part of the leaf treated with ^{64}Cu.

Definitive evidence that appreciable Cu can move in plant phloem has been provided by experiments involving application of Cu to leaves of Cu deficient plants in the field. An early summer treatment with Cu sprays of foliage on some limbs of Cu deficient almond trees resulted in leaves and kernels on unsprayed limbs recovering from Cu deficiency and increasing their Cu concentrations 3 to 15 fold over controls in late summer (Kester et al., 1961).

Analyses of phloem sap also suggest that Cu is phloem mobile. Phloem exudates from several plant species have given values for Cu concentrations ranging from 3 to 140 µM (Table I). As in xylem sap, any Cu in the phloem sap is likely to be complexed with soluble organic N compounds which are usually present in high concentrations and would bind Cu^{2+} even more strongly at the prevailing high pH values of 8 to 8.5. The principal organic N solute or solutes of the phloem sap is often the same as that in the xylem sap: it varies with species and may be an amide, amino acid, or ureide. In addition, phloem sap of most species generally contains a variety of amino acids including relatively more of glutamic and aspartic acids, valine, serine, threonine, glycine, and alanine (Pate, 1976).

There is little information on the extent to which values for Cu concentrations in phloem exudates vary with environmental conditions or with plant development and genotype. One study has reported a 2½ fold diurnal fluctuation from 5 to 12 μM in the Cu concentrations of phloem exudate from young developing pods of *Lupinus angustifolius*: sugar concentrations fluctuated in parallel so that the weight of Cu relative to sugar remained relatively constant, varying from 4 to 6 μg Cu/g sugar (Hocking *et al.*, 1978).

Tammes and Van Die (1966) attempted to relate the capacity of the phloem sap to supply mineral nutrients to inflorescences by comparing concentrations of individual nutrients on a dry weight basis. The concentration of Cu in inflorescence was double and that in leaves 10 times that of Cu in phloem sap. On the other hand, since the phloem sap had 2 μg Cu/g dry matter of sap its concentration of 6 μM Cu could have been sufficient to supply the minimal needs of 1-3 ug/g in the developing organs of most of the plants whose requirements have been measured (see Robson and Reuter, 1981). Most of the other phloem saps of Table I would likewise appear to have had adequate Cu to supply developing organs since they contained between 1.5 and 50 ug Cu/g dry matter. Indeed most of these values would indicate a luxurious or even toxic supply of Cu. However, some of the high values reported could be erroneous due to contamination arising from the considerable difficulties experienced in collecting adequate quantities of sap for analysis (Pate, 1975) and from problems in analytical procedures such as those discussed later in this Symposium by Simmons. But overall the data suggest that phloem sap usually contains plenty of Cu to supply developing organs.

Yet young organs are amongst the first to develop symptoms of Cu deficiency (see Bussler, 1981), resembling in this way the onset of Ca deficiency, a fact generally taken to indicate extreme phloem immobility of this element. Recent studies of changes with time of leaf Cu contents have both confirmed and clarified this paradox. They have shown

that under some conditions Cu is highly mobile in plant
phloem but immobile under others: perhaps the term
"variably mobile" more adequately describes the behaviour of
Cu in phloem than the terms "partially mobile" (Bukovac and
Wittwer, 1957) and "intermediate mobility" (Epstein, 1972).

C. Leaf Cu Contents

Changes with time of leaf Cu contents have been used
to draw inferences about the mobility of Cu in plant phloem.
Such studies suffer from the serious disability that they
measure a *net* value, recording only the differences between
import of Cu into a leaf and its export. They seriously
underestimate or even completely obscure phloem export whenever Cu import into a leaf is appreciable. As a result,
stable or increasing values for leaf Cu contents provide no
information on Cu export. On the other hand declining values
provide an estimate of the minimum quantity of Cu exported
via the phloem. Only by using labelled isotopes can this
problem be avoided. Lack of a suitable Cu radioisotope has
prevented such studies but later in this Symposium, Walker
and Welch describe a method for using mass isotopes of Cu
which might permit critical, long term studies of Cu transport to be undertaken.

While some studies of the Cu content of plant leaves
suggest that Cu is not mobile in plant phloem many others
suggest that it is highly mobile. Thus, in contrast to B
and Ca, the Cu content of leaves generally reaches a peak
from which it declines during senescence (McClung and Lott,
1956; Labanauskas *et al.*, 1959). The loss can be substantial. Leaves of oats lost 25% of their Cu during grain
ripening (Williams and Moore, 1952), onion leaves exported
40% of their Cu during bulb development (Carolus, 1962), and
sugar cane leaves lost a large proportion of their Cu to the
stem during senescence (Mukherjee, 1969). After flowering,
leaves of wheat given a just adequate or luxury supply of Cu
lost 60 and 70% of their Cu respectively and this was sufficient to account for about 60% of the Cu in the grain of each
treatment at maturity (Fig. 1; Loneragan *et al.*, 1980).

By contrast with Cu adequate wheat plants, Cu deficient
plants lost no Cu from their total leaves after the time of
flowering in controls (Fig. I). Since the Cu deficient
plants failed to form stems, flower, or grain the retention
of Cu by leaves may have resulted from the lack of a sink.

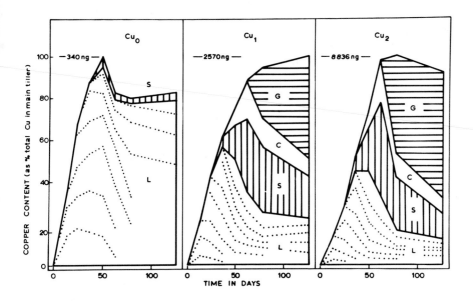

FIGURE 1. *Distribution of Cu among leaves (L), stem (S), inflorescence chaff (C), and grain (G), of the primary shoot of wheat plants at deficient (Cu$_0$), marginal (Cu$_1$) and luxury (Cu$_2$) supplies of Cu. After Loneragan et al., 1980).*

Certainly, individual older leaves lost appreciable Cu to younger leaves. But the loss of Cu was delayed so that, at some harvests, the oldest leaf of severely Cu deficient plants had more Cu and higher Cu concentrations than the same leaf of plants given adequate or luxury supplies of Cu. Senescence and loss of N was also delayed. In all treatments, movement of Cu from old leaves paralleled movement of N, leading to the suggestion that protein hydrolysis is a pre-requisite to the release of Cu from leaves in the phloem sap (Loneragan *et al.*, 1976, 1980).

In a series of experiments, Jillian Hill has shown a very close relationship between the amounts of Cu and N in the oldest leaf of wheat seedlings in response to Cu nutrition and to shading (Hill *et al.*, 1979b, 1979c). Figure 2 shows how in Cu adequate plants, the Cu and N contents of the oldest, unshaded leaf declined at first slowly and then rapidly with senescence and loss of chlorophyll.

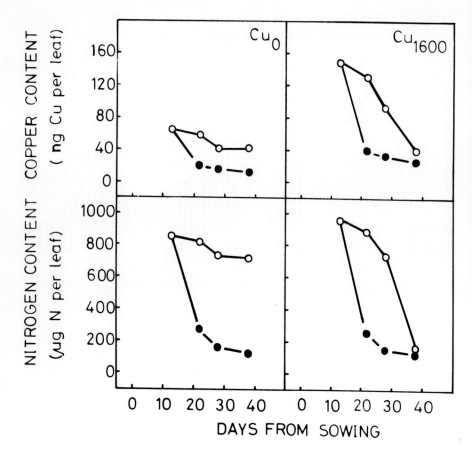

FIGURE 2.   The effects of plant Cu supply and leaf shading on the Cu and N contents of the oldest leaf of wheat plants. Cu_0, deficient Cu supply; Cu_{1600} luxury Cu supply; o, unshaded;   ● shaded.   After Hill et al., 1979b.

In Cu deficient plants, the same leaf did not senesce during the experiment and both Cu and N contents showed little decline.   Shading of the oldest leaf induced its chlorosis and senescence in the Cu deficient plant and accelerated them in the Cu adequate plant: in both Cu treatments, the loss of Cu from the leaf paralleled the accelerated loss of N.   In other experiments in which application of NH_4NO_3 or Cu deficiency has delayed senescence and loss of N it has also delayed a loss of Cu from the oldest leaf of wheat (Hill *et al.*, 1978;   Loneragan *et al.*, 1980).

In the Cu deficient plants, the loss of Cu from the old shaded leaves was accompanied by a transitory and marked increase in growth and in the Cu concentrations and contents of roots, tillers, and young leaves. Shading of old leaves had clearly resulted in mobilisation of Cu from them to developing organs (Hill *et al.*, 1979b).

In contrast to these studies, Wood and Womersley (1946) concluded that no Cu moved from oat leaves during grain development even though almost all their protein was exported. Re-examination of their data suggests an alternative explanation. In their experiment, the Cu content of plants increased progressively to maturity, possibly from contamination in the culture system. Consequently, Cu would have been entering leaves in the xylem sap during senescence and this could have masked any loss of Cu by export in the phloem.

Although most of the Cu in wheat leaves appears to remain in them until they senesce, small quantities may move from them together with small amounts of N prior to senescence. In Cu adequate wheat plants, the oldest leaf attained its maximal Cu and N contents and concentrations 11 days after sowing. Thereafter, its Cu and N decreased in 2 stages. In the first stage, the leaf lost about 10% of its Cu and N slowly over 10 days. In the second stage, which commenced before and continued during the loss of chlorophyll, the oldest leaf rapidly lost a further 70% of its Cu and 80% of its protein and total N. Over the entire period, the rate of change in Cu and N concentrations were remarkably parallel as if the 2 nutrients were closely coupled in their phloem export (Hill *et al.*, 1979c). At full senescence, leaves retained about 20% of their original Cu content so that the amount of Cu in senesced leaves, unlike that of N, varied greatly between treatments.

In the coupling with N of its loss from leaves, Cu resembles organic S (Ergle, 1954; Eaton, 1966; Bouma, 1975). However, Cu differs from S in that it does not appear to change in strict stoichiometric proportions with N. Thus, although the losses from the oldest leaf of Cu and N appeared to proceed in a constant proportion in plants from each of 3 Cu treatments, the proportions in each treatment were markedly different (Loneragan *et al.*, 1980). It is as if all Cu entering the leaf is bound by N compounds in chemical forms or in organelles which retain it against phloem export until protein hydrolysis when they are released and exported in some form coupled with N. As a result of this behaviour, the movement of Cu from leaves in phloem is largely controlled by their senescence.

III. DISTRIBUTION OF Cu IN PLANTS

A. *Between Roots and Shoots*

The distribution of Cu between roots and shoots varies widely with plant species and cultivars and with environmental conditions.

The concentration of Cu in roots is generally higher than that in shoots but the degree of difference varies markedly with the level of Cu supply. Roots of plants grown for long periods at low levels of Cu have concentrations approaching those of tops (Fig. 3; Jarvis, 1978; Graves, 1978; Jarvis and Robson, 1981).

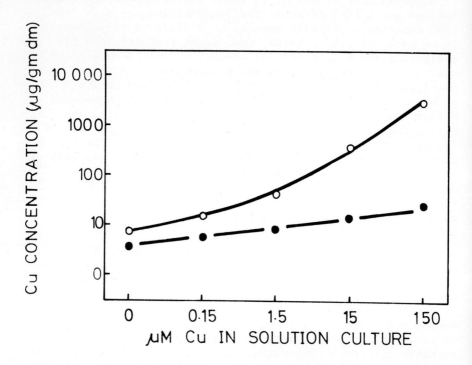

FIGURE 3. *Relationship between Cu concentration in solution culture and Cu concentrations in the roots (o) and shoots (●) of perennial ryegrass. After Jarvis, 1978.*

With increasing concentrations of Cu in solution, the concentration of Cu in roots of most species increases much
more rapidly than that of tops (tobacco (Struckmeyer *et al*.,
1969); orange (Brams and Fiskell, 1971); red clover (Hill,
1973); ryegrass (Jarvis, 1978); chrysanthemum (Graves,
1978). The proportion of the plant's Cu in roots thus
varies greatly with Cu supply. For example, in the ryegrass of Figure 3, the proportion of the plant's Cu in its
roots varied from 39% to 96% as Cu levels added to solutions
varied from 0, when Cu deficiency depressed growth slightly,
to 150 µM when Cu was toxic.

While at high levels of Cu supply most species behave
like ryegrass, some species differ. Later in this Symposium
Woolhouse and Walker present evidence suggesting that many
plant species which are able to tolerate high levels of Cu
in their environment appear to do so by accumulating Cu in
their roots and excluding it from their tops. Other species
in the same environment accumulate very much higher concentrations of Cu in their leaves.

Nitrogen fertilizers frequently change the content of
Cu in plants and its distribution between roots and shoots.
However, these changes are usually accompanied by such profound effects upon plant growth and development that they
obscure any other relationships between Cu and N which
might exist. For example, application of NH_4NO_3 almost
doubled the proportion of Cu present in plant tops relative
to roots but it also increased root growth 5 fold and
shoot growth 10 fold so that it diluted the available Cu
and doubled the weight of tops relative to roots (Chaudhry
and Loneragan, 1970).

Despite the problems of interpretation occasioned by
growth phenomena, there are persistent reports of Cu concentrations in plant shoots correlating with N concentrations. For example, Rasheed and Seeley (1966) reported
that Cu concentrations in the shoots of a wide range of
species from pastures throughout the United Kingdom correlated closely, positively, and linearly with their N
concentrations. Gladstones *et al*. (1975) also observed
that Cu concentrations in the shoots of a number of cereals
and pasture grasses increased linearly with N concentrations:
unlike those of Rasheed and Seeley (1966), legume shoots
behaved differently having higher Cu concentrations but
more variable and lower ratios of Cu to N. Other workers

have also observed close relationships between increasing
Cu and increasing nitrogen levels in leaves of tung trees
(Gilbert, 1951), oats (Cheshire *et al.*, 1967), and the
Rhodesian copper flower *Becium homblei* (Reilly, 1969).
Gilbert (1951) suggested that the relationship resulted from
protein forming strong complexes with Cu and immobilizing
it within plant leaves. Reilly (1969) suggested that
similar complexes between Cu and NH_2 groups of amino acids
and proteins probably account for the observed relationship
as well as the remarkable ability of the Rhodesian copper
flower to tolerate high levels of Cu in the leaf. All of
these observations and suggestions are consistent with the
hypothesis presented earlier that Cu accumulates in leaves
and only moves from them when N levels decline.

B. *Within Roots*

Much of the Cu in the roots of plants is associated with
cell walls. Graham (1981) has already pointed out in the
previous chapter that most of this Cu can be desorbed from
cell walls by Pb^{2+} or Cu^{2+} but not by Ca^{2+} or other cations.
However, little has been done to quantify the ability of
roots to adsorb Cu^{2+} by specific reactions and electrostatic
processes.

In Cu deficient ryegrass almost all of the Cu in roots
was associated with cell walls. The proportion of root
Cu associated with cell walls steadily decreased to 25% with
increasing levels of Cu supply to toxic levels. It is not
clear where or in what form this non-wall Cu was held. But
in ryegrass, whatever its form and location, Cu in the root
appeared to be largely immobilised. Thus when plants grown
for a period at slightly toxic levels of Cu were transferred
to solutions without Cu, their roots retained most of their
Cu over a period of 14 days (Jarvis, 1978). Indeed, even
when grown into moderate or severe Cu deficiency, roots of
red clover and wheat retained most of the Cu which they had
accumulated in an earlier period of adequate supply (Jarvis
and Robson, 1981).

C. *Within Shoots*

The average concentration of Cu in whole shoots of plants
is generally highest in young seedlings and decreases
steadily towards maturity (Gladstones and Loneragan, 1975;
Loneragan *et al.*, 1980; Reuter *et al.*, 1981b). It also

varies with Cu and N supply. Within plant shoots too, Cu
concentrations of organs also vary with their age and with
the level of Cu and N supply.

Within plant shoots, the distribution of Cu among leaves
fits the pattern expected from its postulated variable
mobility tending to follow that of a mobile element in Cu
adequate plants and of an immobile one in Cu deficient plants.
Thus plants supplied with adequate or luxury levels of Cu
generally have relatively high concentrations of Cu in their
shoot tips and young leaves with lower concentrations in
their old leaves. By contrast, Cu deficient plants usually
have lower Cu concentrations in their young leaves than in
their old leaves, as for example in peanuts (Nualsri *et al.*,
1977), chrysanthemum (Graves, 1978), and wheat (Fig. 4;
Loneragan *et al.*, 1980).

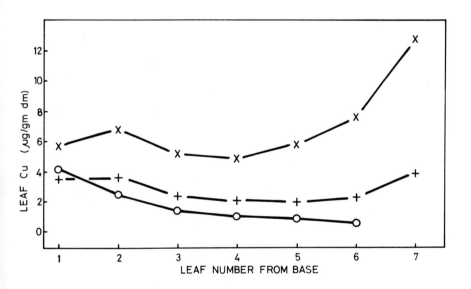

FIGURE 4. *Relationship of Cu concentration in leaf to
position of leaf on the primary stem of wheat plants grown
at deficient (o), just adequate (+), and luxury (x) Cu
supply. From data of Loneragan et al., 1980*

Copper deficient subterranean clover differed from this pattern in having higher Cu concentrations in young than in old leaves. Possibly this different pattern is related to the fact that in subterranean clover,in contrast to peanuts, Cu deficiency limits N fixation in root nodules before affecting host plant growth (Nualsri, 1978; Snowball *et al.*, 1980). However, even in this species the difference in Cu concentrations between young and old leaves was much smaller at deficient than at adequate Cu supply (Reuter *et al.*, 1981b).

As already discussed, concentrations of Cu in particular leaves of wheat (Loneragan *et al.*, 1980) and subterranean clover (Reuter *et al.*, 1981b) were also highest when they were young and decreased with age, especially after senescence.   Stems of wheat followed the same pattern.

Within the leaves of subterranean clover, Cu supply had a marked effect on the ratio of Cu concentrations in blades and petioles (Reuter *et al.*, 1981b). When Cu was adequate, Cu concentrations in blades invariably exceeded those of the petioles subtending them and the differential increased with increasing Cu supply. However, in Cu deficient plants, Cu concentrations in leaf blades were equal to or less than those in their petioles. In severely deficient plants, Cu concentrations of petioles increased appreciably above those of their blades and also above those of the petioles of less deficient plants. Since Cu deficiency also retarded plant development by delaying production of leaves and branches, senescence of plant parts, and flowering it had complex effects upon the distribution of Cu within the plant.

The activities of 5 Cu enzymes in leaves, stems, and petioles of subterranean clover confirmed that young organs were the first to suffer from Cu deficiency. Indeed, in Cu deficient clover plants, enzyme activities in young leaves were up to 30 fold lower than those of Cu adequate plants while the activities in mature green leaves, older leaves in early senescence, and petioles and stems in the same plants showed no differences (Walker and Loneragan, 1981).

The striking effects of Cu deficiency in preventing the movement of Cu to young organs probably accounts for the depression of lateral bud development in chrysanthemum (Graves, 1974), wheat (Hill *et al.*, 1978), and subterranean clover (Reuter *et al.*, 1981a). It may also account for the peculiar sensitivity of pollen to Cu deficiency at anthesis in wheat (Graham, 1975, 1976) and several other species (Dell, 1981).

Some attempts have been made to examine the distribution of Cu within plant cells. However, from what is known of the variability in Cu contents of leaves with Cu and N supply and with age, large differences in the distribution must be expected under varying conditions. In most studies, appreciable quantities of leaf Cu have been found in chloroplasts. Neish (1939) fractionated clover leaves and found 75% of the Cu within chloroplasts where it was tightly bound in organic forms. By contrast Timperley *et al.* (1973) found only 35% of the Cu in tree leaves in chloroplasts: about 50% was in their cytoplasmic fraction, 2-19% in mitochondria and nuclei, and 3-8% in ribosomes.

Within chloroplasts a considerable proportion of the Cu appears to be bound in plastocyanin and possibly also Fraction 1 protein. Katoh *et al.* (1961) found that over half the Cu was present in plastocyanin. Moreover, as Bussler (1981) discusses later, both plastocyanin and Cu decreased markedly in chloroplasts of Cu deficient plants (Baszynski *et al.*, 1978). On fractionation of Cu compounds in mature leaves of Cu adequate plants, Walker (1980) also found a relatively large proportion of leaf Cu to be associated with Fraction 1 protein. Fraction 1 protein consists primarily of one or two proteins with ribulose biphosphate carboxylase-oxygenase activity. These enzymes contain some firmly bound Cu as Walker and Webb discuss in the next chapter. Walker (1980) has suggested that, in addition to this firmly bound Cu, Fraction 1 protein may contain a variable amount of loosely bound Cu: in Cu deficient plants, the amount of Cu associated with Fraction 1 protein decreased markedly while the level of Fraction 1 protein remained relatively unchanged. Such a variable association of Cu with Fraction 1 protein would permit the protein to act both as an agent for removing Cu^{2+} which is highly toxic to the photosynthetic system (Sandmann and Böger, 1980) and as a reserve store of Cu.

While still largely speculative, a variable association of Cu with Fraction 1 protein would provide an explanation for the retention and release of Cu from leaves. Fraction 1 protein increases rapidly in leaves during their expansion. Thereafter, Fraction 1 protein may decline to a small extent

(Kannangara and Woolhouse, 1968; Batt and Woolhouse, 1975;
Woolhouse and Batt, 1976) but the bulk of it in leaves persists with little apparent turnover until leaf senescence
when it is exported rapidly together with most of the other
chloroplast N (Huffaker and Peterson, 1974) which accounts
for some 90% of the leaf N lost at this time (Morita, 1980).
Hill *et al.* (1979c) suggested that the early decline of
Fraction 1 protein observed by Kannangara and Woolhouse,
(1968) might account for the early and small coupled decline
of Cu and N in wheat leaves which has already been discussed.
The loss of Cu after senescence could be related to hydrolysis
and export of Fraction 1 and other chloroplast proteins.

If Fraction 1 protein does not operate in this way,
some other compound or compounds must act to store Cu^{2+} in
non-toxic forms. The recent discovery of metallothionen
in roots of *Agrostis gigantea* (Rauser and Curvetto, 1980)
may be significant. In microorganisms and in animals
metallothionen functions in this way for Cu^{2+} and other
toxic metal ions as Woolhouse and Walker discuss in some
detail in a later chapter of this Symposium.

D. *Within Flowers and Seeds.*

Within the reproductive organs, as in vegetative parts,
Cu concentrations of Cu adequate plants are highest in the
young parts. In flowers, anthers containing pollen and
ovaries had much higher Cu concentrations than filaments,
styles or perianth parts (Table II). In seeds, embryos had
high concentrations relative to other parts. For example,
in cereal grains from several sources, Cu concentrations in
embryos were 2½ to 8 times greater than those in the endo-
sperm of the same grain (Table III). But in contrast to
vegetative organs, Cu concentrations in various parts of
wheat grain had the same relation to each other in Cu
deficient as in Cu adequate plants (Hill *et al.*, 1979a).
At the same time, Cu concentrations in wheat grain and its
parts were 10 fold higher in Cu adequate than in Cu deficient
plants (Wheat 3, Table III).

While under some conditions, Cu concentrations in seeds
and grains respond to applications of Cu fertilizers, under
other conditions they do not. For example, when a wide
range of species and cultivars of annual crop and pasture
plants were grown in the field, their seeds and grains

TABLE II. CONCENTRATIONS OF Cu IN FLORAL AND OTHER PARTS OF 4 PLANT SPECIES[a].

	Plant Species[b]			
	A	D	F	M
	μg Cu per g dry matter			
Anthers[c]	9	41	9	13
Pollen	-	-	-	5
Filaments	1	8	6	-
Styles	5	6	6	6
Ovaries	13	19	12	7
Sepals[d]	-	8	7	-
Petals[e]	4	7	7	-
Leaves	9	7	17	12
Roots	58	-	-	-
Seed[f]	9	-	-	2

[a] Data from Knight et al., 1973

[b] A = antirrhinum, D = daffodil F = fuchsia, M = maize

[c] Includes unshed pollen in all species and whole spikelet with anthers and filament in M.

[d] Sepals includes perianth in D

[e] Petals include corona in D

[f] Seed or grain produced from the flowers analysed.

showed no response to a 3 fold increase in rates of
application of Cu fertilizer (Gladstones, *et al.*, 1975).
In the same experiment, Cu concentrations in seeds and
grains varied widely with species. Small seeded legumes
had Cu concentrations in their seeds up to 6 times higher
than those in grains of wheat, oats, and barley. Large
seeded legumes and rye had intermediate concentrations.

The processes controlling Cu concentrations in seeds and
grains have received little study. In lupins, small quantities of Cu were lost from pods during seed development
(Hocking *et al.*, 1978). But for most species and under
most conditions the dominant processes are probably those
of retranslocation of Cu from vegetative organs after
flowering in combination with those which determine seed
yield. For example, in Figure 1, wheat plants at luxury
Cu supply had as much Cu in their primary shoots at flowering as they had at maturity: the amount of Cu gained by
grain during maturation was matched by that lost from
leaves, stems, and flowers. Wheat plants given marginally
adequate Cu increased their total Cu contents to a small
extent following flowering: but even in this treatment
more than 70% of the Cu in grain at maturity appeared to
have been retranslocated from other parts following
flowering. At flowering, the concentration of Cu in
plants was higher at luxury than at adequate Cu so that
much more Cu was subsequently translocated to grain. Since
plants of both treatments produced identical grain yields,
the Cu concentration in grain was 4 fold higher at luxury
than at adequate Cu. Similarly, the differences among
species in Cu concentrations of seeds and grains in the field
experiment discussed above corresponded generally with Cu
concentrations of whole shoots at flowering.

However, under other conditions the extent to which Cu
moves from leaves to seeds and grain following flowering
may be very much smaller. In the extreme case where no
grain forms as the result of pollen sterility (Graham, 1976)
no Cu moves. Hill *et al.* (1978) have also observed that
an appreciable amount of Cu remained in leaves and stems
of 4 wheat cultivars at maturity and that this quantity
increased with increasing Cu status of the plant. The
supply of N also influenced translocation of Cu from leaves
in at least 3 ways - by increasing growth and inducing Cu
deficiency with the consequences already discussed, by
increasing amount of vegetative tissue and thus increasing
the quantity of Cu retained at maturity, and by delaying

TABLE III. CONCENTRATIONS OF Cu IN PARTS OF SEEDS AND GRAINS OF 2 LUPIN AND 5 CEREAL SPECIES

Seed Part[a]	Plant Species[b]						Wheat		
	Lupinus a	Lupinus b	Barley	Maize	Rye	Rice	1	2	3
	μg Cu per g dry matter								
Whole[a]	11	14		1.5	4	5	-	4	0.6-5
Seed Coat[a]	2	4	14	7	10	10	8	23	-
Endosperm	-	-	3	0.9	2	5	1.2	3	0.5-5
Aleurone	-	-	-	-	-	-	-	12	-
Embryo	14	17	10	7	12	19	11	19	1.8-18

[a] For cereals, "seed" and "seed coat" include pericarp of grain and, in the case of rice only, the aleurone layer.

[b] Lupinus a = albus and l - luteus (Gladstones and Drover, 1962)

Barley, Rye, Wheat 1 (Höfner et al., 1961)

Maize, Rice, Wheat 2 (O'Dell et al., 1972)

Wheat 3: data are those for the 2 out of 8 treatments which had the lowest and highest concentrations of Cu in the whole grain (Hill et al., 1979a).

leaf senescence. Under some conditions, one or more of these
effects of N on translocation of Cu were critical depressing
both Cu concentrations in grains and grain yield.

Other agronomic practices which influence the processes of Cu translocation to developing seeds may also be
expected to have important effects on seed production.
For example, any agronomic practice which delays senescence
following flowering would prevent phloem translocation of Cu
to developing fruits requiring that all of their Cu requirements be met from the xylem sap. Where the store of Cu
available to the xylem sap from roots and stems is limited,
as appears to be the case in wheat, seed development would
depend upon continued absorption of Cu from the soil: it
would then be vulnerable to climatic conditions limiting
the absorption of Cu from the soil, as Grundon discusses
later in this Symposium. Under such conditions, plants
would have their seed production severely depressed by Cu
deficiency and respond to foliar applications of Cu even
though they might have very high concentrations of Cu in
their leaves.

CONCLUSIONS

The work reviewed in this paper suggests that Cu
probably occurs in xylem and phloem saps complexed with
soluble organic N compounds such as amino acids. Once
present in the xylem and phloem saps, these Cu complexes
move freely within the translocatory streams. However, processes operating in both root and leaves appear to control
the distribution and redistribution of Cu within the plant
by governing the release of Cu for excretion into either
xylem or phloem saps.

In roots, high concentrations of Cu may be held
against transport to shoots even under conditions of severe
Cu deficiency. The processes governing this retention of Cu
are not understood.

In shoots, the behaviour of Cu appears to be coupled
with N metabolism or transport. Present evidence suggests
that Cu entering leaves is bound by N compounds such as
proteins which release little if any of it for phloem transport until they are hydrolysed. As a result green leaves
may accumulate very high concentrations of Cu and retain them
against transport to young leaves and developing meristems

even during the development of Cu deficiency. This hypothesis explains why Cu deficiency symptoms develop
characteristically in young leaves and meristems. It also
explains why Cu concentrations of plant shoots may not
correlate with the onset and intensity of Cu deficiency and
why increasing levels of N fertilizers may increase the
plant's apparent requirement for Cu.

If correct, the hypothesis provides a sounder basis for
the interpretation of physiological and biochemical phenomena associated with development of Cu deficiency, for the
development of better procedures for diagnosis of Cu
deficiency, and for prediction of conditions likely to lead
to the development or enhancement of Cu deficiency in
agronomic and horticultural crops.

REFERENCES.

Baszynski, T., Ruszkowska, M., Krol, M., Turkendorf, A.,
    and Wolinska, D. (1978). *Z. Pflphysiol.* 89, 207.
Batt, T. and Woolhouse, H.W. (1975). *J. exp. Bot.* 26, 569.
Bouma, D. (1975). *In* "Sulphur in Australian Agriculture"
    (K.D. McLachlan, ed.) pp.79-86. Sydney Univ. Press,
    Sydney.
Brams, E.A. and Fiskell, J.G. (1971). *Soil Sci. Soc. Amer.
    Proc.* 35, 772.
Bukovac and Wittwer (1957). *Pl. Physiol.* 32, 428.
Bussler, W. (1981). Chapter 10, this Volume.
Carolus, R.L. (1962). Int. Hort. Congr., Rep., 16th
    (Brussels, Belgium). 2, 202.
Chaudhry, F.M. and Loneragan, J.F. (1970). *Aust. J. agric.
    Res.* 21, 865.
Cheshire, M.V., DeKok, P.C., and Inkson, R.H.E. (1967).
    J. Sci. Fd. Agric. 18, 156.
Dell, B. (1981). *Ann. Bot.* (In Press).
Eaton, F.M. (1966). *In* "Diagnostic Criteria for Plants
    and Soils" (H.D. Chapman, ed.) pp.444-475. Univ.
    Calif. Div. agric. Sci., Berkeley, Calif.
Epstein, E. (1972) "Mineral Nutrition of Plants:
    Principles and Perspectives". John Wiley and Sons, Inc.,
    New York.
Ergle, D.R. (1954). *Bot. Gaz.* 115, 225.
Freeman, H.C. and Martin, R.P. (1969). *J. biol. Chem.*
    224, 4823.

Gilbert, S.G. (1951). *Pl. Physiol.* 26, 398.

Gladstones, J.S. and Drover, D.P. (1962). *Aust. J. exp. Agric. Anim. Husb.* 2, 213.

Gladstones, J.S., Loneragan, J.F., and Simmons, W.J. (1975). *Aust. J. agric. Res.* 26, 113.

Graham, R.D. (1975). *Nature* (Lond.) 254, 514.

Graham, R.D. (1976). *J. exp. Bot.* 27, 717.

Graham, R.D. (1981). Chapter 7, this Volume.

Graves, C.J. (1974). *Ann. Bot.* 38, 729.

Graves, C.J. (1978). *Ann. Bot.* 42, 117.

Hill, J.M. (1973). *J. exp. Bot.* 24, 525.

Hill, J., Robson, A.D. and Loneragan, J.F. (1978). *Aust. J. agric. Res.* 29, 925.

Hill, J., Robson, A.D. and Loneragan, J.F. (1979a). *Aust. J. agric. Res.* 30, 233.

Hill, J., Robson, A.D. and Loneragan, J.F. (1979b). *Ann. Bot.* 43, 449.

Hill, J., Robson, A.D. and Loneragan, J.F. (1979c). *Ann. Bot.* 44, 279.

Hocking, P.J., Pate, J.S., Atkins, C.A., and Sharkey, P.J. (1978). *Ann. Bot.* 42, 1277.

Höfner, W., Schaumlöffel, E., and Völker, L. (1961). *Planta* (Berlin) 57, 608.

Huffaker, R.C. and Peterson, L.W. (1974). *Ann. Rev. Pl. Physiol.* 25, 363.

Jarvis, S.C. (1978). *J. Sci. Fd. Agric.* 29, 12.

Jarvis, S.C. and Robson, A.D. (1981). Unpublished manuscript.

Kannangara, C.G. and Woolhouse, H.W. (1968). *New Phytol.* 67, 533.

Katoh, S., Suga, I., Shiratori, I., and Takamiya, A. (1961). *Arch. Biochem. Biophys.* 94, 136.

Kester, D.E., Uriu, K., and Aldrich, T. (1961). *Proc. Amer. Soc. hort. Sci.* 77, 286.

Knight, A.H., Crooke, W.M., and Burridge, J.C. (1973). *Ann. Bot.* 37, 159.

Labanauskas, C.K., Jones, W.W., and Embleton, T.W. (1959). *Proc. Amer. Soc. hort. Sci.* 74, 300.

Loneragan, J.F., Snowball, K., and Robson, A.D. (1976). *In* "Transport and Transfer Processes in Plants" (I.F.Wardlaw and J.B. Passioura, eds.) pp. 463-469. Academic Press, New York.

Loneragan, J.F., Snowball, K., and Robson, A.D. (1980). *Ann. Bot.* 45, 621.

McLung, A.C. and Lott, W.L. (1956). *Proc. Amer. Soc. hort. Sci.* 67, 113.

Morita, K. (1980). *Ann. Bot.* 46, 297.
Mukherjee, K.L. (1969). *J. Indian bot. Soc.* 67, 180.
Neish, A.C. (1939). *Biochem. J.* 33, 300.
Nualsri, L. (1978). Ph.D Thesis. Univ. W. Austral.
Nualsri, L., Robson, A.D. and Loneragan, J.F. (1977). Intern. Soc. Soil Sci. Comm. IV, V. August, 1977. Kuala Lumpur.
O'Dell, B.L., deBoland, A.R., and Koirtyohann, S.R. (1972). *Agric. Fd. Chem.* 20, 718.
Österberg, R. (1974). *Co-ord. Chem. Rev.* 12, 309.
Parker, A.J. (1981). Chapter 1, this Volume.
Pate, J.S. (1975). *In* "Transport in Plants. I. Phloem Transport" (M.H. Zimmermann and J.A. Milburn, eds.) pp. 451-473. Encycl. Pl. Physiol. (N.S.) 1. Springer-Verlag, Berlin.
Pate, J.S. (1976). *In* "Transport and Transfer Processes in Plants" (I.F. Wardlaw and J.B. Passioura, eds.) pp. 253-281. Academic Press, New York.
Pate, J.S., Sharkey, P.J., and Lewis, O.A.M. (1974). *Planta* (Berl.) 120, 229.
Pate, J.S., Sharkey, P.J., and Lewis, O.A.M. (1975). *Planta* (Berl.) 122, 11.
Rasheed, M.A. and Seeley, R.C. (1966). *Nature* (Lond.) 212, 644.
Rauser, W.E., and Curvetto, N.R. (1980). *Nature* (Lond.) 287, 563.
Reilly, C. (1969). *New Phytol.* 68, 1081.
Reuter, D.J., Robson, A.D., Loneragan, J.F., and Tranthim-Fryer, D.J. (1981a). *Aust. J. agric. Res.* (In Press).
Reuter, D.J., Robson, A.D., Loneragan, J.F., and Tranthim-Fryer, D.J. (1981b). *Aust. J. agric. Res.* (In Press).
Robson, A.D. and Reuter, D.J. (1981). Chapter 13, this Volume.
Sandmann, G. and Böger, P. (1980). *Pl. Physiol.* 66, 797.
Sarkar, B. and Kruck, T.P.A. (1966). *In* "The Biochemistry of Copper" (J. Peisach, P. Aisen, and W. Blumberg, eds.) pp. 183-196. Academic Press, New York.
Sillen, L.G. and Martell, A.E. (1971). "Stability Constants" Suppl. No. 1, Chem. Soc. Spec. Publ. No. 25.
Snowball, K., Robson, A.D., and Loneragan, J.F. (1980) *New Phytol.* 85, 63.
Stevenson, F.J. and Fitch, A. (1981). Chapter 4, this Volume.
Struckmeyer, B.E., Peterson, L.A., and Hsi-Mei Tai (1969). *Agron. J.* 61, 932.
Tammes, P.M.L. (1958). *Acta Botan. Neerl.* 7, 233.

Tammes, P.M.L. and Van Die, J. (1964). *Acta Botan. Neerl.* 13, 76.

Tammes, P.M.L. and Van Die, J. (1966). *Proc. Kaninkl. Ned. Akad. Wetenschap.* 69, 655.

Tiffin, L.O. (1972). *In* "Micronutrients in Agriculture" (J.J. Mortvedt, P.M. Giordano, and W.L. Lindsay, eds.) pp. 199-229. Soil Sci. Soc. Am. Inc., Madison.

Tiffin, L.O. (1977a) *In* "Implications of Metals in the Environment". pp. 315-334. Proc. 15th Ann. Hanford Life Sci. Symp., Sept. 29-Oct. 1, 1975, Richland, Wash. Published by ERDA, TIC, CONF. 750929, Oak Ridge, Tenn.

Tiffin, L.O. (1977b). Ann. Rep. ERDA (AEC) Contract No. AT (49-7)-1. Agric. env. Quality Inst., Beltsville agric. Res. Center, U.S.D.A., Beltsville, Md.

Timperley, M.H., Brooks, R.R., and Peterson, P.J. (1973) *J. exp. Bot.* 24, 889.

Walker, C.D. (1980). Ph.D. Thesis, Murdoch University.

Walker, C.D. and Loneragan, J.F. (1981). *Ann. Bot.* (In Press).

Walker, C.D. and Webb, J.M. (1981). Chapter 9, this Volume.

Williams, C.H. and Moore, C.W.E. (1952). *Aust. J. agric. Res.* 3, 343.

Wood, J.G. and Womersley, H.B.S. (1946). *Aust. J. exp. Biol. med. Sci.* 24, 79.

Woolhouse, H.W. and Batt, T. (1976). *In* "Perspectives in Experimental Biology, Vol. 2 - Botany" (N. Sunderland, ed.) pp. 163-175. Pergamon, Oxford.

9

COPPER IN PLANTS: FORMS AND BEHAVIOUR[1]

Colin D. Walker[2]

School of Environmental and Life Sciences
Murdoch University
Perth, Western Australia

John Webb

School of Mathematical and Physical Sciences
Murdoch University
Perth, Western Australia

I.   INTRODUCTION

In plants, new roles and new forms of Cu are regularly
being identified, although many are now known.   Before
1960, known forms of Cu in plants included the proteins
cytochrome oxidase (Caughley *et al.*, 1976; Ericinska and
Wilson, 1978), ascorbate oxidase (Dawson, 1966; Burstein
et al., 1976) phenolase (Henry, 1976; Mayer and Harel,
1979), laccase (Fee, 1975; Mayer and Harel, 1979) and
diamine oxidase (Matsuda and Suzuki, 1977).   Since that
time, plastocyanin (Katoh, 1977), superoxide dismutase
(Mitchelson *et al.*, 1977), Fraction 1 protein (Jensen and
Bahr, 1977), plantacyanin (Aikazyan and Nalbandyan, 1975)
and possibly quinol oxidase (Palmer, 1979) have been identified as Cu proteins of wide occurrence in plants.   Further-

---

[1]*Supported in part by the Australian Meat Research
Committee.*
[2]*Present address: U.S. Plant, Soil and Nutrition
Laboratory, Ithaca, New York.*

more, several 'simple' blue proteins have been isolated from
individual plant sources. Low molecular weight Cu species
have been found in vascular exudates (Tiffin, 1972) and
organic extracts (Bremner and Knight, 1970; Timperley
et al., 1973) of plants. In this chapter we consider the
above Cu forms in relation to normal plant metabolism and,
to a less extent, to metabolism in Cu deficiency. We make
only passing reference to 'non-functional' forms as occur in
Cu accumulator plants (Brooks, 1972; Foy *et al.*, 1978) and
probably also in plants experiencing toxic levels of Cu
(Bremner, 1974; Foy, *et al.*, 1978) and to the Cu apparently
unavailable for retranslocation that remains in normal plant
tissue after senescence (Loneragan *et al.*, 1976).

We further explore the behaviour of these forms of Cu by
considering the effect of Cu deficiency on the enzymatic
activities of these several Cu enzymes. Recent reviews of
trace element forms in plants include those of Nicholas and
Egan (1975) and Tiffin (1977).

II. KNOWN FORMS OF CU IN PLANTS

Many reports of Cu associated with plant compounds exist
in the literature. Foremost of these are enzymic studies
indicating trace but stoichiometric levels of Cu in purified
plant protein, often as an extrinsic feature in the study of
the protein and its properties. Other reports describe the
nature of translocatable Cu (Tiffin, 1972; Hoefner, 1966)
and of Cu compounds recovered in organic solvent extracts
(Bremner and Knight, 1970; Timperley *et al.*, 1973;
Ozolinya and Lapinya, 1976), in digests (Bremner and Knight,
1970), in organelle fractions (Timperley *et al.*, 1973) and in
crude protein fractions (Ozolinya and Lapinya, 1976).

The distribution of Cu among organelles of plant cells
is discussed in the preceding chapter of this volume
(Loneragan, 1981). So has the evidence suggesting that a
substantial proportion of Cu in plants is associated with
nitrogenous compounds. Copper protein forms are widely
distributed and, as will become more evident later, perform
major metabolic functions in the plant. Thus, the correlation and associated behaviour of N and Cu in many plant species
seems a fundamental factor in the behaviour of Cu in the
biology of plants.

A. *Low Molecular Weight Forms*

The nature of low molecular weight Cu species has been examined in extracts of plant materials. Using paper electrophoresis, Hoefner (1966) and Tiffin (1972) detected the presence of anionic forms of Cu in broad bean and tomato xylem exudates respectively, indicating the complexation of Cu ions with negatively charged ligands (most likely to be organic compounds, probably amino acids). Similarly, Bremner and Knight (1970) and Timperley *et al.*, (1973) found several anionic forms of Cu in ethanolic extracts of rye-grass and tree leaves respectively. Thompson and his colleagues (Thompson and Tiffin, 1974; Tiffin and Thompson, 1975; pers. comm.) using a Ni radioisotope, found several anionic species in a number of higher plants that they considered would complex Cu *in vivo*. These ligands were not simple amino acids; they initially appeared in amino acid fractions but have not been completely characterised.

Reilly and his colleagues (Reilly *et al.*, 1970; Reilly, 1972) reported that in extracts of the Cu accumulator plant *Becium homblei*, Cu and amino acids moved together in paper chromatography indicating the possibility of complexation. Farago and Mullen (1979) reported that, in proline rich extracts of Cu accumulator plants of *Armenia maritima*, Cu was found in forms identified by chromatographic behaviour to be a Cu proline complex. The studies of Tiffin and Thompson (1975) indicate that in some cases the Cu ligands may not be the more common (and readily recognised) protein amino acids. Many other plant amines are known (Smith, 1975) and are possible ligands. The possibility of mixed complexes (amino, organic, phenolic acids) should also be considered. The cationic Cu hexaaquo species was not reported in any of the work discussed.

If Cu amino acid species (or similar) comprise the main mobile forms of Cu in vascular tissue of plants (Tiffin, 1972), then an intracellular pool of similar species also seems likely, particularly on the evidence from mammalian systems (Harris and Sass-Korsak, 1967) that Cu moves across membranes as Cu amino acid complexes. Arman and Wain (1963) applied Cu solutions to the surface of apple tree leaves and found increases in Cu uptake when specific stoichiometric proportions (of either one, 2 or 3 molecular equivalents) of anions or organic acid ligands were added to the Cu solutions. These results indicate the transport of Cu into leaves in complexed forms with these ligands.

Low molecular weight complexes of Cu are recognised to have catalytic properties (Mottola *et al.*, 1968). For example, the Cu cysteine species catalyses the autoxidation of ascorbic acid, a property it shares with the Cu protein ascorbate oxidase.   The significance of such properties *in vivo* is hard to determine, but some Cu species may have sufficient catalytic power to account for the extreme toxicity of Cu in biological systems when in excess (Arman and Wain, 1953).

B.   *Copper Proteins*

Copper in proteins has been recognised as occurring in 3 distinguishable types (Broman *et al.*, 1962) on the basis of optical, magnetic, and redox properties as summarized in Table I.   Although this scheme has been suggested to not be universally valid (Mondovi *et al.*, 1976; Beinert, 1977), the information available on Cu in proteins can be examined quite usefully following this terminology.

Type I Cu is that form responsible for the intense blue colour of many Cu proteins.   It may be the only form present in a protein (e.g., plastocyanin, azurin), or it may occur in conjunction with types II and III Cu, (e.g., laccase, ascorbate oxidase).   The coordination geometry of Cu in the type I form has been particularly fascinating to workers as the intense absorption band at about 600 nm (Table I) could not be satisfactorily reproduced in model studies (Amundsen *et al.*, 1977).   More recently, this absorption band has been claimed to be due to a Cu thiolate charge transfer (Karlin and Zubieta, 1979).

The crystallographic examinations of plastocyanin (Colman *et al.*, 1978) and azurin (Adman *et al.*, 1978) showed that the type I Cu ion of these proteins was held in a distorted tetrahedral coordination.   Colman *et al.*, (1978) observed that the four ligand groups involved in Cu coordination are conserved in all plastocyanin and azurin sequences that are known and that "the azurin and plastocyanin families have similar Cu sites in keeping with their similar biological roles as electron transport proteins".   They noted further that the tetrahedral distortion of the preferred square planar coordination of the Cu(II) ions was responsible for a number of the observed properties of 'blue' proteins, viz., the narrow hyperfine splittings of the epr spectra, the possession of redox potentials higher than those of Cu(II) in low molecular weight complexes of square planar structure, and the nature of the electronic, circular dichroism and magnetic circular dichroism spectra.

TABLE I. Properties of Copper Sites in Proteins

Type	Synonym	Optical[a] Behaviour Abspn(nm) (ext coeff)	Redox Behaviour (mV)	EPR Behaviour A (cm^{-1})	Metal Species[b]	Examples
I	blue	600 ($3-11.10^3$)	$184-395^c$ $415-767^d$	0.004-0.009	Cu(II)	Plastocyanin, other 'blue' proteins
II	non-blue	600 ($1-4.10^2$)	440^e	0.012-0.020	Cu(II)	Galactose Oxidase (resting enzyme), also in assoc. with other forms in laccase, ascorbate oxidase
III	EPR-non-detectable	340 (3.10^3)			spin-paired Cu(II)-Cu(II) or Cu(I)	Oxyhemocyanin, with other forms in laccase, ceruloplasmin

[a] Absorption in nm and (extinction coefficient)
[b] Tentative form
[c] Simple 'blue' proteins with no 'oxidase' function
[d] 'blue' oxidases
[e] Value for Galactose Oxidase

Despite some variability in the properties of different Cu species placed in the type II class (Mondovi *et al.*, 1976), these species exhibit similar optical spectra, in that they possess extinction coefficients much lower than those of type I Cu (hence 'non-blue'), and similar magnetic spectra with 'normal' hyperfine splitting (Table I). These properties of the type II Cu resemble those of the Cu(II) ion in many low molecular weight compounds. X-ray crystallography data for superoxide dismutase (Richardson *et al.*, 1975) indicates a square planar coordination of the type II Cu with four histidine residues in this protein, indicating that type II Cu is likely to occur in sites lacking the considerable distortion (away from preferred square planar coordination) observed in type I Cu protein sites.

If the claim that the 'blue' character of type I Cu is due to Cu thiolate charge transfer is correct (Karlin and Zubieta, 1979), then the lack of sulphur ligands (as is the case with superoxide dismutase) is possibly a further distinction that can be made between types I and II Cu.

Type III Cu is epr nondetectable and is the source of the 330-350 nm peak observed in some Cu proteins (Table I). The lack of residual electron spin in this species may be due to spin pairing of two closely coordinated Cu(II) ions or the presence of the diamagnetic Cu(I) ion. At least one case of the former is known (mushroom oxyphenolase: Mason, 1965). Single type III Cu species in which the latter would be the only possible structure in Cu thionein (Weser and Rupp, 1978).

Many Cu proteins are known to occur in higher plants (Table II). Of these, a number are well characterised and have been placed into the functional classes proposed for all Cu proteins (Malkin and Malmström, 1970; Fee, 1975). Thus plant Cu proteins function as terminal oxidases, mono- and di-oxygenases, in superoxide radical catalysis, and in electron transport pathways. The considerable involvement of biological Cu with the biochemistry of oxygen (Hayaishi, 1974; Ochai, 1975) is particularly evident in the plant Cu proteins.

1. Cytochrome Oxidase. Copper is present in cytochrome oxidase, the terminal oxidase of the mitochondrial electron transport chain. This protein has been intensively studied in recent years and many reviews are available (Wharton 1973; Nicholls and Chance, 1974; Fee, 1975; Caughley *et al.*, 1976; Beinert, 1977; Erecinska and Wilson, 1978). The minimum functional unit of cytochrome oxidase

TABLE II. Copper Proteins Of Higher Plants

Protein	Copper Stoichiometry	Copper Form I	II	III	Molecular Weight	Subcellular Localisation
Cytochrome Oxidase	2[a]	1			200,000	Mitochondria
Ascorbate Oxidase	8	2	2	4	140,000	Cytoplasm,Cell Wall
Phenolase	1		1?		40,000	Chloroplast
Laccase	4	1	1	2	110,000	
Quinol Oxidase	?					Mitochondria
Diamine Oxidase	(0.08-0.09%)					Nucleus?
Superoxide Dismutase	2		2		33,200	Chloroplast, Mitochondria
Fraction 1	?		present		650,000	Chloroplast
Plastocyanin	1 or 2		all		10,500 or 21,000	Chloroplast
Rice bran blue protein	1		1		23,000	
Mung bean blue protein	1		1		22,000	
Stellacyanin	1		1		20,000	
Umecyanin	1		1		14,500	
Mavicyanin	1		1		18,000	
Plantacyanin	1		1		8,000	

[a] Cytochrome Oxidase possesses one Cu in a heme-Cu complex

has a molecular weight of 200,000 daltons and consists of
7 poly-peptides subunits with 2 heme groups and 2 Cu atoms.
It spans the inner mitochondrial membrane in a specific
orientation.   One of the Cu atoms interacts with the cytochrome a_3 heme group and is unique in this behaviour, while
the other is tentatively regarded as a type I Cu (Erecinska
and Wilson, 1978;   Karlin and Zubieta, 1979).

2.   *Ascorbate Oxidase*.   Ascorbate oxidase is a 'blue'
oxidase, catalysing the oxidation of ascorbate to L-dehydro-
ascorbate, and is widely found in higher plants (Dawson,
1966;   Burstein *et al.*, 1976;   Dawson *et al.*, 1975).
Halloway *et al.*, (1970) reported its presence in cell walls,
and other workers have found it in the cytoplasm (James and
Cragg, 1943).   Although this oxidase can be considered as
a terminal respiratory oxidase (Brown and Hendricks, 1952;
Perumal and Beattie, 1966), its physiological function may
still be uncertain (Malmström *et al.*, 1975), as plant respiratory requirements may be met by mitochondrial respiration.
Ascorbate oxidase in the cytoplasm may be involved in
anabolic metabolism (Bidwell, 1979), being linked to certain
dehydrogenases in a redox shuttle involving glutathione.

3.   *Phenolase and Laccase*.   The Cu proteins phenolase
(Henry, 1976;   Halliwell, 1978a;   Mayer and Harel, 1979)
and laccase (Fee, 1975;   Meyer and Harel, 1979) catalyse
oxygenation reactions of plant phenols.   Laccase is a 'blue'
Cu protein found in *Rhus* exudates and in fungal species and
has been found not only to oxidise phenols (catechols appear
to be its native substrate), but also to oxidise aromatic
amines and ascorbic acid (Levine, 1966).   Laccase is not
inhibited by carbon monoxide whereas higher plant phenolase
is (Kertesz, 1966).

Phenolase has been reported in mitochondria, but more
frequently in the thylakoid membrane of chloroplasts (Henry,
1976;   Mayer and Harel, 1979).   It has 2 distinct enzyme
functions, namely the mono-oxygenation of monophenols or
'cresolase', and the mono-oxygenation of o-diphenols or
'catecholase' (Brooks and Dawson, 1966).   The former
function may also be referred to as 'tyrosinase' and the
latter as 'polyphenolase' or 'polyphenol oxidase'.   However,
these names are more accurately ascribed to enzymes with
only one of these functions (occurring elsewhere in nature),
and the term 'phenolase' will be used to refer to the 2
associated functions occurring in higher plants.

The properties of the phenolase complex of spinach beet
have been recently reviewed (Butt, 1979).   Purified pheno-

lase is found to have a molecular weight of around 40,000 daltons (Vaughan *et al.*, 1975), and at least one Cu atom per molecule, although several workers have reported less than one Cu per 40,000 dalton unit, possibly through the presence of Cu chelates in the purification method (Kidron *et al.*, 1977).

Phenolic oxidation products produced by the action of phenolase are involved in lignin and alkaloid biosynthesis (Hanson and Harvir, 1979), in the synthesis of the quinone intermediates of electron transport chains of photosynthesis and respiration (Bidwell, 1979), and in the production of brown, melanotic patches that are observed in many plants in response to wounding (Racusen, 1969). They are active as phytoalexins, being inhibitors of invasive spore germination and fungal growth, and as toxins against herbivores. They are also a problem in the preparation of plant extracts (Anderson, 1968).

Phenolase is important in senescing tissue where it is coupled to the oxidation of cellular constituents via its immediate phenolic substrates (Racusen, 1969; Patra and Mishra, 1979).

4. *Diamine Oxidase.* Early work on diamine oxidase isolated from pea seedling (Hill and Mann, 1962) established that this enzyme contained Cu. It has been found that amine oxidases from legume species possess Cu, whereas those of cereals contain flavin adenine dinucleotide (FAD) (Hirosawa and Suzuki, 1975). All the Cu in legume oxidase is divalent and does not change valency during the enzyme reaction (Hill, 1971). The legume and cereal forms oxidize amines by different mechanisms (Smith, 1977; Yasunobu *et al.*, 1977). Both enzymes are at high levels in the seedling stage of the plant, metabolizing the polyamines such as spermidine and spermine that are also present in these tissues (Matsuda and Suzuki, 1977; Smith, 1977). Polyamines are found widely in plant material and are precursors of alkaloids (Yasunobu *et al.*, 1977) and probably also auxin. In germinating pea seedlings, diamine oxidase activity is induced by putrescine, spermidine, and amithine (Srivastava *et al.*, 1977). In fractionated extracts of plant tissue, the enzyme appears to be normally bound to a particulate fraction which may be nuclei (Smith, 1976).

5. *Superoxide Dismutase.* Superoxide dismutases are widely distributed in biological systems (Mitchelson *et al.*, 1977) and are thought to act as a cellular defence against the toxic superoxide radical generated in a wide range of

biochemical reactions (Halliwell, 1978b). Copper-zinc superoxide dismutases are dimeric enzymes with one Cu and one Zn per subunit (Marmocchi *et al.*, 1978). The Cu ion is involved in the mechanism of the disproportionation of the superoxide radical (Malmström *et al.*, 1975). However the function of the Zn ion is less clear. Bovine erythrocyte superoxide dismutase has been crystallised and its structure has been determined by X-ray diffraction (Richardson, *et al.*, 1975).

In plants, Cu-Zn superoxide dismutase has been found in chloroplasts (Asada *et al.*, 1973) where it functions to regulate superoxide radicals generated by photorespiration (Halliwell, 1978b). Several isozymes of superoxide dismutase, some possessing Fe or Mn, rather than Cu and Zn, have been reported in mitochondria, while Fe superoxide dismutases occur in the periplasmic space of simple organisms. Whether the Fe and Mn enzymes of higher plants would have similar intracellular distribution is an open question (Mitchelson *et al.*, 1977).

6.   *Plastocyanin.*   Early work established the presence of Cu in chloroplasts (Neish, 1939). Katoh (1960) isolated and subsequently characterised plastocyanin. Katoh *et al.*, (1961) found that plastocyanin is localised in the chloroplast, where it contains over 50% of the copper present in this organelle. Plastocyanin is widely found in green plants being an intermediate in the electron transport chain of photosynthesis. The work of Smith (1977) on the thylakoid membrane suggests that plastocyanin occupies a position on the outside surface of this chloroplast structure. The properties of plastocyanins from many plant sources have been examined (Ramshaw *et al.*, 1973) and recently reviewed (Katoh, 1977). Purified plastocyanins reported recently (Boulter *et al.*, 1979) represent plant sources from twelve families of flowering plants. The structure of french bean plastocyanin has been examined by X-ray crystallography (Colman *et al.*, 1978).

7.   *Fraction I Protein.*   This is an apparently pure protein which is the major protein of photosynthetic tissues. It has ribulose biphosphate (RuBP) carboxylase activity and frequently also RuBP oxygenase activity.

Whether these two enzyme activities belong to a single protein or to 2 closely associated proteins is a subject of current controversy. So also is the role of Cu in these enzymes. A report that a Cu protein with ribulose biphosphate (RuBP) oxygenase activity may be separated from RuBP

carboxylase activity of Fraction 1 (Brãnden, 1978) has been
denied by other workers (McCurry *et al.*, 1978).   The
indication that RuBP oxygenase is a Cu enzyme may be supported by the observation that chloroplasts of Cu deficient
plants have a higher quantum efficiency (Basynski *et al.*,
1978).   However, much of the earlier work failed to find
any requirement for Cu for RuBP oxygenase, even when Cu
binding was found (see Jensen and Bahr, 1977).   The lack
of a transition metal in the oxygenase mechanism is disturb-
ing to some workers (Lane and Miziorko, 1978), who consider
the question open.   On the other hand, Cu could interfere
with photosynthetic efficiency in the light reaction, since
high levels of free Cu ions inhibit electron transport in
both photosystems, leading to lipoxygenase activity
(Sandmann and Bõger, 1980).

The association of Cu with Fraction 1 is also controv-
ersial.   Some authors (Lorimer *et al.*, 1973;   Chollet *et
al.*, 1975;   Johal and Bourque, 1979) have reported no
detectable Cu in certain Fraction 1 preparations;   others
present clear evidence for the presence of Cu in similar
preparations (Wishnick *et al.*, 1969, 1970;   Daley, *et al.*,
1980).

Contamination from any of the Cu proteins present in
chloroplasts (plastocyanin, superoxide dismutase, phenolase)
might account for some of the reports of Cu in Fraction 1
protein.   However, EDX measurements on the pyrenoid of green
algae, in which none of these enzymes has been reported,
indicate that Fraction 1 protein binds Cu *in vivo* (Salisbury
and Floyd, 1977, 1978).

Nicholson *et al.*, (1980) also found Cu to be associated
with Fraction 1 protein in fractionated extracts of Cu
intoxicated mung plants.   Moreover, the parallel decline of
Cu and N in leaves during the early stage of their senescence
(Hill, *et al.*, 1979;   Loneragan, 1981) when Fraction 1 is
the major declining protein (Wittenbach, 1978) suggests an
association between Cu and Fraction 1.

The apparent conflict in the association of Cu with
Fraction 1 protein may result from a variable capacity of
this protein to bind Cu in response to Cu status and tissue
age.   Walker (1980) gives evidence of such variability in
subterranean clover in which Fraction 1 protein contained
Cu when extracted from old leaves and other tissues of high
Cu plants but contained none when extracted from young
leaves of low Cu plants.

Clearly the role of Cu in Fraction 1 protein needs re-
assessment especially in the light of recent results suggesting that Fraction 1 may have an important role in the control

of Cu metabolism within the plant.

8.   Quinol oxidase.   Palmer (1979), in reviewing the metabolism of plant mitochondria, describes the recent discovery of an 'alternative oxidase' which is cyanide resistant, unlike cytochrome oxidase, and which, in partially purified preparations, contains significant amounts of Cu and probably also a flavin group.   The presence of this protein in mitochondria distinguishes it from higher plant phenolase which is found in the thylakoid membrane of the chloroplast.   The 'alternative' title describes its function in providing a second pathway of ubiquinone oxidation in the mitochondria.   This pathway is thought to transfer electrons directly from the quinol substrate to molecular oxygen, avoiding the cytochrome electron transport chain. It has been proposed that lipoxygenase activity in plant mitochondria operates through the same enzyme but more recent work (Miller and Opendorf, 1981) suggest that this may not be so.   The role of Cu in these processes has not been established.   Similar processes are known in micorbial mitochondria.

9.   Other Plant Copper Proteins.   A number of Cu proteins has been reported from individual plant sources. How widely they are distributed in higher plants is uncertain.   These proteins are all 'blue' proteins and possess only one Cu atom per molecule, this being a type I species. Similar proteins, e.g., azurin, have been found in fungi and algae.   The plant proteins are mung bean blue protein from etiolated seedlings of mung bean (Schichi and Hacket, 1963), rice bran blue protein (Ida and Morita, 1969), *Rhus* stellacyanin (Peisach *et al.*, 1967), umecyanin from horse-radish root (Paul and Stigbrand, 1970), mavicyanin from zucchini squash (Marchesini *et al.*, 1979) and plantacyanin from spinach (Aikazyan and Nalbandyaw, 1975).   Plantacyanin was originally isolated from cucumber skin and termed 'cusacyanin':   it may be fairly widely distributed in higher plants (Aikazyan *et al.*, 1974;   Aikazyan and Nalbandyaw, 1975).

Moore and Williams (1976) and Karlin and Zubieta (1979) have reviewed collectively a number of the 'blue' proteins along with the very similar and widely distributed plasto-cyanin.   The redox potentials of these proteins range from 184 mV (stellacyanin) to 395 mV (azurin).   Plastocyanin has a potential of around 390 mV.

The low potentials and the fact that most of the reduced proteins are not reoxidized by molecular oxygen indicate

that they are unlikely to be oxidases. Four of those
above (mung bean and rice bran blue protein, umecyanin, and
stellacyanin) come from non-photosynthetic tissues pre-
cluding a role in electron-transport in photosynthesis.
However, certain of the 'blue' proteins may have electron
transport functions (Karlin and Zubieta, 1979). The evidence for azurin indicates that it functions as an electron
carrier in the respiratory chain between cytochromes and
cytochrome oxidase, and stellacyanin also appears to be an
electron carrier. Functions for the remaining 'blue'
proteins are unknown.

C. Comments on Plant Cu Forms

Copper plays an active role in enzymes performing key
functions in respiration and photosynthesis and Cu proteins
have been implicated in lignification, anabolic metabolism,
cellular defence mechanisms, and hormone metabolism
(Bussler, 1981). The metabolic significance of certain
blue proteins is unknown. Copper amino acids or similar
low molecular weight compounds appear to be involved in the
transport of Cu in the plant.

However, there is relatively little indication of any
mechanism of control of the distribution of Cu itself
between tissues, cells, organelles, or compounds by the
plant. Thus, no Cu compounds have been ascribed functions
of "Cu storage" or "Cu membrane transport" or "insertion of
Cu into enzymes". Known forms of Cu might be examined for
possible roles relating to these areas. Nor are there
plant hormones known to regulate Cu metabolism in any
specific manner. The suggestion that low molecular weight
species of Cu may be involved in Cu transport within vascular tissue (and possibly in trans-membrane movement) may
provide an opening for further studies of this general area.

III. EFFECTS OF DEFICIENCY ON FORMS OF Cu

It is evident that Cu is involved in many compounds of
both low and high molecular weight which are of functional
importance in the metabolism of higher plants. Consequently growth of plants under conditions of inadequate levels
of available Cu leads to impaired growth. Physiological
aspects of Cu deficiency are considered in the next chapter
of this Volume (Bussler, 1981). In this section, Cu

deficiency is related to depressed levels of the functional forms of Cu.

Of those plant enzymes known to contain Cu (Table II), around half have been studied in Cu deficient plants (Table III). Almost invariably, Cu deficiency depresses the activities of these enzymes. The depression of enzyme levels varies with plant parts (Hill, 1973; Lyszcz *et al.*, 1976; Walker and Loneragan, 1981), plant age (Brown and Hendricks, 1952; Hill, 1973; Lyszcz *et al.*, 1976), species (Brown and Hendricks, 1952; Bar Akiva *et al.*, 1969; Lyszcz *et al.*, 1976; Basynski *et al.*, 1978), and subspecies (Brown and Hendricks, 1952; Graves *et al.*,1979) (Table III). The comparison of enzyme depressions reported in different studies is difficult because of differing culture methods, nutrient supply, plant organs sampled, and plant ages. Some workers have grown plants in solution culture (Bailey and McHargue, 1944; Perumal and Beattie, 1966; Bar Akiva *et al.*, 1969; Cartwright and Hallsworth, 1970; Hill, 1973) while others have used Cu deficient soils (Brown and Hendricks, 1952; Lyszcz *et al.*, 1976; Basynski, *et al.*, 1978; Graves *et al.*, 1979; Walker and Loneragan, 1981). Both substrates have been successful in terms of obtaining plants displaying Cu deficiency symptoms and diminished enzyme levels. Solution culture was found suitable for the study of root nodule enzymes in the work of Cartwright and Hallsworth (1970). Bligny and Douce (1977) studied cytochrome oxidase in a phytostat solution culture of sycamore cells, obtaining cells very deficient (less than 5% normal) in the level of oxidase present.

Activities of Cu enzymes in individual parts of the complete plant and their response to Cu deficiency have also been studied. Hill (1973) and Lyszcz *et al.*, (1976) reported that Cu deficiency resulted in depressed enzyme levels throughout all organs of the plants studied. By contrast, Walker and Loneragan (1981) found that Cu deficiency severely depressed enzyme activities in young plant tissues but had little effect elsewhere: they suggested that old leaves retained their Cu against translocation into the growing shoot tip.

Lyszcz *et al.*, (1976) and Walker and Loneragan (1981) found differences in enzyme levels between old and young leaves of healthy plants. In earlier studies, most workers had also examined leaf tissue, although the age of the tissue was frequently unspecified. This fact may have some small effect on the magnitude of differences observed between treatments.

Proportional differences of enzyme levels between

TABLE III.   Copper Enzymes Affected by Copper Deficiency in
             Plants

Enzyme	Species	% Activities Cu deficient Cu adequate	Reference
Ascorbate	Lupins(3spp.)	5%-74%	b
Oxidase	Soybeans(2cv.)	33%-53%	b
	Wheat	0%-40%	b
	Corn	0%-19%	b
	Apple	5%	c
	Lemon	53%	d
	Lime	69%	d
	Subterranean Clover	67%	e
	Red Clover	19%-93%	f
	Oats	0%-92%	g
	Sunflower	12%-61%	g
	Subterranean Clover	34%	k
Phenolase	Tomato	70%	a
	Red Clover	0%-92%	f
	Sunflower	4%-81%	g
	Chrysanthemum(8cv.)	0%-50%	j
	Subterranean Clover	62%	k
Cytochrome	Subterranean Clover	75%-80%	e,k
Oxidase	Red Clover	15%-113%	f
	Sycamore	5%	h
Diamine	Red Clover	9%-88%	f
Oxidase	Subterranean Clover	45%	k
Plastocyanin	Oats	47%	i
	Spinach	40%	i
Superoxide Dismutase	Subterranean Clover	91%	k

a:Bailey and McHargue, 1944;   g:Lyszcz et al., 1976;
b:Brown and Hendricks, 1952;   h:Bligny and Douce, 1977;
c:Perumal and Beattie, 1966;   i:Basynski et al., 1978;
d:Bar Akiva et al., 1976;   j:Graves et al., 1979;
e:Cartwright and Hallsworth,   k:Walker and Loneragan, 1981.
   1970;
f:Hill, 1973;

treatments generally increased with plant age (Hill, 1973;
Lyszcz *et al.*, 1976). For example, leaves of deficient red
clover seedlings had 70% of normal ascorbate oxidase activity of 9 days, 60% at 15 days, and 20% at 42 days. These
observations follow from the inability of the seedlings to
obtain Cu from their deficient substrate. The author gives
no indication as to when these plants first showed signs of
deficiency (Hill, 1973).

Photosynthesis involves the Cu proteins plastocyanin,
superoxide dismutase, and Fraction 1 protein. The phenolase
complex also occurs in the chloroplast and may be significant in quinone metabolism. Copper deficiency depresses
photosynthate levels and respiratory activity as indicated
for example, by the lowered level of reduction of the redox
dye triphenyltetrazolium chloride (TTC) in wheat plant stems
(Brown and Clarke, 1977).

The photosynthetic carbon dioxide fixation of plants
was found to be depressed by Cu deficiency in the early work
of Loustalot *et al.*, (1945). The level was lower in leaves
displaying symptoms of deficiency (20% normal level) than
in those which appeared as normal (50% normal level).
Subsequently, Spencer and Possingham (1960) established a
positive correlation between Cu supply of plants and the
rate of the Hill reaction, now recognised to be the non-
cyclic photoreduction of redox indicators by illuminated
chloroplasts. This work was soon followed by the discovery
of the Cu electron transport protein, plastocyanin (Katoh,
1960). Bishop (1964) presented further evidence of the
presence of a Cu compound (i.e., plastocyanin) in photoreduction in the alga *Scenedesmus obliquus* and its depression as the principal effect of Cu deficiency.

Recent reports (Lyszcz *et al.*, 1976; Basynski *et al.*,
1978) confirm the inhibiting effect of Cu deficiency on
photosynthesis. Carbon dioxide fixation was again depressed and this could not always be attributed to depressed
ribulose diphosphate carboxylase activity. Basynski *et al.*,
(1978) reported that in chloroplasts of Cu deficient plants
the levels of plastocyanin and Cu were around 50% of normal
levels. Photosynthetic electron transport activities were
similarly depressed, as were levels of plastic pigments when
measured on a leaf weight basis. The depression of the
latter may result from an effect on phenolase levels or
from the lack of photosynthate in the plant. The higher
quantum efficiency of chloroplasts from Cu deficient plants
(Basynski *et al.*, 1978) may be attributed to a lowered rate
of photorespiration resulting from a diminished level of
ribulose diphosphate oxygenase.

The evidence suggests that photosynthesis in Cu deficient plants operates at a depressed rate because of the
lack of the electron transport protein, plastocyanin. While
further limitations through the depression of levels of
other Cu proteins are also possible, their effects have not
been reported in the Cu deficient chloroplast.

Respiration in plants, when defined vary broadly as the
consumption of oxygen, may be considered to involve many
Cu proteins, i.e., all oxidases and oxygenases.   In the
generation of adenosine triphosphate (ATP) through oxidative phosphorylation, the widely recognised role of a Cu
protein is that of cytochrome oxidase, the terminal oxidase
of the electron-transport chain of the mitochondria. Superoxide dismutase and quinol oxidase also occur in the mitochondria and have auxiliary functions in the latter process.
The active cytoplasmic ascorbate oxidase and chloroplast
phenolase and ribulose biphosphate oxygenase may also contribute substantially to the oxygen consumption in other
sections of the plant cell.

Cartwright and Hallsworth (1970) found that Cu deficiency in nodulated subterranean clover led to a depression of
leaf and nodule respiration, particularly marked in younger
leaf tissue. Lyszcz *et al.*, (1976) found a depression of
respiration in Cu deficient oat and sunflower plants.
Bligny and Douce (1977) found that Cu deficiency depressed
levels of cytochrome oxidase during growth of sycamore cell
cultures. The respiration of such cells was more severely
inhibited by the cyanide ion (an inhibitor of cytochrome
oxidase). Thus, depression of mitochondrial respiration
occurred in this case.

The immediate repercussions of depressed oxidative
phosphorylation are likely to be a depression of levels of
ATP and of rates of energy dependent processes utilising
ATP and thus involves many areas of plant metabolism.

The initial work recognising functional roles of Cu in
plants was soon followed by suggestions that such forms
might be monitored as an assessment of nutrient status.
This concept has been continued to be raised over the years,
most notably by Bar-Akiva *et al.*, (1969, 1971). Bar-Akiva
et al., (1969) found that ascorbate oxidase activity in
leaf tissue of Cu deficient citrus trees could be re-
elevated to normal levels by incubation of the tissue with
Cu solution. The 'replacement' of enzyme activity by such
a technique was proposed as a method of diagnosing Cu
deficiency in the field as Robson and Reuter discuss later
in this Symposium. How low an enzyme activity needs to be
depressed before the lack of function limits metabolism

remains to be discovered.   This level is likely to vary
considerably within different organs of the plant.   However,
the existence of such a level is well recognised, at least
in terms of the whole plant, as the 'functional nutrient
requirement' of the plant (Loneragan, 1968).

IV.   CONCLUDING REMARKS

The nature of the chemical forms of Cu in plants is
being revealed gradually by chemical, biochemical, and
nutritional studies.   Complexes of Cu with organic compounds
of low molecular weight and with proteins both occur.   Of
the low molecular weight Cu species which have been detected
in plants few have been characterised.   By contrast, a
large number of Cu proteins have been identified in plant
materials.   They include proteins with no known function
and enzymes performing key functions in a wide range of plant
processes.   But so far, no Cu compound has been ascribed
the functions of "Cu storage", "Cu membrane transport",
"insertion of Cu into enzymes", or "Cu transport in vascular
tissues".   Much therefore remains to be done before the
behaviour of Cu in plants can be understood.
Copper deficiency has been shown to depress the levels
and enzyme activities of many Cu compounds.   In a single
plant, the effect may vary between organs and with age of a
particular organ.   The fact that many compounds may be
affected in a single organ suggests that Cu deficiency may
limit plant metabolism through several enzymes.   However
the relationship between depression of enzyme activity *in
vitro* and plant metabolism *in vivo* is not simple.   For
example, the *in vivo* sensitivity of enzyme function to Cu
deficiency can be expected to relate to factors such as
relative binding strength of the enzyme for Cu, available
Cu at the site of enzyme activity, and the extent to which
the normal level of enzyme activity may be depressed before
it limits metabolism.   These and other problems of the
relationships among Cu deficiency, activity of Cu enzymes,
and physiological processes are developed further in the
next chapter (Bussler, 1981).

ACKNOWLEDGEMENT

Professor J.F. Loneragan is thanked for continuing
encouragement and stimulating discussions.

REFERENCES

Adman, E.T., Stenkamp, R.E., Sieker, L.C., and Jensen, L.H. (1978). *J. Mol. Biol. 123*, 35.

Aikazyan, V.T., Markosyan, K.A., and Nalbandyan, R.M. (1974). *Biokhimiya 39*, 1184.

Aikazyan, V.T., and Nalbandyan, R.M. (1975). *FEBS Lett. 55*, 272.

Amundsen, A.R., Whelan, J., and Bosnich, B. (1977). *J. Am. Chem. Soc. 99*, 6730.

Anderson, J.W. (1968). *Phytochem. 7*, 1973.

Arman, P., and Wain, R.L. (1963). *Ann. Appl. Biol. 51*, 439.

Asada, K., Urano, M., and Takahashi, M. (1973). *Eur. J. Biochem. 36*, 257.

Bailey, L.F., and McHargue, J.S. (1944). *Pl. Physiol. 19*, 105.

Bar Akiva, A. (1971). *In* "Recent Advances in Plant Nutrition". (R.M. Samish, ed.). p.115. Gordon and Breach, New York.

Bar Akiva, A., Lavon, R., and Sagiv, J. (1969). *Agrochimica 14*, 47.

Basynski, T., Ruskowska, M., Król, M., Turkendorf, A., and Wolińska, D. (1978). *Z. Pfl. Physiol. 89*, 207.

Beinert, H. (1977). *Coord. Chem. Rev. 23*, 119.

Bidwell, R.G.S. (1979). "Plant Physiology". 2nd Ed. Macmillan, New York.

Bishop, N.I. (1964). *Nature (London) 204*, 401.

Bligny, R., and Douce, R. (1977). *Pl. Physiol. 60*, 675.

Boulter, D., Peacock, D., Guise, A., Gleaves, J.T., and Estabrook, G. (1979). *Phytochem. 18*, 603.

Bränden, R. (1978). *Biochem. Biophys. Res. Commun. 81*, 539.

Bremner, I. (1974). *Quart. Reve. Biophys. 7*, 75.

Bremner, I., and Knight, A.H. (1970). *Br. J. Nutr. 24*, 279.

Broman, L., Malmström, B.G., Hasa, R., and Vännagard, T. (1962). *J. Mol. Biol. 5*, 301.

Brooks, D.W., and Dawson, C.R. (1966). *In* "The Biochemistry of Copper". (J. Peisach, P. Aisen, and W.E. Blumberg, eds.) p.343. Academic Press, New York.

Brooks, R.R. (1972). "Geobotany and Biogeochemistry in Mineral Exploration". Harper and Row, New York.

Brown, J.C., and Clarke, R.B. (1977). *Pl.Soil 48*, 509.

Brown, J.C., and Hendricks, S.B. (1952). *Pl. Physiol. 27*, 651.

Burstein, S.R., Gerwin, B., Taylor, H., and Westley, J. (1976). *In* "Iron and Copper Proteins". (K.T.Yasunobu, H.F. Mower, and O. Hayaishi, eds.) p.472.Plenum Press,

New York.

Bussler, W., 1981. Chapter 10, this volume.

Butt, V.S. (1979), *Recent. Adv. Phytochem. 12,* 433.

Cartwright, B., and Hallsworth, E.G. (1970). *Pl. Soil 33,* 685.

Caughley, W.S., Wallace, W.J., Volpe, J.H., and Yoshikawa, S. (1976). *In* "The Enzymes", Vol. 13, 3rd Ed. (P.D. Boyer, ed). p.299. Academic Press, New York.

Chollet, R., Anderson, L.L., and Horsepian, L.C. (1975). *Biochem. Biophys. Res. Commun. 64,* 97.

Colman, P.M., Freeman, H.C., Guss, J.M., Murata, M., Norris, V.A., Ramshaw, J.A.M., and Venkatappa, M.P. (1978). *Nature (London) 272,* 319.

Daley, L.S., Wendel, C., Talent, J., and Theriot, L.J. (1980). *Pl. Physiol. 65,* S517.

Dawson, C.R. (1966). *In* "The Biochemistry of Copper". (J. Peisach, P. Aisen, and W.E. Blumberg, eds.) p.305. Academic Press, New York.

Dawson, C.R., Strothkamp, K.G., and Krul, K.G. (1975). *Ann. N.Y. Acad. Sci. 258,* 209.

Erecinska, M., and Wilson, D.F. (1978). *Arch. Biochem. Biophys. 188,* 1.

Farago, M.E., and Mullen, W.A. (1979). *Inorg. Chim. Acta 32,* L93.

Fee, J.A. (1975). *Struct. Bonding (Berlin) 23,* 1.

Foy, C.D., Chaney, R.L., and White, M.C. (1978). *Ann. Rev. Pl. Physiol. 29,* 571.

Graves, C.J., Adams, P., and Windsor, G.W. (1979). *J. Sci. Fd. Agric. 30,* 751.

Halliwell, B. (1978a). *Prog. Biophys. Mol. Biol. 33,* 1.

Halliwell, B. (1978b). *Cell Biol. Int. Rep. 2,* 113.

Halloway, M., Phethean, P.D., and Laggart, J. (1970). *Phytochem. 9,* 935.

Hanson, K.R., and Harvir, E.A. (1979). *Recent Adv. Phytochem. 12,* 91.

Harris, D.J.M., and Sass-Kortsak, A. (1967). *J. Clin. Invest. 46,* 659.

Hayaishi, O. (1974). "Molecular Mechanisms of Oxygen Activation". Academic Press, New York.

Henry, E.W. (1976). *Z. Pfl. Physiol. 78,* 446.

Hill, J., Robson, A.D., and Loneragan, J.F. (1978). *Aust. J. agric. Res. 29,* 925.

Hill, J., Robson, A.D., and Loneragan, J.F. (1979). *Ann. Bot. 44,* 279.

Hill, J.M. (1971). *In* "Methods in Enzymology". Vol. 17. (H. Tabor, and C.W.Tabor, eds.) p.730. Academic Press, New York.

Hill, J.M. (1973). *J. exp. Bot. 24,* 525.
Hill, J.M., and Mann,P.J.G. (1962). *Biochem. J. 85,* 198.
Hirosawa. E., and Suzuki, Y. (1975). *Phytochem. 14,* 99.
Hoefner, W. (1966). *Symp. Int. Agrochim. 6,* 94.
Ida, S., and Morita, Y. (1969). *Rep. Agric. Biol. Chem. 33,*
    10.
James, W.O., and Cragg, J.M. (1943). *New Phytol. 42,* 28.
Jensen, R.G., and Bahr, J.T. (1977). *Ann. Rev. Pl. Physiol.*
    28, 379.
Johal, S., and Bourque,D.P. (1979), *Science 204,* 75.
Karlin, K.D., and Zubieta, J. (1979). *Inorg. Persp. Biol.*
    Med. 2, 129.
Katoh, S. (1960). *Nature (London) 186,* 533.
Katoh, S. (1977). *In* "Photosynthesis I: Photosynthetic
    Electron Transport and Photophosphorylation". Encyclo-
    paedia of Plant Physiology, Vol. 5. (A. Trebst, and
    M. Avron, eds.) p.247. Springer, Berlin.
Katoh, S., Suga, I., Shiratori, I., and Takamiya, A. (1961).
    Arch. Biochem. Biophys. 94, 136.
Kertesz, D. (1966). *In* "The Biochemistry of Copper". (J.
    Peisach, P. Aisen, and W.E. Blumberg, eds.) p.339.
    Academic Press, New York.
Kidron, M., Harel, E., and Mayer, A.M. (1977). *Phytochem.*
    16, 1050.
Lane, M.D., and Miziorko, H.M. (1978). *In* "Photosynthetic
    Carbon Assimilation." (H.W. Siegelman, and G. Hind,
    eds.) p. 19. Plenum Press, New York.
Levine, W.G. (1966). *In* "The Biochemistry of Copper".
    (J. Peisach, P. Aisen, and W.E. Blumberg, eds.) p.371.
    Academic Press, New York.
Loneragan, J.F. (1968). *Nature (London) 220,* 1307.
Loneragan, J.F. Chapter 8, this Volume.
Lorimer, G.H., Andrews, T.J., and Tolbert, N.E. (1973).
    Biochem. 12, 18.
Loustalot, A.J., Burrows, F.W., Gilbert, S.G., and Nason, A.
    (1945). *Pl. Physiol. 20,* 283.
Lyszcz, S.T., Ruskowska, M., Wejcieska, U., and Zinkiewicz,
    E. (1976). *Acta Agrobot. 29,* 99.
Malkin, R., and Malmström, B.G. (1970). *In* "Advances in
    Enzymology and Related Areas of Molecular Biology",
    Vol. 33. (A. Meisler, ed.) p. 177. Wiley, New York.
Malmström, B.G., Andreasson, L. E., and Reinhammer, B.
    (1975). *In* "The Enzymes". Vol. 12., 3rd Ed. (P. D. Boyer,
    ed.) p. 507. Academic Press, New York.
Marchesini, A., Minelli, M., Merkle, H., and Kroneck, P.M.H.
    (1979). *Eur. J. Biochem. 101,* 77.
Marmocchi, F., Venardi, G., Bossa, F., Rigo, A., and Rotilo,

G. (1978).   *FEBS Lett. 94,* l09.

Mason, H.S. (1965). *Ann. Rev. Biochem. 34,* 595.

Matsuda, H., and Suzuki, Y. (1977). *Pl. Cell Physiol. 18,* 1131.

Mayer, A.M., and Harel, E. (1979). *Phytochem. 18,* 193.

McCurry, S.D., Hall, N.P., Pierce, J., Paech, C., and Tolbert, N.E. (1978). *Biochem. Biophys. Res. Commun. 84,* 895.

Miller, M.G., and Obendorf, R.L. (1981). *Pl. Physiol.* (In Press).

Mills, C.F. (1958). *Soil Sci. 85,* l00.

Mitchelson, A.M., McCord, J.M., and Fridovich, I. (1977). "Superoxide and Superoxide Dismutases." Academic Press, New York.

Mondovi, B., Morpurgo, L., Rotilio, G., and Finazzi-Argo, A. (1976). *In* "Iron and Copper Proteins". (K.T. Yasunobu, H.F. Mower, and O. Hayaishi, eds.) p.424. Plenum Press, New York.

Moore, G.R., and Williams, R.J.P. (1976). *Coord. Chem. Rev. 18,*125.

Mottola, H.A., Haro, M.S., and Freiser, H. (1968). *Anal. Chem. 40,* 1263.

Neish, A.C. (1939). *Biochem. J. 33,* 300.

Nicholls, P., and Chance, B. (1974). *In* "Molecular Mechanisms of Oxygen Activation". (O. Hayaishi, ed.) p.479. Academic Press, New York.

Nicholas, D.J.D., and Egan, A.R. (1975). (eds.) "Trace Elements in Soil-Plant-Animal Systems". Academic Press, New York.

Nicholson, C., Stein, J., and Wilson, K.A. (1980). *Pl.Physiol. 66,* 272.

Ochai, E.I. (1975). *J. Inorg. Nucl. Chem. 37,* 1503.

Ozolinya, G.R., and Lapinya, L.P. (1976). *Sov. Pl. Physiol. 23,* 953.

Palmer, J.M. (1979). *Biochem. Soc. Trans. 7,* 246.

Patra, H.K., and Mishra, D. (1979). *Pl. Physiol. 63,* 318.

Paul, K.G., and Stigbrand, T. (1970). *Biochim. Biophys. Acta, 221,* 255.

Peisach, J., Levine, W.G., and Blumberg, W.E. (1967). *J. Biol. Chem. 242,* 2847.

Perumal, A., and Beattie, J.M. (1966). *Proc. Am. Soc. Hortic. Sci. 88,* 41.

Racusen, D. (1969). *Can. J. Bot. 48,* l029.

Ramshaw, J.A.M., Brown, R.H., Scawen, M.D., and Boulter, D. (1973). *Biochim. Biophys. Acta, 303,* 269.

Reilly, C. (1972). *Z. Pfl. Physiol. 66,* 294.

Reilly, C., Rowel, J., and Stone, J. (1970). *New Phytol. 69,*

993.

Richardson, J.S., Thomas, K.A., Rubin, B.H., and Richardson, D.C. (1975). *Proc. Natl. Acad. Sci. U.S.A. 72*, 1349.

Robson, A.D., and Reuter, D.R. (1981). Chapter 13 this Volume.

Salisbury, J.L., and Floyd, G.L. (1977). *Pl. Physiol. 59*, S111.

Salisbury, J.L., and Floyd, G.L. (1978). *J. Phycol. 14*, 362.

Sandmann, G., and Boger, P. (1980). *Pl. Physiol. 66*, 797.

Schichi, H., and Hackett, D.P. (1963). *Arch. Biochem. Biophys. 100*, 185.

Smith, D.D. (1977). *Biochim. Biophys. Acta 459*, 468.

Smith, T.A. (1975). *Phytochem. 14*, 865.

Smith, T.A. (1976). *Phytochem. 15*, 633.

Smith, T.A. (1977). *Phytochem. 16*, 1647.

Spencer, D., and Possingham, J.V. (1960). *Aust. J. biol.Sci. 13*, 441.

Srivastava, S.K., Prakash, V., and Naik, B.I. (1977). *Phytochem. 16*, 185.

Thompson, J.F.,and Tiffin, L.O. (1974). *Pl. Physiol. 54*,S24.

Tiffin, L.O. (1972). *In* "Micronutrients in Agriculature". (J.J.Mortvedt, P.M. Giordano, and W.L. Lindsay, eds.) p.119.S.S.S.A., Madison.

Tiffin, L.O. (1977). *In* "Implications of Metals in the Environment", Proc. 15th Annu. Hartford Life Sci. Symp. p.315. E.R.D.A., Oak Ridge.

Tiffin, L.O., and Thompson, J.F. (1975). *Pl. Physiol. 56*,19.

Timperley, M.H., Brooks, R.R., and Peterson, P.J. (1973), *J. exp. Bot. 24*, 889.

Vaughan, P.F.T., Eason, R., Paton, T.Y., and Rictie, G.A. (1975). *Phytochem. 14*, 2383.

Walker, C.D. (1980). "Copper Compounds in Subterranean Clover". Ph.D. Thesis, Murdoch University.

Walker, C.D., and Loneragan, J.F. (1981). *Ann. Bot.* (In Press).

Weser, U., and Rupp, H. (1978). *In* "Trace Element Metabolism of Man and Animals -3" (W.Kirchgessner, ed.) p.40 A.T.W., Freising-Weihenstephan.

Wharton, D.C. (1973). *In* "Inorganic Biochemistry", Vol. 2. (G.L. Eichhorn, ed.) p.955. Elsevier, Amsterdam.

Wishnick, N., Lane, M.D., and Scrutton, M.C. (1970). *J.Biol. Chem. 245*, 4939.

Wishnick, N., Lane, M.D., Scrutton, M.C. and Mildvan, A.S. (1969). *J. Biol. Chem. 244*, 5761.

Wittenbach, V.A. (1978). *In* "Photosynthetic Carbon Assimilation". (H.W.Siegelman, and G. Hind, eds.) p.416. Plenum Press, New York.

Yasunobu, K.T., Ishizaki, H., and Minamuir, N. (1977). *Mol. Cell. Biochem. 13*, 3.

10

PHYSIOLOGICAL FUNCTIONS AND UTILIZATION OF COPPER

Wolfgang Bussler

Department of Plant Nutrition,
Faculty of International Agri-
cultural Development, Technical
University,   Berlin

I. INTRODUCTION

Copper is an essential nutrient for plants and therefore must have at least one of the following characteristics
for all higher plant and varieties:

(1) The element must be present in the plant tissue
(in sufficient amounts and in balance with other nutrients).
Without it the plant cannot complete its life cycle
(Epstein, 1965).

(2) It is a part of the molecule of an essential
plant constituent or metabolite (Epstein, 1965).

(3) It must be directly involved in the metabolism of
the plant (Arnon and Stout, 1939) in a specific role.

(4) It is needed during the complete life cycle for
metabolic reactions or for maintaining the cytoplasmic
structure on which metabolism is dependent.

It seems to me important to remember this definition.

Since we know from the experiments by Anna L. Sommer
(1931) that Cu is essential we are able to predict the
behaviour of a plant if Cu is not available in sufficient
amounts relative to the other growth conditions.

COPPER IN SOILS AND PLANTS

All reactions of a plant to Cu and all physiologic
data are characterised by the general optimum curve.

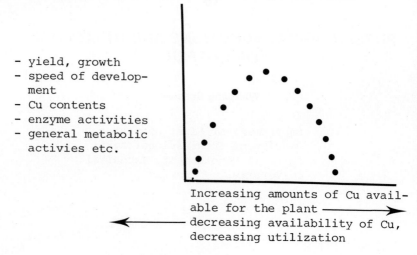

- yield, growth
- speed of develop-
  ment
- Cu contents
- enzyme activities
- general metabolic
  activies etc.

Increasing amounts of Cu available for the plant ⟶
⟵ decreasing availability of Cu,
decreasing utilization

All factors which increase the rate of general development of a plant induce a higher need for Cu since Cu is
involved in many basic physiological reactions involving
synthesis of metabolites, physiological steps in development,
and differentiation as well as in catabolism and turn over.

All factors which depress development induce lower
needs for Cu.

If Cu is the limiting factor in the presence of all
other nutrients, the plant cannot complete its life cycle;
yield and growth are progressively depressed as the availability of Cu decreases, becoming nil in its absence. Under
such limiting conditions the plant follows the principle of
Le Chatelier. The speed of growth diminishes, typical enzyme
activities and general metabolic activities decrease. The
plants do not develop according to their genetic possibilities. They stop growth and development. Their metabolism
is interrupted in distinct reaction chains or moved into
another direction. The plant develops symptoms of deficiency.

We know that a nutrient inside the plant may have
different and many functions. The effect of a nutrient is
dependent on its binding form and compartmentation. One Cu
ion in the plant may exert during its stay in the cell many
different functions.   Their importance for metabolism may
vary in degree but not in principle. The amount of Cu

deposited as slag in cell structures excluded from turnover
seems to be very limited.

It is to be expected that at the beginning of deficiency
the water soluble or easily bound forms of Cu will diminish
first and induce the first Cu deficiency symptoms. The
general metabolism has not a sudden breakdown. The metabolism
is delayed step by step. During this development of
increasing deficiency we cannot expect to find Cu always in
the same binding form, in the same fraction of an acidic,
alkaline, complexing, or exchanging solvent. The fractions
may change according to the development of the plant and
according to the degree of Cu deficiency. This subject
has been reviewed in the preceding chapter (Walker and Webb,
1981), and the need for further investigations has been
emphasised.

The principle problem in clarifying the physiological
functions of Cu or in finding the primary effect of Cu on
metabolism lies in the fact that two mechanisms become
operative under Cu deficiency conditions:

(1) Some reactions are directly delayed by Cu
deficiency: there is a disproportion between Cu-dependent
reactions and available Cu.

(2) Low amounts of available Cu are depressed in their
activity by antagonisms of disproportionally high amounts
of other bivalent metals; e.g. the 'normal' contents of Fe
and Mn became too high for a Cu-deficiency plant.

It is impossible to induce only one of these mechanisms.

Copper deficiency can never be limited to Cu deficiency
alone. It is always connected with an imbalance of other
growth factors. Plants with a recognisable Cu deficiency
have always more than one factor which differentiates them
from the controls (less Cu in the organs, changed ion ratios,
more ash, etc.). Changes in the dry matter influence the
calculations about Cu contents per gram dry matter even if
they have no effect on the Cu contents.

Other difficulties arise from the type of experimentation:

(1) If we induce a severe Cu deficiency in a
highly purified environment we can observe depressions of
growth and development and at a late stage a general
disturbance of metabolism. Under such conditions, applications of Cu will also improve reactions in which Cu is not
directly involved so that it seems that Cu influences all
physiological reactions. As deficiency progresses, there is
a general breakdown of metabolism and bound Cu becomes available again even from the older leaves in which its movement
appears to be related to the movement of N (Hill, Robson, and
Loneragan, 1978, 1979).

(2) Plants with their leaf Cu contents below optimal
levels are maintained under conditions of mild copper
deficiency, and the severity of the deficiency symptoms may
decrease to the stage of a latent deficiency where typical
symptoms are hidden and the deficiency is only detectable
by biochemical methods. This is often the case with Cu
deficiency in the field or in pot experiments with a Cu
deficient soil. Pissarek (1974) has also found only few
symptoms in experiments with nutrient solutions of low Cu
content.

(3) If plants are grown initially in a complete
nutrient solution and then transferred to a Cu-deficient
solution at a stage when they are growing rapidly, severe Cu
deficiency symptoms may develop rapidly: they may develop
more quickly and appear more intense than they do under the
experimental conditions described in type (1). Since the
roots of these plants are intact, the developing symptoms
are likely to be closer to those of simple Cu deficiency
and not be complicated by other metabolic disturbances
induced by poor rooting.

The physiology of Cu can only be understood in connection with the general physiology of the living organism. Cu
alone is toxic. Cu in excess (decreasing part of the optimum
curve) also inhibits the general metabolic activity, inducing
deficiencies in other trace nutrients or exerting a direct
toxic influence on the living system.

For optimal development the plant must have an optimal
amount of reactive Cu in its cells. This optimal amount
depends on plant species or variety (Rademacher, 1937) and on
all other growth factors. Among the latter, the balance of
heavy metal nutrients is important. This arises because Cu
functions mainly in enzymes (Peisach *et al.*, 1965) and

because, in enzymes, Cu with its variable valency ($Cu^+ \rightleftharpoons Cu^{2+}$) behaves similarly to Fe and Mn. The variable valency, the similar ion diameter and atom weight, the affinity to proteins and proteids leads to antagonisms between Cu and Fe and Mn.

It is therefore important to establish in the plant cell a balanced ratio of heavy metal nutrients. The balanced ratio can be found in connection with enzyme activities, the amount of synthesized lignin, and in fertilizer experiments with the method of systematic variations (Hòmes, 1974). The ratio may vary from variety to variety.

For these reasons, the "sufficient amount" of Cu which plants require for optimal development cannot be defined as an absolute norm. For example, the Cu content in the grains of 16 wheat varieties grown in the same soil was found to range between 5.6 and 16.7 ppm (Schütte, 1964). Nambiar (1976) also demonstrated that genetic differences influence the Cu nutrition of cereals, genotypes with high protein contents in the grain being more sensitive to Cu deficiency than genotypes with low grain protein. According to Graham and Anderson (1979), ryes and triticales have proved to be relatively Cu-efficient and wheats Cu-inefficient.

Because of its role in a large number of key enzymes (Peisach *et al.*, 1965; Walker and Webb, 1981), Cu influences nearly all metabolic pathways.

II. THE INFLUENCE OF CU ON PHYSIOLOGICAL PROCESSES

A. *Carbohydrate Metabolism.*

1. *Photosynthesis*

Copper deficiency has been reported to inhibit photosynthetic activity and to induce chlorosis and structural malformations in leaves. Since chlorosis involves the breakdown of chloroplasts which are the organelles for photosynthesis, symptoms of chlorosis can be expected to lead to a decrease in the rate of photosynthetic activity.

Chlorosis is an early and typical symptom of some deficiency diseases (N, S, Mg, Fe, Mn, Zn) as well as of some toxicities (SO_2, O_3, alkali). Similar effects might be expected for Cu deficiency since plant chloroplasts contain appreciable Cu. For example, in clover, 75% of total Cu

was found in the chloroplasts where most of it was organically
bound (Neish, 1939). However in Cu deficiency, development
of chlorosis is very variable making Reuther's comment in
1957 very apt :

"The symptoms of Cu deficiency in green plants vary considerably with the species and perhaps other complicating
factors. No general description of visual Cu deficiency
symptoms can therefore be made".

For example, in our experiments, chlorosis has appeared late
in Cu deficient plants as secondary, atypical symptoms accompanied by the general disturbance of the metabolism (Rahimi,
1972). This is in agreement with observations of Cu
deficiency symptoms described by Bergmann and Neubert (1976)
in crop plants and by Penningsfeld (1978) in ornamental
plants. By contrast, other authors have described chlorosis
as a typical symptom for Cu deficiency (Purvis and Carolus,
1964; Ruiter, 1969). Baszynski *et al.* (1978) have also
measured a decrease in the chlorophyll contents of oats and
spinach in Cu deficiency while Snowball *et al.* (1980)
observed a general paleness of colour extending to all leaves
of subterranean clover.

In the case of subterranean clover, the chlorosis
appeared to result from a secondary effect of Cu deficiency
on N fixation since the chlorosis disappeared when N fertilizers were applied to Cu deficient plants (Snowball *et
al.*, 1980). In other experiments where Cu deficiency has
induced chlorosis, the leaves develop first as green leaves
and only later become chlorotic. Moreover, some authors have
reported severe structural aberrations in cells of Cu
deficient leaves which are still green.

Reed (1941) found palisade cells separated beneath the
stomata, large cavities, shrinkage, lysis of the contents,
loss of dark pigments and plastids clumped together. Rahimi
(1972) described the palisade cells of sunflower leaves as
short, with smaller and less chloroplasts compared with the
control. But these Cu deficient plants were green, even
though growth had already stopped and leaves were distorted.
Such late development of chlorosis points to an indirect
role for Cu: the later symptoms appear in the sequence of
symptom development the less likelihood there is for the
symptom to have a direct dependence on the function of the
deficient Cu.

The evidence thus suggests that damage of the chloroplasts

is a late step in developing Cu deficiency. Hence if Cu deficiency depresses the rate of photosynthesis directly it must do so through some other disturbance. In addition to depressing the contents of chlorophyll a and b in leaves of oats and spinach, Cu deficiency depressed β-carotene, lutein, neoxanthine, violaxanthine, plastoquinone A, plastohydro-quinone A, αtocopherol, αtocopherylquinone, and Vitamin K_1 (Baszynski *et al.*, 1978).

According to Baszynski *et al.* (1978) the synthesis of all thylakoid prenyl lipids as well as of chlorophylls was lower during Cu deficiency. The formation of plastoquinones (C_{45} chain) was much more strongly inhibited than that of phytol derivatives and carotenoids. The rates of photo-synthesis, expressed as O_2 evolved per unit leaf area, were lower in the deficient than in the control plants. However, on the basis of chlorophyll, the O_2 evolution was slightly higher in deficient plants. Plastocyanin and total Cu con-tent in deficient chloroplasts decreased over 50% in com-parison with the control plants. Electron micrographs point to disturbances mainly within the stroma lamellae.

Compared with those of control plants, the green cells of Cu deficient plants had the following characteristics:

fewer chloroplasts
less Cu in the chloroplasts,
less plastocyanin,
less lipids,
disturbed membrane systems.

Since the plants had a slightly higher oxygen evolution relative to chlorophyll, plastocyanin could not have been limiting for photosynthetic reactions. Plastocyanin is a non-autoxidizable Cu containing enzyme which mediates the photosynthetic electron transport for photosystem I and appears to have no function in the absorbancy of light in the primary act of photosynthesis (Levine, 1969; Crofts and Wood, 1978). The fact that plastocyanin was not limiting to photosynthesis under conditions of Cu deficiency could indicate preferential binding for Cu in this enzyme: within it, Cu may be bound more strongly and its turnover may be slower than in other compounds.

Copper may also have a secondary role in the maintenance of the stability of thylakoid membranes. Here it may be replaced by other divalent metals, since, even under strong

conditions of Cu deficiency the breakdown of chloroplasts
and their membrane systems is delayed for a long time.

2. *Respiration*

Since mitochondrial cytochrome-c-oxidase contains Cu, Cu
must be involved in respiration. However, microscopic examinations do not show any accumulation of assimilates in Cu-
deficient cells (see also 3 below), as would be expected if
respiration was limited. It seems that the Cu in the cytochrome-c-oxidase Cu complex has a high degree of stability
and cannot be removed under Cu deficiency conditions.
Respiration and catabolism continue even under the strongest
and longest lasting Cu deficiency conditions. Dead Cu-
deficient leaves were never found to contain starch or other
physiologically usable energy resources. The deficient cells
were optically nearly empty.

3. *The distribution of carbohydrates*

For a long time it was thought that a decreased fruit
setting which has frequently been observed under Cu
deficiency conditions (see E.2 below) resulted from a
depression in the amounts of soluble carbohydrates available
to the fruit. This assumption was often supported by
observed chloroses in Cu deficient plants and the belief
that Cu is involved in photosynthesis. However, in a recent
paper, Graham (1980a) has demonstrated that Cu deficient
wheat plants accumulate soluble carbohydrates. This accumulation leads to high concentrations and excretions of honey-
like, globular exudates from the stems of senescing and
deficient plants. This indicates more a surplus than a
deficiency of carbohydrates. Graham attributes this
behaviour to the absence of a sink for carbohydrates
resulting from failure of flower set due to the sterility of
pollen in Cu deficient plants (see E.1 below). These
findings are supported by the facts that Cu deficient leaves
remain green or dark blueish green for a long time, that
plastocyanin is not the limiting factor in photosynthesis,
and that enough carbohydrates are present for excessive
tillering. That the carbohydrate distribution is not
limited by Cu deficiency was most distinctly seen in the
roots of Graham's experiments which had no endogenous supply
of carbohydrates but were dependent on translocation: 18 to
20 weeks after emergence, the roots of Cu deficient plants
contained about 100 mg. of soluble carbohydrates g^{-1} dry

weight, whereas the controls had almost none. These
findings do not suggest that Cu is involved in carbohydrate
translocation, either in Cu adequate or in Cu deficient
plants. Rather, the carbohydrate distribution depends on
the presence or absence of a sink.

B. *Nitrogen metabolism*

1. *Nitrogen reduction and fixation*

Asano (1959, 1960) has reported the separation of an
enzymatic nitrite reducing system from a micrococcus strain
into a soluble component which is activated by Cu^+ and Cu^{2+}
ions, and a particulate component which is enhanced by Fe^{2+}
and Fe^{3+}.

An improvement of growth and N content by Cu additions
to low Cu plants of *Alnus* and *Casuarina* was reported by Bond
and Hewitt (1967). Copper has also an indirect effect on
nodule formation in the legume subterranean clover (Cartwright and Hallsworth, 1970): Cu deficiency was found to
have a marked effect on nodule development and N fixation
while leaf growth was reduced only slightly. Cartwright
and Hallsworth suggested that Cu is critical in maintaining
cytochrome oxidase in the nodule. If the activity is reduced
then oxygen levels could increase in the nodule, which in
turn would restrict N fixation. The more the total environment in the N fixing system is reduced, the greater is the
rate of reduction of atmospheric N.

In a recent paper, Snowball *et al.* (1980) confirmed that
Cu deficiency decreased N fixation in subterranean clover.
They used evidence from three approaches to conclude that Cu
is specifically required for symbiotic N fixation. Firstly,
while both Cu and N application increased the growth of the
legume, the interaction between Cu and N on growth was negative. Secondly, the application of Cu increased the concentrations of both total and protein N in the plant. Lastly,
N fixation as measured by the acetylene reduction assay
increased with Cu application to the soil. Conversely, N
fixation per plant diminished with decreasing Cu supply along
with yield and N content in the tops.

If Cu is directly involved in N fixation then it is
difficult to understand why application of NH_4NO_3 could
induce Cu deficiency symptoms on the leaves on Cu deficient
soils where Cu was not applied, since the Cu in the N

fixation system could be used elsewhere. These symptoms were
not observed (except for Lancelin soil) where N was not
applied. On 2 of 3 soils, application of NH_4NO_3 increased
the growth of the clover only at low levels of Cu application
but not at high ones. This could have been an effect of a
disturbed ratio of trace element metals, possibly of Mo:Cu.
Here we have one of the difficulties mentioned in the introduction. More than one process was influenced by the experiment. Nitrogen application hindered the activity of nodules
to fix N. By applying NH_4NO_3 there was no need for any N
fixation. Protein and N concentrations increased in the tops.
Copper without NH_4NO_3 application increased the activity of
the nodules to fix N. Protein and N concentrations increased
in the tops. Application of both nutrients influenced both
reactions.

2. *Protein metabolism*

Protein metabolism delivers the organic raw products for
the synthesis of structures which are able to form together
with Cu the large number of Cu enzymes in plants. Many of
these Cu enzymes show decreasing activity with decreasing Cu
contents in the tissue and this is restored again if Cu is
given to the system (Walker and Webb, 1981). We know that Cu
may be the limiting factor for this enzyme activity. All
these enzymes are also influenced by the general conditions
of the protoplasmic medium such as pH, temperature swelling
conditions (mainly due to Na, K, Ca, Mg) and O_2 and CO. The
activity of Cu proteins is also often inhibited by various
chelating agents (Goldstein, 1966). That enzyme activity
depends on other factors and not only on Cu should be
remembered.

Where Cu deficiency decreases the activities of Cu
enzymes there is no evidence that the protein component of
the enzymes was ever limiting. Nevertheless, like deficiencies of other catalytically important elements, Cu
deficiency interferes with protein synthesis and causes an
increase in soluble N compounds (Brown *et al.*, 1958). The
pattern of amino acids also changes: Cu-deficient tomatoes
have more aspartic acid, glutamic acid, asparagine, β-
alanine, arginine, proline, and less histidine, lysine,
phenylalanine, oil, starch and reducing sugars than plants
with normal Cu content (Possingham, 1956).

By while Cu deficiency has profound effects on protein

metabolism, there is no evidence that this results from a direct effect of the deficiency. A short time after the first effects of Cu deficiency, metabolism suffers from secondary induced imbalances not directly connected with Cu. All these imbalances together induce the development of micro and macro symptoms. These symptoms are an expression of a disturbed metabolism. Copper is the limiting factor for growth and development. But not all metabolic reactions are stopped. The nutrient uptake of other ions continues, the assimilation of the "dark or blueish green" leaves is not interrupted, the translocation of assimilates is hindered because of the stopped growth, and there is no or only a small dilution of metabolites by growth.

The present evidence suggests that the influence of Cu on protein metabolism is indirect, resulting from a secondary effect of a long lasting Cu deficiency. Whether the Cu proteins ever become limiting to the growth of Cu deficient plants is not known (Price *et al.*, 1972; see C below).

c. *Cell wall metabolism - lignin synthesis*

Cell wall composition changes with age. At the earliest stages of development pectins predominate. Later cellulose and hemi-celluloses appear. At these stages of a young cell, the wall is extensible and the cell is able to enlarge. When the cell reaches its final size, lignin, salts, and other substances are also encrusted in the structure of the wall. The wall becomes rigid as typified by cell walls in old plant organs, especially wood. The turnover of cell wall constituents diminishes and the wall becomes a dead product which is not or only to a limited degree involved in metabolism. Copper is important in cell wall metabolism having its most spectacular effects as the latest stages of stabilisation during lignification. Hagin (1959), Schütte and Mathews (1969), Rahimi and Bussler (1974), Pissarek (1974), and Graham (1976a) have all described a decrease in lignification under Cu deficiency conditions. This is in agreement with observed Cu deficiency symptoms.

Early investigators (Lipman and McKinney, 1931; Sommer 1930, 1931), observed that the leaves of various plants such as barley and lin, sunflower, and tomatoes all developed symptoms of bending, distortion, twisting, loss of turgor, withering and death with the onset of Cu deficiency. Stems became soft, weak, curved, and bent. Onion bulbs had a lack of firmness in texture (Sauchelli, 1969). Oat plants died in a Cu deficient nutrient solution during the development of the fourth leaf (Hagin, 1959). We are convinced today that bending, weakness, and arrested development have their causal origin in delay or inhibition of lignification ('delay' if copper is available in insufficient amounts, 'inhibition' if copper is almost completely absent). The degree of lignification depends on the amount of physiologically active Cu. The less Cu is active, the more lignification is hindered. Under different conditions of Cu deficiency different degrees of lignification are found. Under extreme conditions of Cu deficiency all stages of lignification can be demonstrated in one and the same plant. The older organs show a complete lignification made possible at least from the seed Cu, the younger organs show no lignification at all because of limited Cu translocation and absence of Cu uptake. Organs between these stages show different degrees of lignification. This explains also the different results obtained by many investigators.

In mildly Cu deficient sunflowers and oats Pissarek (1974) observed a decrease in lignification mainly in the sclerenchyma. In severely Cu deficient plants Rahimi (1972) found a decrease in lignification of xylem elements and thinner walls in the cells of the epidermis, collenchyma, sclerenchyma, phloem, and xylem. The number of xylem cells also decreased. The Cu deficient stems and petioles were gummy when cut. Pissarek (1974) observed that where cells had no or insufficient lignified walls they collapsed in very young plants due, he suggested, to force of suction during transpiration. Rahimi (1972) found furthermore, a deficient formation of xylem cells and more xylem cells with annular or helical thickening which were extensible as in young tissue. The delayed lignification is detectable before outer symptoms of Cu deficiency appear, a fact which I have been able to use in diagnosing the status of Cu in a citrus plantation in Iran (Bussler, unpublished data). In this case, the Cu status had a clear bearing on the potential value of irrigation.

The mechanical weakness of cell walls resulting from poor lignification leads to bending and distortion of leaves and stems. Rademacher (1940), Schropp and Koronowski (1969) and Pissarek (1974) suggest that this is the cause of the disturbed negative geotropism which occurs in Cu deficiency. I believe that bending is only caused by weakness and distortions by irregular, not symmetric disturbances of lignification. Not all the bundles of the same cross section of

an organ are uniformly influenced by Cu deficiency. That Cu
deficiency affects sclerenchyma before xylem may result from
its greater distance from the Cu distribution system. In
current investigations, we have also observed this difference
between xylem and sclerenchyma in tomatoes.

If we compare the physiological functions of Cu with the
deficiency symptoms in plants only one function seems to be
related to the developing symptoms. This is the function of
Cu in the phenoloxidase enzymes which are involved in the
biosynthesis of lignin. Lignin is a mixture of condensation
polymers of phenylpropane like aromatic rings with a 3-carbon
side chain and substituted phenylpropane compounds. Lignin
stabilises different plant tissues such as xylem and sclerenchyma. If the phenoloxidase enzymes are not active, as for
example in Cu deficient tissues, less lignin is formed and
these tissues remain weak, allowing the organs to bend from
this loss of rigidity.

The formation of lignin in plant cells can easily be
examined by the addition to tissue sections in microscopic
preparations of a drop of a specific reagent such as:

(a)   cotton brown (the lignified tissue turns yellow to brown,
non-lignified tissues remain colourless (Hagin, 1959), or,
even better

(b)   phloroglucinol-HCl. (Dissolve 8 g phloroglucinol in
100 ml abs. ethanol: to the clear solution add 20 ml fumi-
HCl: mix thoroughly. To obtain an even better contrast add
one drop of conc. HCl to the preparation. The lignified
tissue becomes red, non-lignified tissues remain colourless
(Rahimi and Bussler, 1974).

Figure 1 shows the use of this latter technique to illus-
grate how lignification of xylem cells of tomato plants
respond to the level of Cu in the nutrient solution. Lignin
only formed when Cu was present in the nutrient solution.
In the absence of Cu, lignin did not form and this process
was not reversed when the plants were resupplied with Cu.

The fact that lignin synthesis responds so readily to
the level of plant Cu supply suggests that the binding of Cu
to the enzymes in this process is weaker than binding by
other enzyme systems such as plastocyanin.

Fig. 1: Shows a cross section of a tomato stem from left to
        right:

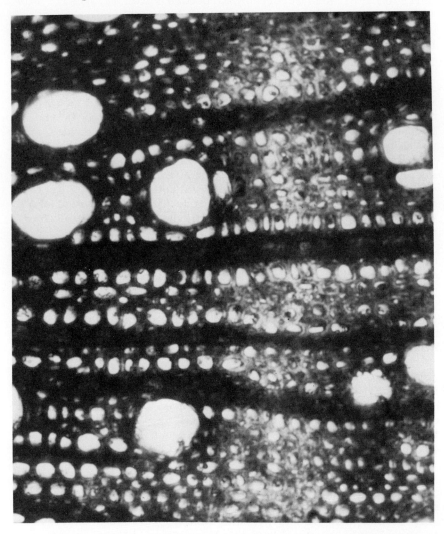

- o,oo1 ppm Cu in the nutrient solution, a weak lignification is indicated by a weak red coloration.
- plant in a copper-free solution, no lignification at
  all, this process is irreversible.
- lignification after addition of o,o5 ppm Cu,
  lignification is normal. (Foto by Rahimi, 1972).

D. *Water relations*

In sunflower and oats, Pissarek (1974) found a higher water consumption in mildly Cu deficient plants than in Cu adequate plants. At the same time plants wilted. Collapse of xylem vessels also disturbed water distribution and contributed to the wilting symptom. By contrast, Graham (1976a) observed an increased leaf diffusive resistance and, at the same time, increased leaf water potential in severely Cu deficient wheat relative to plants receiving Cu.

These results seem to be contradictory. But, in my opinion, the experiments are not comparable. The authors used different plant species at different levels of Cu supply; they also used different methods for analysing the water relations at different times. Moreover, as Graham demonstrates, both the stage of physiological development and the degree of Cu deficiency are of paramount importance in the plant's response to Cu. In his experiments, the weekly water use in the first 14 weeks of growth was higher in Cu adequate than in Cu deficient plants. Later, the water use of Cu deficient plants was higher than the water use of the controls. This was true for high and low water supply. This crossover of the water use curves occurred after anthesis. At this time, both low water supply and low Cu supply caused partial stomatal closure but the interaction of water and Cu treatments was small and not significant.

To find a physiological explanation for these results the experiments should be repeated in a different way. The water relations should not be measured for the plant as a whole but for individual leaves, since the leaves of a Cu deficient plant may have different characteristics in their water relations depending on the degree of Cu deficiency.

Both authors agree with the importance of structural aspects for the water movement. In Cu deficient plants, decreased lignification of xylem vessels led to their collapse and in this way blocked water transport. Thus wilting resulted from decreased lignification and was not related to plant water potential or to stomatal closure (Graham, 1976a). Decreased lignification was also observed by Schütte and Mathews (1969). So the role of Cu in the water status of the plant seems not to be a direct one. Copper influences the lignification, the amount and the rigidity of xylem vessels and perhaps also, although here we have no experimental results, their permeability for water to the surrounding tissue.

E.   *Reproduction*

   1. *Pollen sterility*.   The first Cu deficiency disease
of economic importance was Cu deficiency in barley as described by Lipman and McKinney (1931).   They observed that in
a Cu free nutrient solution barley was unable to develop
seeds.   The disease appeared also in different cereals in Cu
deficient soils or Cu fixing humic sands and was later described as "reclamation disease, wither tip, yellow tip, and
white tip".   In extreme cases no seeds at all developed,
leading to the suggestion that Cu had its main physiological
function in the generative phase of development.   However,
Graham (1975) found that Cu deficiency depressed grain numbers
in wheat plants through pollen sterility.   This result is
in agreement with the findings by Hill *et al*. (1979) that the
Cu concentration of the individual wheat grain is not
causally related to the individual grain development: Cu
deficiency decreased the number of grains without decreasing
the weight of individual grains.   This suggests that the
function of Cu in the process of grain development is limited
to its effects on pollen.

   In Graham's (1975) Cu deficient wheat plants, the flower
and its parts were smaller: the ovule was slightly smaller
and the anthers much smaller, while the filaments did not
elongate.   The pollen grains were smaller and did not stain
with I_2-K solution as starch was absent.   Pollen was not
viable accounting for the sterility of cereals under Cu
deficient conditions.   By contrast with pollen grains, ovules
were much less sensitive to Cu deficiency.   Indeed, ovules
of Cu deficient plants were able to set grain when treated
with pollen from Cu sufficient plants (Graham, 1975).   To
explain why Cu deficiency influenced the reproductive phase
more than the vegetative one Graham (1975) has suggested
that "synchronized meiotic divisions in a large number of
pollen mother cells may produce a localized demand for Cu
exceeding that which the deficient plant can supply".   In his
experiments, application of Cu to Cu deficient wheat overcame
the Cu deficiency provided that it was supplied by the early
boot stage, not later than 11 weeks after sowing.   Application of Cu to the soil or by foliar spray were both effective
(Graham, 1976b).   Graham's (1976b) results indicate that
Cu deficiency may interrupt microsporongenesis at a specific
stage of development, at or near meiosis.

In current experiments I have found that Cu deficiency
affects tomato pollen in the same manner as wheat (fewer
pollen grains, mis-shaped, smaller, without starch: Bussler,
unpublished data). Dell (1981) has also found that lignified
wall thickenings were reduced or absent in the endothecia
of anthers of Cu deficient plants.

All these facts suggest that the nuclear material is
readily affected by Cu deficiency (see Ozolina, 1963). The
production of DNA and RNA decreased at copper deficiency.
The peculiar sensitivity to Cu deficiency of pollen in
Graham's experiments is a characteristic of a relatively
mild Cu deficiency. Under conditions of severe Cu deficiency,
the general development and differentiation of the flower is
affected long before single organs. Copper deficiency
affects the formation of flowers at their initiation before
anthers and ovules are differentiated from the embryonic
tissue. Copper deficient flowers or ears are formed in a
diminished number and are of a smaller size. But even this
reduced flowering capacity cannot be satisfied under Cu
deficient conditions. The Cu dependent mechanisms which
influence the productive organs in their primary development
from undifferentiated embryonic tissue to differentiated
organs are not known. Here copper may have a primary role.

2. Seed production. Symptoms of Cu deficiency primarily appear on the younger parts of a plant. Young leaves
or the growing point were principally affected in wheat, oat,
broad bean, and lucerne (Hewitt *et al.*, 1954). Sometimes Cu
deficiency has also been described as developing first in
older leaves of other plant species. In some of our own
experiments we have also observed Cu deficiency symptoms on
older leaves while the young leaves were sound but this
proved to be due to impurities of later used irrigation
water: it was an experimental accident having nothing to do
with the mobility of Cu inside the plant.

Recent results indicate that Cu deficiency always starts
to develop in younger tissues (Rahimi and Bussler, 1974).
The development of symptoms in young organs has led to the
conclusion that Cu is not easily translocated from older
to younger plant organs during the development of Cu
deficiency. Hill *et al.* (1978) have also shown that Cu
translocation in the phloem out of older leaves is dependent
on Cu content and leaf senescence (Loneragan, 1981).

The anthers are the youngest parts of a plant at the

time of pollen formation. If pollen sterility as an effect
of Cu deficiency is to be generalised for all flowering
plants, then seed production must decrease.

Reuter *et al.* (1981) found that Cu deficiency depressed
seed yield of Seaton Park subterranean clover: it also
depressed top and root growth (weight and number of organs),
retarded phasic development, decreased internode elongation,
and decreased pollen fertility. Plant senescence and
maturity were also delayed. The decreased seed yield was
influenced by all these effects of Cu deficiency but
especially by the pollen sterility. The small pollen grains
lacked starch at maturity and their cytoplasm was degenerated.
Where seeds formed, their weights and germination were not
affected. This is in agreement with the finding by Graham
(1975) that ovules of Cu deficient plants are not as much
influenced by Cu deficiency as pollen grains. A primary
role of Cu in seed formation could not be deduced from these
results.

Seed and fruit production also decrease if flowers form
an abscission layer and drop at ripe or unripe stages. We
have found this in tomatoes with severe Cu deficiency: but
it occurs at a late stage of development and a secondary
influence of other disturbed growth factors cannot be
excluded (Bussler, unpublished data).

F. *Disease resistance*

This section discusses the relationship between the
plant's internal Cu status and its resistance to disease:
the effects of Cu as a fungicide are not discussed. Graham
(1980b) found that wheat plants deficient in Cu were particularly susceptible to powdery mildew. The severity of
infection of leaves and stems decreased with increasing
level of Cu supply. This was evident from the 7th week after
emergence to harvest. One explanation for the higher infection of Cu deficient plants is the poor lignification since
a well lignified sclerenchyma can stop the entry of mycelium.
Non-lignified tissue together with high concentrations of
carbohydrates could create conditions favourable for fungal
growth. Such conditions exist for senescing Cu deficient
wheat plants in which Graham (1980a) has described an
accumulation of soluble carbohydrates as a source of honey-
like, globular exudates from the stems but this mechanism
cannot be invoked to explain the greater incidence of disease

before anthesis when Cu deficient plants were relatively low in soluble carbohydrates.

Other possibilities for a higher or lower resistance exist in the activity of Cu containing enzymes through their roles in metabolism of phenolic compounds or of the amines and polyamines which would act as a nutrient medium for the fungus. Causal connnections are not known.

It seems that the mechanisms of disease resistance are manyfold and differ with the age of the plant and with the degree of Cu deficiency. The structure of the plant as well as the physiological composition may be involved in this behaviour. A luxurious consumption of Cu might provide an additional mechanism of disease resistance by providing an accumulation of copper near the plant surface and may have the double effect of internal and 'external' copper.

III.   SUMMARY

Copper deficiency affects a multitude of physiological processes in the plant - carbohydrate metabolism (photosynthesis, respiration, and carbohydrate distribution), N metabolism (N fixation and protein synthesis and degradation), cell wall metabolism (especially lignin synthesis), water relations, seed production (especially through pollen viability) and disease resistance. It may also affect ion uptake, plant differentiation in early development phases, and perhaps other steps of the metabolism of secondary plant substances.

Copper must function in many of these processes since many of the operative enzymes within them contain Cu. But the evidence indicates that in most of the processes affected by Cu deficiency, the shortage of Cu operates indirectly. Any change in the concentration of the physiologically active Cu in direction of deficiency or toxicity influences plant metabolism generally and many symptoms - in the widest sense of the word - of a disturbed Cu nutrition may not be directly related to Cu.

Lignin synthesis and pollen viability, which are especially sensitive to Cu deficiency, may be exceptions to this generalisation. In these processes, Cu deficiency may act reasonably directly. If so, then the activities of the Cu enzymes in these processes would appear to be more sensitive to the level of Cu in the cell than the Cu enzymes of other processes. Alternatively, these Cu sensitive processes are most active in developing regions of the plant relying on the transpiration stream for their supply of Cu.

Whatever the reasons it is clear that if we wish to understand the physiology of Cu we must be able to understand the physiology of the plant as a whole.

REFERENCES

Arnon, D.I., and Stout, P.R. (1939). *Pl. Physiol.* 14, 371.
Asano, A. (1959). *J. Biochem.*(Tokyo) 46, 1235.
Asano, A. (1960). *J. Biochem.*(Tokyo) 47, 678.
Baszynski, T., Ruszkowska, M., Krol, M., Tukendorf, H., and Wolinska, D. (1978). *Z. Pf. Physiol.* 89, 207.
Bergmann, W. und Neubert, P. (1976) "Pflanzendiagnose und Pflanzenanalyse". p.145. VEB Gustav Fischer Verlag, Jena.
Bond, G. and Hewitt, E.J. (1967). *Pl. Soil* 27, 447.
Brown, J.C., Tiffin, L.O., and Holmes, R.S. (1958). *Pl. Physiol.* 33, 38.
Cartwright, B. and Hallsworth, E.G. (1970). *Pl. Soil* 33, 685.
Crofts, A.R. and Wood, P.M. (1978) *In* "Current Topics in Bioenergetics, 7". (D.R. Sanadi and L.P. Vernon, eds.). p.175. Academic Press, New York.
Dell, B. (1981). *Ann. Bot.* (In press).
Epstein, E. (1965). *In* "Plant Biochemistry" (J. Bonner and J.E. Varner, eds.). p.438. Academic Press, New York.
Goldstein, M. (1966). cited by Peisach *et al.* (1978).
Graham, R.D. (1975). *Nature* (London) 254, 514.
Graham, R.D. (1976a). *Austr. J. Pl. Physiol.* 3, 229.
Graham, R.D. (1976b). *J. exp. Bot.* 27, 717.
Graham, R.D., and Anderson, G.D. (1979). Int. Symp. on Trace Element Stress in Plants, Univ. of Calif., Los Angeles. *J. Pl. Nutr.* (In press).
Graham, R.D. (1980a). *Z. Pfl. Ernahr. Bodenk.*, 143, 161.
Graham, R.D. (1980b). *Pl. Soil*.
Hagin, M. (1959). Kupfermangel bei Hafer unter Berücksichtigung anatomischer Veränderungen. *Diss. TUB, D 83, Nr.* 104.
Hewitt, E.J., Bolle-Jones, E.W., and Miles, P. (1954). *Pl. Soil* 5, 205.

Hill, J., Robson, A.D., and Loneragan, J.F. (1978). *Austr. J. agric. Res.* 29, 925.
Hill, J., Robson, A.D., and Loneragan, J.F. (1979). *Austr. J. agric. Res.* 30, 233.
Homès, M.V.L. *et* Germaine H.J. Homès-van Schoor (1974-9). Academie Royale de Belgique, Bulletin de la Classe des Sciences, 5e serie, Tome LX, p.1012-1028.
Levine, R.P. (1969). *Ann. Rev. Physiol.* 20, 523.
Lipman, C.B. and McKinney, G. (1931). *Pl. Physiol.* 6, 593.
Loneragan, J.F. (1981). Chapter 8, this Volume.
Nambiar, E.K.S. (1976). *Austr. J. agric. Res.* 27, 453.
Neish, A. (1939). *Biochem. J.* 33, 300.
Ozolina, G. (1963). *Chem. Abstr.* 62, 1042 d.
Peisach, J., Aisen, P., and Blumberg, W.E. (1966). *In* "Copper in Biological Systems". p.588. Academic Press, New York.
Penningsfeld, F. (1978). Diaserie: Mikronährstoffmangelerscheinungen an Zierpflanzen. Ed. Metalldünger Jost GmbH, Postfach 224, D 5860 Iserlohn.
Pissarek, H.P. (1974). *Z. Pfl. Ernähr., Bodenk.* 137, 224.
Possingham, J.V. (1956). *Austr. J. biol. Sci.* 9, 539.
Price, C.A., Clark, H.E. and Funkhouser, E.A. (1972) *In* "Micronutrients in Agriculture". (J.J. Mortvedt, P.M. Giordano, and W.L. Lindsay, eds.). p. 231-239. Soil Sci. Soc. of Amer., Madison
Purvis, E.R.- and Carolus, R.L. (1964). "Hunger signs in crops". David McKay, New York.
Rademacher, B. (1937). *Z. Pfl. Krank. Pfl. Schutz* 47, 545.
Rademacher, B. (1940). *Mitt. Forstw. Forstwiss.* Hannover 11, 335.
Rahimi, A. (1972). Kupfermangel symptome und ihre Entwicklung bei höheren Pflanzen. *Diss. TUB, D 83, Nr.* 14.
Rahimi, A., und Bussler, W. (1974). *Landw. Forsch. SH* 30/2, 101.
Reed, H.S. (1941). *Am. J. Bot.* 28, 10.
Reuter, D.J., Robson, A.D., Loneragan, J.F., and Tranthim-Fryer, D.J. (1981). *Aust. J. agric. Res.* 32 (In press).
Reuther, W. (1957). *In* "Yearbook of Agriculture". p.1288. USDA, Washington, D.C.
Ruiter, J.H. (1969). *Pl. Soil* 31, 197.
Sauchelli, V. (1969). "Trace elements in Agriculture". Van Nostrand Reinhold, New York.
Schropp, W., und Koronowski, P. (1969). *In* "Handbuch der Pflanzenkrankheiten" (P. Sorauer, ed.). p.170. Parey, Berlin, Hamburg.
Schütte, K.H. (1964). "The biology of trace elements; their role in nutrition". Crosby Lockwood and Son Ltd.,London.

Schütte, K.H. and Mathews, M. (1969). Trans. roy. Soc. S.
    Afr. 38, 183.
Snowball, K., Robson, A.D., and Loneragan, J.F. (1980). *New
    Phytol*. 85, 63.
Sommer, Anna L. (1930). *Am. Fert*. 72, 15
Sommer, Anna L. (1931). *Pl. Physiol*. 6, 339.
Vierke, G., und Struckmeier, P. (1977). *Z. Naturforsch*.
    32 c, 605.
Walker, C.D., and Webb, J. (1981). Chapter 9, this Volume.

11

THE PHYSIOLOGICAL BASIS OF COPPER TOXICITY AND COPPER TOLERANCE IN HIGHER PLANTS

H.W. Woolhouse

S. Walker[1]

John Innes Institute
Norwich, Norfolk, U.K.

INTRODUCTION

A. *A Note concerning the Chemistry*

In this paper the term 'heavy metals' is studiously
avoided because we are firmly committed to the importance
of adopting a new nomenclature as advocated by Nieboer and
Richardson (1980) arising from the work of Ahrland *et al.*
(1958). These authors based their system on empirical
thermodynamic data, i.e. trends in magnitude of equilibrium
constants describing the formation of metal-ion ligand complexes. The formative reaction may be written: $M + L \rightleftharpoons ML$:
where M represents the metal ion, L the ligand, and ML the
metal-ligand complex. The corresponding equilibrium constant,

K_{ML}, is given by:
$$K_{ML} = \frac{[ML]}{[M] \times [L]}$$

In this convention, three classes of metals are recognised:
Class A: in which the metal ions are sometimes referred
to as 'hard acids'. These metals have the ligand or donor

[1] *Work supported by NERC Studentship*

atom preference sequence for ligands: $F^- > Cl^- > Br^- > I^-$ and for metal-binding donor atoms in ligands:

$$O > S \simeq Se$$

$$N > As$$

$$O > N > S$$

Class B: in which the metal ions may be referred to as 'soft acids'.   Ions of these metals exhibit the opposite preference sequences:

$I^- > Br^- > Cl^- > F^-$ and

$$Se \simeq S > O$$

$$As > N$$

$$S > N > O$$

Class Borderline: in which the metal ions are clearly distinct from those of class A but the distinction between class B and the borderline group is less clear.   Among the borderline metal ions, class B character increases in the order:

$$Mn^{2+} < Zn^{2+} < Ni^{2+} < Fe^{2+} \simeq Co^{2+} < Cd^{2+} < Cu^{2+} < Pb^{2+}$$

Copper in the form Cu^+ belongs to class B but Cu^{2+} falls into the borderline group albeit towards the end of the sequence of metals in this group showing most class B character.   For the purposes of the present work we may suppose that we are concerned almost entirely with Cu^{2+} since the Cu^+ ion readily disproportionates in the presence of water.   Calculations based on the appropriate oxidation potentials show that the maximum concentration of Cu^+ is unlikely to exceed 10^{-2}M.

$$Cu^+ + e \longrightarrow Cu^o \qquad E^{\ominus} = 0.52V$$

$$Cu^{2+} + e \longrightarrow Cu^+ \qquad E^{\ominus} = 0.15V$$

$$2Cu^+ \longrightarrow Cu^o + Cu^{2+} \qquad E^{\ominus} = 0.37V$$

The free energy for this reaction, $\Delta G = -nFE$ but the equilibrium constant, $Keq = -\log \dfrac{\Delta G}{RT}$,

$$\text{where Keq} = \frac{\left[Cu^{2+}\right]}{\left[Cu^{+}\right]^2}$$

so that $Keq = 10^6$. Hence, even if $Cu^{2+} = 10^{-2}M$ then $[Cu^+] = 10^{-4}M$.

Thus we may suppose that for disproportionation will be such that we are for the most part dealing with the Cu^{2+} ion.

B. *The Biological Context*

Copper is an essential element for all organisms, being a constituent of enzymes which catalyse oxidative reactions in a variety of metabolic pathways. At high concentrations Cu is toxic to most organisms, an attribute which has led to its use in a variety of fungicides and marine antifouling compounds.

In areas where toxic concentrations of Cu are encountered, as in cupriferous outcrops, waste heaps from copper mining and smelting operations, and on copper-treated surfaces, there are to be found organisms which have evolved mechanisms that enable them to tolerate the high concentration of copper. Examples of Cu-tolerance are found in many of the major groups of organisms (Table I) although the evidence suggests that the mechanisms of tolerance differ widely. Indeed even within the angiosperms, with which we shall be exclusively concerned in this article, there appear to be several mechanisms of Cu-tolerance.

Copper tolerance in angiosperms was first recognised in *Melandrium silvestre* (Prat, 1934); the phenomenon has subsequently been studied from the standpoints of plant geography, evolution, genetics and physiological mechanisms.

I. PLANT GEOGRAPHY

It has been known in Europe for centuries that outcrops containing Cu and other metals can be recognised from the distinctive species and anomalous growth and colouration of the plants associated with them (Agricola, 1556). The most extensive and spectacular flora of this kind is that of the copper belt in southern Africa and it appears that early prospectors in the region soon became familiar with the distinctive botanical features (Wilderman, 1921; Robyns, 1932). Detailed botanical descriptions of the copper belt have subsequently been published (Duvingneaud, 1958; Duvigneaud and Denaeyer de Smet, 1963; Horscroft, 1961;

TABLE 1. Examples of Cu-tolerance in Some Major Groups of Organisms

Group	Organism	Reference
Flowering plants	Melandrium silvestre Silene inflata	Prat (1934) Wachsmann (1961) Broker (1963)
	Mimulus guttatus	Allen and Sheppard (1971)
Bryophytes	Several species from a copper contaminated area in Austria	Url (1956)
Liverworts	Cephaloziella spp.	Koch (1932) Lambinon and Auquier (1964)
Algae	Ectocarpus spp. Scenedesmus acutiformis Chlorella fusca	Russel and Morris (1970) Stokes (1975)
Lichens	Genera: Acarospora and Lecanora	Lange and Ziegler (1963)
Fungi	Brown Rot Fungi (many spp.)	Young (1961) Da Costa and Kerruish (1964)
Bacteria	Thiobacillus ferro-oxidans Thiobacillus thio-oxidans Ferrobacillus ferro-oxidans	Bryner et al (1954)
Annelids (Polychaetes)	Nereis diversicolor	Bryan and Hummerstone (1971)

Wild, 1968, 1978). The general feature of the cupriferous
outcrops of this region is that only herbaceous species occur
on the most contaminated soils; these are surrounded by an
intermediate zone of stunted woody plants which form a transition to the prevailing woodland of the area. Ernst (1974)
has developed a detailed phytosociological classification for
this region but this is of dubious value in a situation in
which the vegetation is manifestly a continuum closely allied
to gradients of Cu in the soil (Wild, 1968; Shewry *et al.*,
1979).

II. EVOLUTION

Duvigneaud and Denaeyer de Smet (1963) recognised at
least 26 species or sub-species as endemic to the copper
outcrops of Katanga (Haut Shaba). Endemism of this kind is
less evident in other cupriferous areas of southern Africa
but over four hundred species have been described which have
given rise to Cu-tolerant populations. There has been some
speculation concerning the evolution of these metal-tolerant
communities (Wild and Bradshaw, 1977) but in the absence of
knowledge concerning the climatic history of the region or
of the relevant cytological and hybridisation work there is
little that can be said concerning the origins or antiquity
of these endemic species or Cu-tolerant populations.

Four hundred species exhibiting Cu-tolerance suggests
a common phenomenon, but it must be remembered that this is
only a tiny percentage of the flora over a vast and
floristically rich area. There is indeed a sense in which
it may seem the more remarkable that some of the species
which abound in the region have not given rise to Cu-tolerant
forms. Thus, for example, dominants in the flora of the
region are deciduous trees of such genera as *Isoberlinia,*
Brachystegia and *Julbernardia* but for some unexplained
reason no arborescent species appear to have evolved a high
degree of Cu-tolerance. The grasses form the most con-
spicuous vegetation of the heavily contaminated soils
including species of *Loudetia, Eragrostis, Shizachyrium* and
Danthoniopsis but species of *Hyparrhaenia,* perhaps the most
abundant genus of grasses in the region, are virtually
absent from the Cu outcrops.

The restricted propensity for evolving Cu-tolerance
is emphasised by consideration of the situation in Britain
where the flora is comparatively depauperate and the majority
of the Cu-contaminated soils are of relatively recent origin,
deriving from the industrial revolution. Indeed it may be

said that in Britain there are only about 8 species which occur with any frequency on soils contaminated with Cu; these are the grasses *Festuca ovina*, *F. rubra*, *Agrostis stolonifera*, *A. tenuis*, *Deschampsia caespitosa* and the dicotyledons *Silene maritima*, *Armeria maritima* and *Calluna vulgaris*.

Gartside and McNeilly (1974) examined the potential for evolution of Cu-tolerance in nine common species of the British flora by screening 10,000 seeds of each on a range of soils containing differing amounts of Cu. In two species, *Agrostis tenuis* (selected as a control from a Cu-contaminated site) and *Dactylis glomerata*, it proved possible to obtain fully Cu-tolerant plants from a single cycle of selection. Four species, *Lolium perenne*, *Poa trivialis*, *Cynosurus cristatus* and *Arrhenatherum elatius* yielded some plants with a modest degree of Cu-tolerance whilst *Trifolium repens*, *Anthoxanthum odoratum* and *Plantago lanceolata* did not show any evidence of Cu-tolerance. It is notable that *Dactylis glomerata*, though giving rise to highly Cu-tolerant genotypes in a single cycle of selection, is not found on Cu-outcrops under natural conditions, a fact which may serve as a reminder that there may be other factors, such as hydrology and nutrient status of the contaminated soil, which may preclude the establishment or subsequent growth of the species on such sites even if the Cu-tolerance mechanisms is present. Likewise, in the case of the four species that revealed a moderate degree of tolerance, it was estimated that 14-26 generations would be required under severe selection to achieve full tolerance, but notwithstanding the passage of sufficient time, these species do not appear to have colonised cupriferous sites. Of the species showing no tolerance to Cu, *Anthoxanthum odoratum* and *Plantago lanceolata* are notable in that they do give rise to Zn/Pb-tolerant genotypes. The relative abundance of species on Zn/Pb contaminated sites suggests that genes for Zn/Pb-tolerance may be more widespread than for Cu-tolerance although, in the light of the example of *Dactylis*, such inferences should not be too freely drawn since there may be other edaphic factors which are less restrictive on Zn/Pb than on Cu-contaminated sites.

III. GENETICS OF COPPER TOLERANCE

Through much of the work on tolerance to Cu in plants there is the supposition that the phenomenon is genetically determined. This is based on the observation that there is variation between individual plants and that this is not

TABLE II. Comparisons of Indices of Tolerance and Total Soil-Cu Contents for Populations of *Mimulus guttatus*

Location	Index of Tolerance in $5 \times 10^{-6} g$ Cu $g^{-1} H_2O$	Total soil copper g Cu g^{-1} dry soil
Copperopolis 4	43.2	9.57×10^{-3}
Copperopolis 2	36.0	7.02×10^{-3}
Penn 2	25.7	1.1×10^{-3}
Location 1	13.4	3.1×10^{-5}
Location 2	13.4	8.9×10^{-4}
Location 5	3.2	7.0×10^{-5}
Location 3	1.3	4.3×10^{-5}
Location 4	1.3	9.6×10^{-5}
Lake Berryessa	1.6	2.1×10^{-5}
Putah Creek	0.9	2.0×10^{-5}
Bodega Bay	0	3.5×10^{-5}

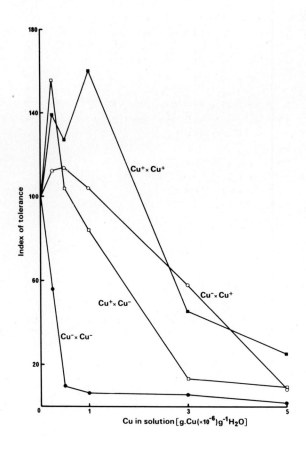

FIGURE 1. The root growth response of the F_1 progeny in copper solutions. Non-tolerant parents are indicated by Cu^- and tolerant parents by Cu^+ with the female plant being indicated first in the cross.

lost in cultivation. Gartside and McNeilly (1974) made crude estimates of the heritability of Cu-tolerance in *A. tenuis, Lolium perenne, Arrhenatherum elatius* and *Dactylis glomerata,* using a polycross technique. As mentioned by the authors, the results are not conclusive in that the basic requirements of the technique were not met, *viz* the selected genotypes were not a random sample from a randomly mating population in respect of the *loci* controlling Cu-tolerance, and it is unlikely that pollination within the polycross was strictly random.

Allen and Sheppard (1971) studied the genetics of Cu-tolerance in *Mimulus guttatus* from copper mine sites in California in which the index of tolerance of the plants broadly correlated with the total Cu content of the soils (Table II); a high degree of heritability was demonstrated (Figure 1). A curious fact to emerge from this work is that when F_1 progeny were tested from crosses involving Cu-tolerant and non-tolerant parents, Cu-tolerance was dominant at low Cu concentrations, intermediate at intermediate concentrations and recessive at high Cu concentrations.

IV. PHYSIOLOGY OF COPPER TOLERANCE

A. *Multiplicity of Mechanisms*

The work of Duvigneaud and Denaeyer de Smet (1963) emphasises the important point that there is more than one mechanism operating for the achievement of tolerance to copper. Table III, extracted from their data, shows the great differences in amounts of Cu present in the leaves of species growing on extremely contaminated soils, i.e. containing more than 5×10^{-4} g Cu g^{-1} dry soil. Evidently some species avoid Cu-toxicity by excluding the metal (Table III, Group I); some species accumulate the metal to very high concentrations (Table III, Group III) whilst others (Group II) occupy an intermediate position. It is of special interest that two species from Group III, *Triumfetta dikuluwensis* (1.23×10^{-4} g Cu g^{-1} dry wt.) and *Becium aureoviride* spp *lupotoense* (2.1×10^{-4} g Cu g^{-1} dry wt.) appear to detoxify copper by accumulation whilst two closely related species from Group I, *Triumfetta cupricola* (2.9×10^{-5} g Cu g^{-1} dry wt.) and *Becium homblei* (3.7×10^{-5} g Cu g^{-1} dry wt.) avoid Cu-toxicity by exclusion of the metal. Standard errors are not given with these analyses and the possibility of contamination with surface dust cannot be dismissed since the species *T. dikuluwensis* and *B. aureoviride* spp *lupotoense*

TABLE III.   List of Plant Species from Copper
             Contaminated Sites in Katanga
             Found Growing on Soils Containing
             $> 5 \times 10^{-4} g$ Cu g^{-1} dry soil

Group I

   Species $< 3.8 \times 10^{-5} g$ Cu g^{-1} dry wt.

        Olax obtusifolia
        Protea hirta
        Cryptosepalum maraviense \
        Andropogon filifolius
        Triumfetta cupricola
        Becium homblei
        Lapeyrousia sandersonii
        Tristachya helenae

Group II

   Species $> 5 \times 10^{-5} g$ Cu g^{-1} dry wt.
        $< 1 \times 10^{-4} g$ Cu g^{-1} dry wt.

        Becium homblei
        Becium aureoviride spp. *lupotoense*
        Gladiolus tshombeanus
        Dissotis derriksiana
        Xerophyta barbarae ssp. *cuprophila*
        Notholaena inaequalis
        Eragrostis boehmii

Group III

   Species $> 1 \times 10^{-4} g$ Cu g^{-1} dry wt.

        Triumfetta dikuluwensis
        Becium aureoviride ssp. *Lupotoense*
        Pandiaka metallorum
        Ascolepis metallorum
        Silene cobalticola
        Haumaniastrum robertii

TABLE IV. Selection of Papers Reporting
Specificity of Metal Tolerance
Mechanisms

Reference	Conclusions
Gregory and Bradshaw (1965)	In populations of A.tenuis tolerance to Zn is always accompanied by tolerance to Ni
Turner (1969)	Multiple tolerance may occur in clones from habitats contaminated by several heavy metals
Allen and Sheppard (1971)	Selection for tolerance in Mimulus guttatus to Cu led also to increased Zn and Pb-tolerance
Walley et al (1974)	Selection for tolerance in A.tenuis to Cu did not lead to increased Zn-tolerance
Cox and Hutchinson (1980)	In Deschampsia caespitosa selection on soils enriched in Cu and Ni gave rise to plants showing increased tolerance to Zn, Pb, Cd and Al

came from sites extremely rich in copper (3.2×10^{-3} and 1.125×10^{-2} g Cu g^{-1} dry soil respectively). If, however, the results are taken at face value then it would seem that there may be both exclusion and accumulation mechanisms evolving in different species of the same genus.

B.   *Specificity of the Metal Tolerance Mechanism*

Over the past thirty years, a succession of papers reporting the screening of plant populations for metal tolerance have raised the question of specificity. A limited selection of the findings are summarised in Table IV. From this assortment of results it appears that there are no widely applicable generalisations available. For example, in *Mimulus guttatus* selection for tolerance on Cu-enriched soils led to some increase in tolerance to Zn and Pb as well as to Cu, although these elements were not present at high concentrations (Allen and Sheppard, 1971). In contrast, similar selection with *A. tenuis* does not lead to increased Zn-tolerance (Walley *et al.*, 1974). Similarly, in *Deschampsia caespitosa* selection on soils enriched in Cu and Ni gave rise to plants which also showed increased tolerance to Zn, Pb, Cd and Al (Cox and Hutchinson, 1980). *Anthoxanthum* shows Pb and Zn tolerance in the absence of Cu-tolerance but this is not the same situation as in *D. caespitosa*; the strictly comparable proposition is, if it had been possible to select for tolerance to Cu in *Anthoxanthum*, would such genotypes have shown an increased tolerance to Zn and Pb? It would then be possible to explore the intriguing question of whether the mechanism of Cu-tolerance was the same in those strains in which it existed independent of Pb- and Zn-tolerance and those in which tolerance to all three metals was elicited.

Clearly, if one wishes to approach critically the physiology of metal tolerance there are three facts from this body of work which cannot be ignored:

(i)   that in some species tolerance to Cu can be selected and inherited independently of tolerance to other metals,

(ii) that in some species selection for Cu-tolerance alone gives rise also to increased tolerance to other metals, and   (iii) that in some species genetically determined resistance may be found to certain metals but not to Cu: as we shall see later recent work suggests mechanisms by which these different patterns of response might arise.

A critical reading of the papers listed in Table IV emphasises that the methods used in both the soil science and physiology are too approximate to support the kind of debate concerning the specificity or otherwise of metal tolerance which some authors have sought to raise (Cox and Hutchinson, 1980). The toxic agent in most forms of Cu-poisoning is the Cu^{2+} ion but little of the published work takes any cognisance of the activity of the relevant ionic species in the soil solution or culture solutions nor to the modifying effects of calcium and other ions.

C. Copper Toxicity and Physiology Processes.

1. Root Growth. It is a curious fact that although measurements of root elongation are the almost universal method for assessing degrees of tolerance to toxic metals, very few studies have addressed the question of what is happening when root growth is blocked. Root growth involves mitotic division to generate new cells which then elongate and increase in volume; mitosis predominates in the apical zone behind the root cap, and elongation in the sub-apical region. In *Agrostis tenuis* it is found that cell division is confined to the apical 0.9 mm with some cell elongation in this region. The growth of the penultimate 0.9 mm segment is entirely by cell elongation (Wainwright and Wool-house, 1975, 1977). Figure 2 shows the effect of Cu^{2+} on the growth of the elongation zone in three clones of *A. tenuis* during a 24h period. It is clear that Cu^{2+} is inhibitory to root elongation at all of the concentrations used but much less so to the Cu-tolerant clone: the experiment is limited in that it is confined to a single time of measurement after 24 hours of incubation in the test solutions and so gives no indication of the time-course of inhibition. The mechanism of cell elongation is complex and offers many possible sites for the inhibition to operate, amongst which we may note the hormonal system, the process of cell wall biosynthesis and assembly and the maintenance of a positive cell turgor.

2. Effects of Copper on Membrane Permeability. It has been suggested that the primary toxic action of metal ions takes place at the cell surface: alterations in cell membrane properties causing leakage of K^+ and other ions and solutes (Danielli and Davies, 1951; Lowry *et al.*, 1957; Woolhouse, 1969). Rothstein (1959) described the effects of metal ions on various functions of the cell membrane.

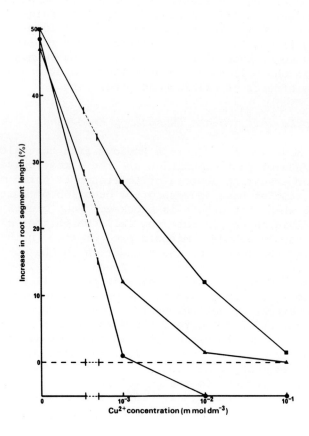

FIGURE 2. Effect of Cu^{2+} ions on the growth
of excised segments from the elongation zone of
the roots of three clones of Agrostis tenuis.
Results are expressed as increase in length of
the 0.9 mm. root segment in 24h as a percentage
of the initial length.   ▲, Zn-tolerant;
■, Cu-tolerant; ●, pasture (non-tolerant) clone.

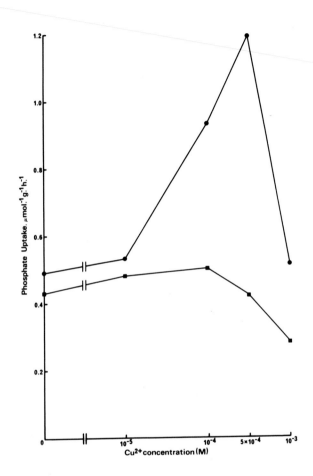

FIGURE 3. The effects of copper on the uptake
of phosphate into roots of tolerant and non-
tolerant tillers of Agrostis tenuis.
●, Cu-tolerant clone; ■, pasture (non-tolerant)
clone.

McBrien and Hassal (1965) showed that Cu^{2+} ions cause loss of potassium from *Chlorella vulgaris* cells.

Wainwright and Woolhouse (1977) studied the effects of temperature and Cu^{2+} concentration on leakage of K^+ from roots of Cu-tolerant; Zn-tolerant; and a non-contaminated pasture clone of *A. tenuis*. In the absence of Cu, K^+ leakage from roots of all three clones was some three times greater at $0^{\circ}C$ than at $25^{\circ}C$. This is consistent with the findings of Mengel and Pfluger (1972) with *Zea mays* who concluded that an energy-dependent K^+ pump is functioning to actively transport K^+ ions into the cell, the action of which is inhibited at low temperatures. Potassium leakage from roots of *A. tenuis* used as an index of membrane damage, increased with increasing Cu concentration over the range 0-1.0 mM Cu: Cu-tolerant clones were less sensitive to Cu-induced damage than either non-tolerant or Zn-tolerant clones (Wainwright and Woolhouse, 1977). In addition to increased resistance of root cell membranes to Cu-induced leakiness, Cu-tolerant plants may also possess a K^+ pump which is less sensitive to Cu than that of non-tolerant plants. Cu-induced K^+ leakage was greater at $25^{\circ}C$ than at $0^{\circ}C$ which may be due to a general increase in membrane permeability at the higher temperature, or to the effect of Cu^{2+} on the K^+ pump being greater because the pump is more active at the higher temperature. Alternatively, Cu^{2+} may be acting to reverse the action of the K^+ pump so that an active efflux of K^+ ions is being observed at $25^{\circ}C$ and a purely diffusive efflux is occurring at $0^{\circ}C$.

Copper caused a 140% stimulation in Pi uptake by the roots of a Cu-tolerant clone of *A. tenuis* over the range 0-0.5 mM and even 1.0 mM Cu^{2+} caused a slight stimulation of uptake (Thompson and Woolhouse 1981). For a non-tolerant strain however 1.0 mM Cu^{2+} caused a 35% inhibition of Pi uptake (Figure 3). This phenomenon was examined in greater detail by pre-loading the roots with ^{32}Pi and then treating them with Cu^{2+} in the presence of unlabelled Pi and measuring the rate of efflux of ^{32}Pi: the experiment was conducted at 10 and $20^{\circ}C$ (Table V). The apparent inhibition of phosphate uptake when the non-tolerant strain is treated with Cu^{2+} is probably accountable in terms of an increased rate of phosphate leakage rather than any direct effect on the uptake process. The Q_{10} for leakage between 10 and $20^{\circ}C$ (3.8) may on the other hand be interpreted as suggesting a metabolic component in the leakage process, as for example a Cu^{2+}-mediated reversal of the Pi pump. Whilst this may seem a curious inference we should note that a direct effect of Cu on the phosphate pump may also be inferred from the

stimulation of Pi uptake in the Cu-tolerant clone (Figure 3).
When Cu^{2+} was supplied to roots of Cu-tolerant and non-
tolerant clones of *A. tenuis* previously incubated with
mannose, leakage of both mannose-6-phosphate and Pi was
observed from non-tolerant plants whilst only Pi was lost
from roots of tolerant plants, a further indication of a
differing extent of membrane damage between the two clones.

TABLE V. The Rates of Leakage of Phosphate
from Roots of Non-tolerant Agrostis
tenuis at Differing Temperatures in
Presence and Absence of Copper

The roots of tillers were pre-loaded in a $^{32}PO_4$
solution ($10^{-6}M$, 185 Bq per plant) for 45 min-
utes at 15°C. Leakage of labelled phosphate was
determined over a 60 minute period and regression
lines drawn to determine initial rates of leakage
at 10°C and 20°C in the presence and absence of
copper at $5 \times 10^{-4}M$. (All solutions contained
10^{-6} $CaSO_4$)

	Treatment	
Temperature		
	Control	Copper
$10°C$	9.5	22.6
$20°C$	17.0	82.6
Q_{10}	1.8	3.8

3. Mechanisms for the compartmentation and Immobilisation of Cu. Experimental evidence is scanty concerning the role of compartmentation and chemical immobilisation processes in the mechanism of Cu tolerance in plants; the principal mechanisms for which some evidence has been gathered are immobilisation in the cell walls, the formation of soluble complexes compartmented in the cell vacuoles and immobilisation in the form of non-diffusible metal-protein complexes.

a. Immobilisation of Cu in cell walls. Cu and Zn bind to the cell walls of roots of *A. tenuis* (Turner, 1970). The amount of Zn bound by the walls of a Zn-tolerant clone increased linearly up to 1.5 mM Zn, suggesting that the walls may function as a metal-immobilising compartment. Jones *et al.* (1971) were able to prepare a Zn-complex from walls of a Zn-tolerant clone of *A. tenuis*, but this was not characterised chemically. The potential of the cell wall as a significant detoxification compartment in higher plants is called in question because of the problem of accounting for specificity of metal tolerance and by the finding of differential responses to Cu^{2+} in elongating segments of roots of Cu-tolerant and non-tolerant clones of *A. tenuis* under conditions in which the Cu-binding capacity of the cell walls is entirely swamped (Wainwright and Woolhouse, 1977).

b. Compartmentation of Cu in soluble vacuolar complexes. The absorption of Cu^{2+} by animal cells does not appear to be an active process but is enhanced by amino acids; Cu is probably transported across the plasma membrane as complexes of the type $[\,Cu\,II\,(AA)_2]^+$, where AA represents an amino acid. The essential acidic L-amino acids give greatest enhancement of Cu-uptake; once in the blood stream the Cu is rapidly transferred to its transport protein ceruloplasmin, though a small fraction persists in a ternary $(Cu^{2+}$-histidine-threonine) complex. There is likewise no necessity to postulate any active mechanism for Cu-uptake into plant cells and the evidence suggests transport through the plant in the form of organic complexes which remain to be characteristed (Graham, 1979). There is no evidence at the present time as to whether soluble Cu-complexes with amino acids or other ligands can be compartmented in soluble form in the cell vacuoles of Cu-tolerant genotypes which accumulate the metal. The recently developed procedures for the isolation of plant cell vacuoles affords an excellent opportunity to check this possibility.

It has been suggested that malate and mustard oil
glucosides may play an important part as soluble ligands for
the vacuolar compartmentation of Zn in Zn-tolerant genotypes
of some species (Mathys, 1977) but it must be emphasised that
this is on the basis of correlations between Zn and malate
content of the tissues and that causality has not been demonstrated; more rigorous techniques of analysis are needed
(Foy *et al.*, 1978).

  c. Compartmentation of Cu in non-diffusible complexes.
Margoshes and Vallee (1957) isolated a Cd and Zn-containing
protein from equine kidney, an organ now known to accumulate
Cd (Vallee, 1979). This material was further purified and
characterised and shown to be a low molecular weight, metal-
binding, cystein-rich protein for which the term metallothionein was coined (Kagi and Vallee, 1960, 1961). Piscator
(1964) found metallothionein present in increased quantity in
livers of rabbits exposed to Cd; he suggested that the biosynthesis of this protein was induced by Cd and had a role
in detoxification.

  Mammalian metallothioneins have a number of features:
they contain an extremely high proportion of cystein
(approximately 33% of the residues); a relatively high proportion of serine (14% of the residues), lysine and arginine
(13% of the residues); possess a single methionine residue;
lack any aromatic amino acids and histidine and contain no
disulphide or free sulphydryl groups. Furthermore all
cysteinyl side chains are bound to the metal ions in mercap-
tide complexes.

  Amino acid sequences are now known for metallothioneins
from horse, man, rabbit, mouse, and *Neurospora crassa*. In
all mammalian metallothioneins the polypeptide chain contains
61 residues with N-acetylmethionine and alanine at the
amino and carboxyl termini respectively. The most noticeable
feature is the predominance of -Cysteine-X-Cysteine- sequences
(where X stands for an amino acid other than cysteine).
Lysine, arginine and serine tend to be located adjacently to
the cysteinyl residues.

  The -Cysteine-X-Cysteine- sequences have been suggested to constitute the primary metal binding sites (Kojima
et al., 1976). A third cysteinyl residue located elsewhere
in the molecule is thought to interact with metal ions
chelated by these bi-dentate sites. The resulting negatively
charged trimercaptide complexes, [$Metal^{2+}$ $(Cystein)_3^-$] , are
responsible for the overall negative charge of the protein
(Kagi, 1970). Copper is bound much more firmly than either
cadmium or zinc (Bremner and Marshall, 1974; Rupp and Weser,

1974;   Rupp *et al.*, 1975;   Rupp and Wester, 1978;   Weser and Rupp, 1979).

Cu-containing proteins with the properties of metallothionein have been isolated from Cu-tolerant strains of yeast (Prinz and Weser, 1975;   Naiki and Yamagata, 1976) and from *Neurospora crassa* (Ammer *et al.*, 1978;   Lerch, 1979). Cd-binding proteins have been isolated from *Anacystis nidulans* (Maclean *et al.*, 1972), soybean (Casterline and Barnett, 1977), tomato (Bartolf, Brennan and Price, 1980), and *Phaseolus vulgaris* (Weigel and Jager, 1980).   In *Phaseolus* subcellular fractionation of Cd-containing tissues showed that more than 70% of the element was localised in the cytoplasmic fraction.

Metallothionein synthesis has recently been demonstrated *in vigro* with isolated   polysomes in the presence of messenger RNA from the liver of rats previously dosed with cadmium or zinc (Squibb and Cousins, 1977;   Andersen and Weser, 1978;   Shapiro *et al.*, 1978).   Results suggest that Zn and Cd are able to induce the synthesis of metallothionein through changes in the intracellular concentration of metallothionein messenger RNA.

Copper may induce synthesis of a Cu-containing protein that is not metallothionein (Premakumar *et al.*, 1975a, b; Winge *et al.*, 1975;   Day *et al.*, 1978) which has been termed copper chelatin.   Cu-binding proteins with low molecular weights and different amino acid compositions to those of metallothionein have been isolated by other investigators (Riordan and Gower, 1975a, b;   Evans and LeBlanc, 1976; Irons and Smith, 1977).

The wide occurrence, and the similarities of primary structure of metallothioneins in animals through microorganisms to plants, suggests that strong evolutionary constraints are acting to conserve this class of proteins. It may be that metal-binding proteins have a role in metal detoxification by binding to and hence reducing the availability of diffusible metal ions within cells.   They may also have a function in regulation of absorption and homeostasis of metal ions.

Different forms of metal-binding proteins have been found in the rat.   Of these, the metallothioneins inducible by Hg, Ag, or Zn and capable of also binding Cu; and the copper-chelatins induced by Cu are of particular interest. If a similar phenomenon is found to occur in higher plants it may provide an explanation of the puzzling pattern of cross-tolerances referred to in section IV*B*;   we are currently exploring this problem.

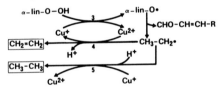

I Formation of starting radical

II Chain initiation

$$\alpha - lin + OH\cdot \longrightarrow \alpha - lin\cdot + H_2O$$

III Radical chain reaction

$$\alpha - lin\cdot + O_2 \longrightarrow \alpha - lin - O - O\cdot$$

$$\alpha - lin - O - O\cdot + \alpha - lin \longrightarrow \alpha - lin - O - OH + \alpha - lin\cdot$$

$$O_2$$

IV Degradation of peroxo compound

FIGURE 4. Copper-mediated lipid peroxidation in the intact cell. In I, the influence of Cu ions on electron transport in the region of PSII and PSI is indicated, as well as the influence on formation of OH· radical from superoxide anion (reactions 1 and 2). Parts II to IV represent the peroxidative reaction steps. Here, Cu ions are necessary for the degredation steps of part IV (reactions 3, 4, and 5). α-lin: α-linolenate; α-lin-O-O·: α-linolenate peroxy radical (ω-3-peroxy radical); α-lin-O·: ω-3-alkoxy radical.

TABLE VI.   *Comparison of Ki Values for Copper*
Inhibition of Root Cell-Wall Acid
Phosphatase Activity in Agrostis
tenuis

Ecotype	Ki (Cu) M
Normal pasture	4.33×10^{-5}
Zinc tolerant	4.25×10^{-5}
Copper tolerant	1.23×10^{-4}

Double reciprocal plots showed that the inhibition of
the cell-wall phosphatase by copper was non-competitive
and that the Kms for the enzyme from the copper-tolerant
plants and from the copper-susceptible plants was the same.
Values of Ki were calculated from the double reciprocal
plots using a modified rate equation derived by Westley
(1969):

$$Ki \ = \ \frac{[I]}{c. \ Vmax \ - \ 1}$$

where I = inhibitor concentration

   c = the intercept on the ordinate when
      1/V is plotted against 1/S

d. Enzyme adaptations. If the primary defences against Cu-toxicity in higher plants involve compartmentation or exclusion of the metal then there should be no significant selection pressure towards Cu-tolerance at the enzyme level. A notable exception to this proposition would be those enzymes which are contained in the extracellular or apparent free space of the root systems and those enzymes carried on the outer surface of the plasma membranes, where there is freedom of access for the toxic ions. In this context it is significant that an acid phosphatase covalently linked to the cell walls of *A. tenuis*, which may with reasonable certainty be regarded as a normal component of the extra-cellular compartment, was shown to have a greater Cu^{2+}-inhibitor constant in preparations from a Cu-tolerant clone when compared to the enzyme from a Zn-tolerant or normal pasture clone (Wainwright, 1975; Wainwright and Woolhouse, 1975) (Table VI). Enzymes internal to the cell membrane on the other hand have not been found to show altered properties in metal-tolerant clones (Cox *et al.*, 1976; Ernst, 1976; Mathys, 1975).

A study of polysaccharide synthesis in preparations of plasma membranes from metal-tolerant and non-tolerant clones would provide a further test of the hypothesis that there must be a selection pressure for metal tolerance for enzymes of the extra-cellular compartment.

e. Tissue damage. There have been innumerable studies of metabolic changes in plant tissues following treatment with toxic concentrations of Cu^{2+}. We may note for example the effects of Cu^{2+} on respiration in the roots of *A. tenuis* (Wu *et al.*, 1975). In this and many comparable investigations however, the time allowed for the treatments is so long that one is involved with a melange of secondary effects which defy interpretation in terms of basic mechanisms.

From the abundance of studies of copper damage the recent work of Sandmann and Boger (1980a and b) may be singled out as of particular interest in that it affords an explanation of the commonly observed symptoms of chlorosis in leaves of plants on cupriferous sites (Shewry *et al.*, 1978). It was found that Cu^{+} and Cu^{2+} ions block photosynthetic electron transport in *Scenedesmus* and in isolated chloroplasts of spinach. Measurements of absorption spectra of the cytochromes and fluorescence emission data were interpreted in terms of two sites of inhibition: on the oxidising side of PSII and the reducing side of PSI. Copper ions may also be responsible for acceleration of peroxidative

degradation of the lipids of the chloroplast membranes
(Figure 4). Sandmann and Boger (1980b) provide several
lines of evidence that reactions of the kind shown in
Figure 4 can occur in chloroplasts *in vitro* but it is not
yet demonstrated that *in vivo* Cu ever accumulates in sufficient quantities within the chloroplast to bring about
reactions of this kind. Clearly, if Cu does accumulate to
these levels, then chlorosis through breakdown of the thyla-
koids could rapidly occur. In any event, where such reactions
do occur under Cu toxicity, they are of an essentially
secondary character arising at a later stage in the demise
of the plant when primary barriers to the exclusion or
compartmentation of Cu in other organs have broken down.

CONCLUSIONS: *Towards a Molecular Biology of Copper
Tolerance?*

If we take a broad perspective view of the study of
Cu-tolerance in plants we are led to conclude that the plant
geography has provided some interesting pointers; the
genetics has not progressed very far; the physiology reveals
a spectrum of very complicated effects on cell growth and
membrane properties and the biochemistry, though very inconclusive, is offering some exciting avenues for further
investigation.
      At the physiological level the studies of membrane
properties offer a difficult challenge, and in those species
in which a central facet of the evolution of Cu-tolerance
has involved alterations of membrane characteristics it will
probably prove extremely difficult to resolve both the
genetics and the details of membrane structure.
      The formation of soluble Cu-complexes awaits the appli-
cation of the new methods for isolation and purification of
cell vacuoles and attendant chemical work. The demonstration
of altered enzyme characteristics in the extracellular compartment should also prove rewarding.
      The rapidly developing evidence of a role for metal-
binding proteins, such as metallothioneins, in the resistance
mechanism is an intriguing finding which may afford an
explanation of some of the hitherto conflicting results
concerning specificity of metal tolerance. It also provides
exciting possibilities for isolation of the genes for these
proteins, hence probing the deeper questions concerning the
nature of their regulation and, through the use of DNA
cloning techniques, the study of their distribution in Cu-
tolerant and non-tolerant genotypes.

REFERENCES

Ahrland, S., Chatt, J., and Davies, N.R. (1958).
    Q. Rev. Chem. Soc. 12, 265.
Agricola, G. (1556). De Re Metallica. English translation
    (1912). *In* "Mining magazine", London (Hoover and
    Hoover, ed.).
Allen, W.R., and Sheppard, P.M. (1971). *Proc. Roy. Soc.*
    Lond. B. 177, 177.
Ammer, D., Budry, R., Lerch, K. (1978). *Experienta* 34, 905.
Anderson, R.D., and Weser, U. (1978). *Biochem. J.* 175, 841.
Bartolf, M., Brennan, E., and Price, C.A. (1980). *Plant*
    Physiol. 66, 438.
Bremner, I., and Marshall, R.B. (1974). *Br. J. Nutr.* 32,
    293.
Broker, W. (1963). *Flora Jena, B.* 153, 122.
Bryan, G.W., and Hummerstone, L.G. (1971). *J. Mar. Biol.*
    Ass. U.K. 845.
Bryner, L.C., Beck, J.V., Davis, D.B., and Wilson, D.G.
    (195-). *Ind. Engng. Chem. Ind. (Int.)* 46, 2587.
Casterline, J.L., and Barnett, N.M. (1977). *Plant Physiol.*
    59, S-124.
Cox, R.M., and Hutchinson, T.C. (1980). *New Phytol.* 64,
    631.
Cox, R.M., Thurman, D.A., and Brett, M. (1976). *New Phytol.*
    77, 547.
Da Costa, E.W.B. : and Kerruish, R.M. (1964). *Forest Prod.*
    J. 14, 196.
Danielli, J.F., and Davies, J.T. (1951). *Adv. Enz.* 11, 35.
Day, F.A., Coles, B.J., and Brady, F.O. (1978). *Bioinorg.*
    Chem. 8, 93.
Duvigneaud, P. (1958). *Bull. Soc. Royale Bot. Belg.* 90,
    127.
Duvigneaud, P., and Denaeyer de Smet, S. (1963). *Bull.*
    Soc. Royale Bot. Belg. 96, 93.
Ernst, W. (1974). *In:* "Schwermetallvegetation der Erde"
    Fischer, Stuttgart.
Ernst, W.H.O. (1977). *In: Symp. Proc. Int. Conf. Heavy*
    Metals Environ., Toronto, 27-31 Oct., 1975. 2 (1), 121.
Evans, G.W., and LeBlanc, F.N. (1976). *Nutr. Rep. Int.* 14,
    281.
Foy, C.D., Chaney, R.L., and White, M.C. (1978). *Ann. Rev.*
    Plant Physiol. 29, 511.
Gartside, D.W., and McNeilly, T. (1974). *Heredity* 32 (3),
    335.
Graham, R.D. (1979). *Plant Cell and Environment* 2, 139.

Gregory, R.P.G., and Bradshaw, A.D. (1965). *New Phytol.* 64, 131.

Horscroft, F.D.M. (1961). *In:* "The geology of the Northern Rhodesian copper belt", p.73. Macdonald, London.

Iron, R.D. ,and Smith, J.C. (1977). *Chem. Biol. Interact.* 18, 83.

Jones, R.G.W.: Sutcliffe, M., and Marshall, C. (1971). *In* "Recent Advances in Plant Nutrition", ed. 2, p.575. Gordon and Breach, New York.

Kagi, J.H.R. (1970). *Int. Congr. Biochem. Abstr. 8th,* p.130, Switzerland, September, 1970.

Kagi, J.H.R., and Vallee, B.L. (1960). *J. Biol. Chem.* 235, 3460.

Kagi, J.H.R. : and Vallee, B.L. (1961). *J. Biol. Chem.* 236, 2435.

Koch, K. (1932). Wissenschaftliche Arbeiten des Bezerkskomitces für Naturdenk-malpflage und Heimatschutz in Osnabruk. 1, 91.

Kojima, Y., Berger, C., Vallee, B.L., and Kagi, J.H.R. (1976). *Proc. Natl. Acad. Sci. USA* 73, 3413.

Lambinon, J., and Auquier, P. (1964). *Natura mosana* 16, 4.

Lange, O.L., and Ziegler, H. (1963). *Mitt. flor. soz. Arbeitsgem. N.F.* 10, 156.

Lerch, K. (1979). *Experientia Supplementum* 34, 173.

Lowry, R.J., Sussman, A.S., and von Boventer, B. (1957). *Mycologia* 49, 609.

Maclean, F.I., Lucis, O.J., Shaikh, Z.A., and Jansz, E.R. (1972). *Fed. Proc. Fed. Am. Soc. Exp. Biol.* 31, 699.

Margoshes, M., and Vallee, B.L. (1957). *J. Am. Chem. Soc.* 79, 4813.

Mathys, W. (1975). *Physiol. Plant* 33, 161.

Mathys, W. (1977). *Physiol. Plant* 40, 130.

McBrien, D.C.H., and Hassal, K.A. (1965). *Physiol. Planta* 18, 1059.

Mengel, K., and Pflüger, R. (1972). *Plant Physiol.* 49, 16.

Naiki, N., and Yamagata, S. (1976). *Plant Cell Physiol.* 17, 1281.

Nieboer, E., and Richardson, D.H.S. (1980). *Environmental Pollution (Series B).* 1, 3.

Piscator, M. (1964). *Nord. Hyg. Tidskr.* 45, 76.

Prat, S. (1934). *Ber. Dt. Bot. Ges.* 52, 65.

Premakumar, R., Winge, D.R., Wiley, R.D.,and Rajagopalan, K.V. (1975a). *Arch. Biochem. Biophys.* 170, 267.

Premakumar, R., Winge, D.R., Wiley, R.D., and Rajagopalan, K.V. (1975b). *Arch. Biochem. Biophys.* 170, 278.

Prinz, R., and Weser, U. (1975). *Hoppe-Seyler's Z. Physiol. Chem.* 356, 767.

Riordan, J.R., and Gower, I. (1975a). *Biochim. Biophys. Acta.* 411, 393.

Riordan, J.R., and Gower, I. (1975b). *Biochim. Biophys. Res. Commun.* 66, 678.

Robyns, W. (1932). *Natuurwet. Tijdschr.* 14, 101.

Rothstein, A. (1959). *Fed. Proc. Fed. Am. Soc. Exp. Biol.* 18, 1026.

Rupp, H., Voelter, W., and Weser, U. (1975). *Hoppe-Seyler's Z. Physiol. Chem.* 356, 755.

Rupp, H., and Weser, U. (1974). *FEBS Lett.* 44, 293.

Rupp, H., and Weser, U. (1978). *Biochim. Biophys. Acta.* 533, 209.

Russel, G., and Morris, O.P. (1970). *Nature (Lond.)* 228, 288.

Sandmann, G., and Boger, P. (1980a). *Z. Pflanzenphysiol.* 98, 53.

Sandmann, G., and Boger, P. (1980b). *Plant Physiol.* 66, 797.

Shapiro, S.G., Squibb, K.S., Markowitz, L.A. and Cousins, R.J. (1978). *Biochem. J.* 175, 833.

Shewry, P.R., Woolhouse, H.W., and Thompson, K. (1979). *Bot. J. Linn. Soc.* 79, 1.

Squibb, K.S., and Cousins, R.J. (1977). *Biochem. Biophys. Res. Commun.* 75, 806.

Stokes, P.M. (1975). *In: Symp. Proc. Int. Conf. Heavy Metals Environ.* Toronto. p.137.

Thompson, D.K., and Woolhouse, H.W. (1981). In Press

Turner, R.G. (1969). *In:* "Ecological Aspects of the Mineral Nutrition of Plants" (I.H. Rorison, ed.), p.399. Blackwell, Oxford and Edinburgh.

Turner, R.G. (1970). *New Phytol.* 69, 725.

Url, W. (1956). *Protoplasma.* 46, 693.

Vallee, B.L. (1979). *Experientia Supplementum* 34, 19.

Wachsmann, C. (1961). *Wilhelms-Universitat zu Munster in Referaten.* 19, 35.

Wainwright, S.J. (1975). PhD Thesis. The University of Leeds, Department of Plant Sciences.

Wainwright, S.J. and Woolhouse, H.W. (1975). *In:* "The Ecology of Resource Degradation and Renewal". The 15th Symposium of the British Ecological Society (Chadwick and Goodman, ed.). Blackwell, Oxford.

Wainwright, S.J., and Woolhouse, H.W. (1977). *J. Exp. Bot.* 28 (105), 1029.

Walley, K.A., Khan, M.S.I., and Bradshaw, A.D. (1974). *Heredity* 32 (3), 309.

Weigel, H.J., and Jager, H.J. (1980). *Plant Physiol.* 65, 480.

Weser, U., and Rupp, H. (1979). *Experientia Supplementum* 34, 221.

Westley, J. (1969). "Enzymic catalysis". Harper and Row, New York, Evanston and London.

Wild, H. (1968). *Kirkia* 7, 1.

Wild, H. (1978). *In:* "Biogeography and Ecology of Southern Africa" (Werger, ed.). p.1301.

Wild, H., and Bradshaw, A.D. (1977). *Evolution* 31 (2), 282.

Wildeman, E. (1921). Comite Special du Katanga, Bruxelles.

Winge, D.R., Premakumar, R., Wiley, R.D., and Rajagopalan, K.V. (1975). *Arch. Biochem. Biophys.* 170, 253.

Woolhouse, H.W. (1969). *In:* "Ecological aspects of the mineral nutrition of plants". British Ecological Society Symposium No. 9. (I.H. Rorison, ed.), Blackwell, Oxford.

Wu, L., Thurman, D.A., and Bradshaw, A.D. (1975). *New Phytol.* 75, 225.

Young, G.Y. (1961). *U.S. Dep. Agr. Forest. Serv. Forest Prod. Lab. Dept.* No. 2223, 7.

COPPER IN SOIL - PLANT SYSTEMS

12

COPPER CONCENTRATIONS IN PLANTS AND THEIR RELATIONSHIP TO SOIL PROPERTIES

S.C. Jarvis

The Grassland Research Institute
Hurley, Maidenhead

I. INTRODUCTION

Detailed investigations of the movement of Cu from soil through plants have, in the main, been restricted to those areas where there has been deficiency in crops. In comparison with some of the other trace elements, the extent of Cu deficiency is, on a global basis, not widespread. There is therefore, for most soils, only limited information about the effects of variation of their properties on the absorption of Cu by, and its distribution in, plants. However, as well as concern over effects of plant growth, there is a growing concern over the supply of Cu to livestock (particularly cattle) for forage crops; this has stimulated greater interest in the factors that may affect the movement of Cu along the pathway soil-plant-animal. Concern over dietary levels for cattle arises, not only because higher concentrations of Cu than those previously defined (A.R.C. 1965) are now thought to be required (Mills, 1975), but also because of the interactions that may limit absorption of Cu from the gut. These interactions can take place at lower concentrations of the interacting elements, i.e. Mo and S, than was previously thought (Suttle, 1978). Such effects have led to a widespread occurrence of a chronic form of hypocupraemia in the U.K. (Mills, 1975). A greater understanding of the factors affecting Cu concentration in plants, even when it is within the range thought to be adequate, is therefore desirable.

Although there are often significant differences between

COPPER IN SOILS AND PLANTS

species, and between varieties, of plants, under most circumstances the range of Cu concentrations in shoots is small. In most species Cu concentrations seldom exceed 30 µg Cu g^{-1} dry matter even when very large additions of Cu have been made to either soils or solution cultures (Jarvis, 1978). In this respect the behaviour of Cu contrasts markedly with that of other trace elements, notably B, Fe, Mn and Mo, when these are added to soils. A few plant species can accumulate very much higher concentrations of Cu in their shoots when there are high concentrations in the soil (Reilly, 1969; Brooks *et al.*, 1978; Woolhouse, this Symposium). In most plants, however, the range of concentration is from 5-15 µg g^{-1} (Bowen, 1979), and in crop plants the usual range is from 5-20 µg g^{-1} (Knezek and Ellis, 1980; Bowen, 1979); in forage plants in the U.K., it is from 2-15 µg g^{-1} (Whitehead, 1966). The concentration at which deficiency occurs varies with the species and the plant part that is sampled, but commonly an effect on plant growth is associated with a concentration of less than 2-3 µg g^{-1} in the dry matter (Gupta, 1979; Robson and Reuter, this Symposium).

The final concentration of Cu in shoots is a reflection of:

(1) the extent to which Cu has been absorbed by roots from the labile reserves in the soil,

(2) the mobility of Cu within the plant, and

(3) the environmental conditions, and the rate and the stage of growth of the plant.

In this review, the relationships between soil properties and the concentration of Cu in shoots, and the extent to which such relationships may be modified by plant factors, are examined.

II. PROPERTIES AFFECTING SUPPLY TO THE SOIL SOLUTION

A. *Parent Material, Weathering and Total Contents*

Copper is a relatively immobile element in soils. For example, in one study Navrot and Singer (1976) have observed that, as basalt rock weathered to form soil, there was little change in concentration (52 and 57 µg g^{-1} in rock and soil, respectively). In contrast, concentrations of the very mobile elements decreased markedly (e.g. Ca decreased from 61,000 to 2,600 µg g^{-1}) while those of the very immobile elements increased (e.g. Al increased from 60,000 to

102,000 µg g^{-1}). Copper deficiency in plants is therefore
often strongly related to geochemistry and tends to be
associated with soils derived from specific parent materials.
Thus, soils derived from coarse-grained materials (i.e. sands
and sandstones) or from acid igneous rocks contain lower concentrations of Cu than those developed from fine-grained
sedimentary rocks (i.e. clays and shales) or from basic
igneous rocks. In temperate regions, where the majority of
soils are relatively young, parent material is therefore
often the dominant factor in determining the total Cu content of the soil. Where weathering processes are more
intense (e.g. tropical regions), or have continued for longer
periods (e.g. in the more mature land forms of Australia),
this relationship may be modified by mobilization and
secondary distribution of the soil Cu.

In England and Wales, Cu deficiency is largely
restricted to crops grown either on organic or peaty soils,
on excessively leached and podzolic sands developed from a
variety of parent materials, or on shallow rendzina soils
developed on chalk (Caldwell, 1971a, 1971b). Caldwell (1971a)
gives further details of Cu-deficiency soils in England and
Wales. In Australia deficiency has occurred on a variety of
soil types, but especially on the extensive areas of calcareous sands and on lateritic podzolic soils (Donald and
Prescott, 1975). Thornton (1979) has recently reviewed the
information on soil types which are likely to produce Cu-
deficient crops in the U.K. and Australia as well as in other
countries.

Other than to identify soils on which plants may be Cu-
deficient, there has been little investigation of the
relationship between soil type and the contents of Cu in the
shoots of plants. A range of 2-250 (median 30) µg g^{-1} has
been suggested for the total Cu contents of all soils (Bowen,
1979), whilst the range in most agricultural soils from a
number of countries is from 5-100 µg g^{-1} (Thornton, 1979).
Most of 751 agricultural soils from 226 farms in England and
Wales contain between 5 and 50 µg g^{-1}, although the complete
range is from 2 to 195 (mean 20) µg g^{-1} (Archer, 1980).
Although it has been suggested that the total contents of
trace elements in soils will give a reasonable indication of
their content in plants (Mitchell, 1964), this is not the
case with Cu. The total concentration in Scottish soils, for
example, gave little guide to the amounts that were available
to plants (Mitchell, 1974). Nevertheless, total concentrations have been used to identify those soils on which
deficiency in crops is likely to occur (Reuther and
Labanauskas, 1966; Davies *et al.*, 1971; Purves and Ragg,

1962). These studies indicated that crops grown on mineral soils with Cu contents of less than 4-6 µg g^{-1}, or on organic soils with less than 20-30 µg g^{-1}, were those most likely to suffer from Cu deficiency. It is likely, however, that these values will vary considerably between different soil types within each of the two broad categories of soils. Over a wider range of total contents in a range of soils there is little relationship between concentrations in soils and those in plants.

B.   *Control of equilibria between solid and solution phases*

Solutions displaced from 20 calcareous soils had concentrations of 0.1 to 0.5 µM Cu (Hodgson *et al.*, 1966) and those from 4 acidic (pH 5.2 to 5.8), Irish soils 0.14 to 0.40 µM Cu (Benians *et al.*, 1977), whilst saturation extracts from 68 soils (Bradford *et al.*, 1971) had 0.2 to 4.0 µM Cu. In most soil solutions, the concentration of Cu is much lower than that determined by the solubility of inorganic precipitates of hydroxides, carbonates or phosphates (Lindsay, 1972; 1979). Precipitates of low solubility may, exceptionally, control the mobility of Cu in highly calcareous soils (Misra and Tiwari, 1966). In soils with intensively reducing conditions (i.e. < 200 mV) sulphide may precipitate from the soil solution and control mobility (Reddy and Patrick, 1977). However, in most soils the concentration in solution is controlled by the sorption/desorption phenomena associated with the various soil colloidal materials (McBride, this Symposium). Studies, with both simple and soil-based systems, have indicated that clays, organic matter and hydrous oxides (particularly those of Mn) all provide strong sorption sites for Cu.

Even although sorption by inorganic and organic colloids may control the concentration of Cu in the soil solution, there is little evidence to show that this process affects absorption by, and contents in, plants. The gradients of sorption isotherms at appropriately low concentrations of Cu in soil and solution phases, should provide some measure of a soil's ability to buffer changes in concentration in solution, and therefore a measure of its ability to maintain supplies to roots. However, neither the gradients of such isotherms, nor their 'equilibrium' concentrations (i.e. the concentrations in solution at which no change in soil Cu takes place) in a range of soils, were related to either the concentration or the total contents in ryegrass when this was grown in pots (Jarvis, 1981a). This finding contrasts

with others for Zn, in which variation between soils of both
of these sorption characteristics could be used to account
for major proportions of the variation in concentration and
total contents in subterranean clover and in wheat (Tiller
et al., 1972).

1. *Association with Mineral Colloids*. There have been
few investigations of the effects of distribution of Cu
amongst the mineral constituents of soils on the contents of
Cu in plants. Both fractionation studies (Le Riche and Weir,
1963; McLaren and Crawford, 1973a; Shuman, 1979) and
laboratory sorption studies (McBride, this Symposium) have
indicated that considerable proportions of the Cu in soils
may be associated with the oxides of Al, Fe and Mn. Laboratory studies have shown that Cu has a high affinity (second
only to Pb) for both natural and synthetic Mn oxides (Davis
and Leckie, 1978; McKenzie, 1980). It might therefore be
expected that such oxides could play an important role in
limiting the mobility of Cu in soils and hence its availability to plants. However, the addition of either birnessite
or a cryptomelane to soils (to give up to 3000 µg manganese
oxide per g soil) had no effect on the extractability of Cu
or its absorption by plants (McKenzie, 1978). This contrasts
markedly with Pb, with which both extractability and absorption by plants were reduced by the addition of the Mn oxide.
There is little other information on the effects of
distribution of Cu amongst the mineral components of soils
on plant contents. Although only very small proportions of
the total Cu in soils are isotopically exchangeable (Kline
and Rust, 1966), it might be expected that clay content,
through its contribution to the C.E.C. of the soil, will
also affect supply to plants. However, inclusion of these
two properties in equations relating extractable Cu to plant
contents does not often improve these relationships (Martens,
1968; Osiname *et al.*, 1973).

2. *Association with Organic Colloids*. As well as providing normal coulombic exchange sites for cations, colloidal
organic matter in soils provides specific 'sorption' sites
for Cu and other metals through other means (Stevenson and
Fitch, this Symposium). The mobility of Cu in biological
systems is determined to a very large extent by its coordination chemistry; copper often forms stronger complexes with
many organic ligands than other divalent transition metal
ions (Parker, this Symposium). Fractionation studies have
shown that the largest fraction of the soil Cu is that
associated with organic matter (Kline and Rust, 1966;

McLaren and Crawford, 1973a; Shuman, 1979), and it has been
calculated (Nriagu, 1979) that, on a global basis, 36 per
cent of the total content of Cu in soils is in this form.
To a large extent the association involves phenolic, hydroxyl
and/or carboxyl groups (Davies *et al.*, 1969). These groups
provide comparatively weak linkages, and Goodman and Cheshire
(1976) have suggested that the more stable, permanent fixation
of Cu may result from complexing with porphyrin groups.

On the basis of the fractionation studies, and on
multiple correlations (Martens, 1968), it has been suggested
that since the bulk of the soil Cu reserves is associated
with organic matter, this provides the main source for plants.
The supply of Cu, as assessed by extractability, often
initially increases when organic matter is added to, and
incubated with soils (Gupta, 1971), but this may be followed
by a decrease in extractability (Elgala and Schlichting,
1976). Even although supply may be temporarily enhanced by
increasing organic matter contents, only that Cu held by the
weaker sorption sites on humic materials can be absorbed by
plants (Petruzzelli and Guidi, 1976). Decreases in plant
contents may then occur as the result of a more stable and
permanent fixation. Thus, there was a decrease in concentration in ryegrass shoots of approximately 2 µg g^{-1} as organic
matter contents of 70 humic, sandy, Dutch soils increased
from 2 to 4 per cent (van Luit and Henkins, 1967). Although
the effect of organic matter content is variable, predictions
of plant contents based on extractable Cu are sometimes
improved when organic matter content is taken into account
(Martens, 1968; Dolar and Keeney, 1971).

3. *Effects of soil pH*. One of the most marked causes
of changes in the contents of many trace elements in plants
is variation of soil pH. This may act either through its
effects on the mobility of the element in soils, or by
affecting its absorption by roots. There is little evidence
to suggest that the mechanisms for the absorption of Cu by
roots are affected by changes in pH. However, laboratory
studies show that, as pH is adjusted towards neutrality, the
mobility of Cu in solution is reduced as the result of
increased sorption on to mineral colloid surfaces (McLaren
and Crawford, 1973b; Forbes *et al.*, 1976; McKenzie, 1980).
Neutral soils adsorb greater quantities of added Cu than acid
soils (Cavallaro and McBride, 1978, 1980), and there are
marked increases or decreases, respectively, in the gradients
of sorption isotherms as the pH of a soil is raised or
lowered (McBride and Blasiak, 1979; Jarvis, 1981a). It is
perhaps worth noting that at low concentrations of Cu, the

gradients of linear sorption isotherms are highly correlated
with pH, but not with some of the other soil properties
thought to be of importance in controlling mobility (i.e.
'free' Fe_2O_3, easily reducible Mn, clay and organic matter
contents or C.E.C.) (Jarvis, 1981a).

Despite the evidence that mobility is greater than low
pH, the effects on concentrations in plants are variable, and
increases are not always recorded. This was demonstrated
over 30 years ago by Piper and Beckwith (1949) who found that
concentration in a number of species (e.g. *Erodium cygnorum*,
Hordeum leporinum), was almost entirely unaffected by a range
of soil pH from 4.5 to 7.5. Similarly, some early Swedish
studies (Lundblad *et al.*, 1949) concluded that the lime content and pH of soils did not affect the availability of Cu as
measured by plant uptake. Liming pasture soils in the U.K.
to raise pH from 4.7 to 6.0 (Archer, 1971) had no effect on
Cu concentration in white clover and timothy; nor did liming
two sandy loam soils (from pH 5.6 to 7.7) on the contents of
peas and barley (Gupta, 1979). In many cases where changes
in concentration have been recorded they are often small
(Hodgson, 1963) and may sometimes be the result of improved
growth as pH is adjusted towards neutrality (Gupta, 1972).
Other examples do, however, provide evidence that supply is
reduced as pH is increased. Van Luit and Henkens (1967)
found maximum concentrations in ryegrass when soil pH was
between 4.5 and 5.0, and Younts and Patterson (1964) found
that concentration in wheat was reduced when pH was raised
above 5.1. Heavy liming (Henkens, 1957) has increased Cu
deficiency in oats, and liming a very acid peat soil
decreased the concentration in cereals at low and extremely
high rates of addition, but increased it at intermediate
rates (Caldwell, 1971b). The variable nature of the effect
of pH was also demonstrated when a dressing of 7.5 t ha^{-1}
$CaCO_3$ was applied to a Scottish soil (Mitchell *et al.*, 1957);
there were marked increases in Cu concentration in red clover
but not in ryegrass.

4. *Effects of Soil Drainage.* Unlike those trace elements with which there are considerable changes in plant
contents with changes in drainage status, Cu is not subject
to oxidation/reduction reactions under changing drainage
conditions. Its mobility and supply to plants, however, may
be indirectly affected by the reduction of hydrous oxides of
Fe and Mn, and the breakdown of organic matter as changes in
drainage occur. Increases in the concentration of Cu in the
soil solution have been found under poorly drained conditions
(Kubota *et al.*, 1963) but this had little effect on contents

or concentration in alsike clover. Similarly, EDTA-extractable Cu was increased when soil in pots was flooded, but the
absorption of Cu by rice was reduced (Beckwith *et al.*, 1975).
Over ten-fold increases in extractable Cu have been found in
some Scottish soils under poor drainage conditions, but
although there were increases in concentration in red clover,
there were not in ryegrass (Table I).

TABLE I.   *Effect of Soil Drainage on Acetic Acid
Extractable Cu in Soils and Concentrations
in Plants.*

Soil parent	Drainage status	Extractable Cu $(µg\ g^{-1})$	Concentration in plant $(µg\ g^{-1})$	
			Ryegrass	Red clover
Argillaceous schist[a]	Good	0.15	3.6	8.4
	Poor	1.05	3.0	10.2
Slate till[b]	Good	0.17	4.0	7.9
	Poor	1.10	3.4	10.3

[a] *Data taken from Mitchell et al., 1957*
[b] *Data taken from Mitchell, 1971.*

In contrast, other field experiments with improved drainage
have resulted in increased concentrations of Cu in plants
(Schwab *et al.*, 1966; Lal and Taylor, 1970). The more usual
pattern of moisture changes that soils undergo, i.e. cycles
of wetting and drying, has little effect on the absorption of
Cu by ryegrass (Nambiar, 1975, 1977a). This species was also
able, if some of its roots had access to water in the sub-
soil, to absorb significant quantities of Cu from nearly dry
topsoil (Nambiar, 1977b).

C. Extractable Cu Contents

The mobility of Cu in soils is affected by a number of
soil properties. Attempts have been made to relate extractable Cu contents, as measures of the Cu which is mobile and

therefore, it is argued, in a form that could be supplied to plants, to plant contents. The extractants used include H_2O, neutral salts, dilute acids and chelating agents (see Gupta, 1979, for recent review). Although widely used, such tests have been neither as precise, nor as reliable, as those for other trace elements, and none of them is generally effective over a range of soils (Mortvedt, 1977). Similarly, on highly contaminated soils there is little relationship between extractable Cu and plant contents (Haq *et al.*, 1980). On the other hand, such measurements have been used to identify some soils on which deficiency in plants is likely to occur (Reith, 1968; Caldwell, 1976). They do not, however, provide an accurate estimate of the Cu contents or concentrations in plants grown on a wide range of soils (Table II). Concentrations in shoots have been related to extractable levels in some soils in Scotland but the relationship held for a single species only (Mitchell *et al.*, 1957).

Recently, chelating agents such as EDTA and DTPA have been used to an increasing extent both in laboratory studies (Borggaard, 1976; Lindsay and Norvell, 1978) and to provide routine assessments of soil Cu for advisory purposes (Caldwell, 1971a, 1976). Their use is based on the marked ability of Cu, and other metals, to form complexes with the organic ligands naturally present in both solid and solution phases of the soil, as well as those added in solution as extractants. The amount of metal that is brought into solution during extraction is a function of both the activity of the metal in the soil solution (i.e. an 'intensity' factor), as well as of the ability of the solid phase to maintain the activity (i.e. a 'quantity' factor) (Lindsay and Norvell, 1978). Assessments of Cu supply based on chelating agents do often provide better predictions of plant contents than those based on other extractants (Dolar and Keeney, 1971; Osiname *et al.*, 1973). Taking variation in other soil properties, e.g. pH, cation exchange capacity, contents of organic matter, hydrous oxides and silt plus clay into account sometimes improves the prediction of plant contents from measurements of extractability (Martens, 1968; Dolar and Keeney, 1971; Osiname *et al.*, 1973). However, the effects are variable and large proportions of the variation in the Cu contents of plants are often unaccounted for by these predictive techniques.

Although neither the total contents of Cu in plants, nor concentration in shoots is strongly related to extractable Cu, absorption per unit of root can be predicted from these measurements of soil Cu (Jarvis and Whitehead, 1981) (Table II). Thus, absorption by ryegrass roots grown in pots on a range of 21 soils is related to EDTA-extractable Cu by the equation:

$$y = 1.20x + 23.78 \ (r = 0.907***)$$

where x = total EDTA-extractable Cu (mg Cu pot^{-1}), and y = total Cu content in the plant per unit of root (µg Cu g^{-1}dry root). This indicates that assessments of soil Cu are unlikely to be successful in providing predictions of plant contents unless other, internal plant factors are also taken into account.

TABLE 11.   *Correlation Coefficients[a] between EDTA Extractable Cu in 21 Soils and Cu Contents of Ryegrass[b]*

Plant Cu[c]	Extractable Cu	
	µg g^{-1}	mg pot^{-1}
Mean concentration (µg g^{-1})	0.344	0.313
Total content, shoots (µg pot^{-1})	0.344	0.255
Total content, shoots + roots (µg pot^{-1})	0.326	0.195
Absorption (µg g^{-1} dry root)	0.766	0.907

[a]$P<0.05$ or 0.001 when $r > 0.433$ or 0.665, respectively
[b]Data taken from Jarvis and Whitehead, 1981.
[c]Data are means or totals for six harvests for shoots and for the final (6th) harvest for roots.

III. PROPERTIES AFFECTING REACTIONS AT THE SOIL/ROOT INTERFACE

The annual removal of a productive grass crop containing 10 µg Cu g^{-1} dry matter and with a yield of 7 tonnes ha^{-1} removes 70 g Cu from the soil. Assuming that the soil has an average content of 20 µg g^{-1} Cu and weighs 2×10^6 kg to a depth of 15 cm, there is sufficient total Cu to supply comparable crops for over 570 years. It has been calculated (Kline and Rust, 1966), on the basis of thermal neutron irradiation studies, that from 35 to 75 per cent of the Cu in soils is in a freely diffusible state and therefore in a

mobile form that could be supplied to plants. The above
soil therefore contains a mobile supply of Cu adequate for
the grass crop for some 200 to 430 years. In practice,
there are considerable controls on the absorption of such Cu
by plants. Even although mobility, and apparent supply to
plants may be modified by changes in soil conditions, it is
evident that concomitant changes in the contents of Cu in
plant shoots do not always result. This, in part, is due to
soil properties affecting absorption by roots.

A. *Effects on Root Interception*

The supply of Cu and other nutrients in the soil
solution to roots is determined to a large extent by soil
characteristics, and takes place through one or more of
three processes. These are:

(1) mass flow with the soil solution.
(2) diffusion through the soil solution and
(3) root interception.

Mass flow and root interception take place under all circumstances, but diffusion to the root only occurs when a
diffusion gradient develops after roots absorb a nutrient
faster than it is supplied by mass flow. Because of its
relative immobility, the major proportion of the Cu supplied
to plants is through root interception (Oliver and Barber,
1966). Thus, on average, 87 per cent of the Cu absorbed
from soil in pots by soybeans was supplied through this
means (Table III). This contrasts with Fe, Mn and Zn which
were largely supplied through diffusion, and with B which
was supplied by mass flow. Other recent studies of the
absorption of Cu by perennial ryegrass from a range of 21
soils (Jarvis and Whitehead, 1981), and of the placement of
Cu fertilizers in soils (Gilkes and Sadlier, 1979) have
also shown that the movement of Cu into plants is, to a
large extent, dependent upon the exploitation of soil by
roots. Soil physical and chemical properties which may
influence the form and distribution of roots, may therefore also indirectly influence the absorption of Cu by
plants.

TABLE 111.   The Supply of Cu to Roots of Soybean[a]

	Soil		Soil/Sand (3:2)	
	Transpiration		Transpiration	
	Low	High	Low	High
Absorption, µg Cu pot^{-1}	13	29	15	27
% Absorption supplied by:				
(i) mass flow	9	26	0	17
(ii) diffusion	0	0	0	0
(iii)root interception	91	74	100	83

[a]Data taken from Oliver and Barber, 1966

B.   Interactions Affecting Absorption

   The mechanisms by which trace elements such as Cu are
absorbed by roots from the soil solution have not been well
defined.   Although absorption is, at least in part, concen-
tration-dependent (Veltrup, 1976;   Jarvis, 1978), plants can
absorb sufficient Cu for growth from extremely dilute solutions (Loneragan, 1975;   Jarvis and Robson, 1981;   Graham,
this Symposium).   Absorption of Cu from the soil solution,
however, has to take place in the presence of a complex
matrix which includes soluble organic, as well as mineral,
constituents.   Both groups of constituents will vary in type,
as well as in quantity, between soils, and both may influence
absorption by roots.

   1.   Soluble Organic Matter.   It has been shown (Hodgson
et al., 1966) that 98 to 98.5 per cent of the Cu in displaced solutions from 20 calcareous soils was complexed by
organic matter, and that the remainder was mostly complexed
with bicarbonate anions.   The limited amount of information
available on the effects of soluble organic ligands from
soils indicates that absorption by intact plants is strongly
influenced by the size of the ligand (Table IV).   Studies
with excised roots have shown that absorption is also
dependent upon the overall charge of the complex (Coombes
et al., 1977).   Large proportions of the Cu taken up by

plants must, however, originate from soluble organic complexes. Copper supplied to roots as a highly stable complex with EDTA is, before it is absorbed, first dissociated from the organic ligand (Goodman and Linehan, 1979). The ease with which the dissociation and subsequent absorption of Cu by roots takes place will therefore depend upon the complexing ligand and this, in turn, will depend on the particular soil on which the plant is growing, as well as conditions within that soil.

TABLE IV. Effect of Soluble Soil Organic Matter on the Absorption of Cu by Barley Seedlings[a]

	Control	*Molecular weight of organic matter*		
		200–400	*1400–1500*	*>5000*
Absorption (μg Cu g^{-1} root)	*41*	*29*	*13*	*6*
% transported to shoots	*1.3*	*2.3*	*2.9*	*3.7*

[a]*Data taken from Mercer and Richmond, 1972.*

Plants themselves may be responsible for changes in the form of Cu in the soil solution. For example, as barley growing on a calcareous peat soil approached ear emergence, there was a significant increase in Cu concentration in the soil solution. This coincided with an increase in soluble organic matter which probably resulted from root exudates (Nielson, 1976a, b). Furthermore, the influence of other soil properties (e.g. pH, water status) on the absorption of Cu may, partly, be indirectly through their effects on soluble organic matter. The variable effects of these properties on the concentration of Cu in plants may therefore be through such changes. For example, as soil pH changes, not only the quantity and the nature of the ligands present in solution change, but also the degree of complexing may be changed. Thus, at pH 4.5 only 35 per cent of the Cu in a 10^{-4}M solution of $Cu(No_3)_2$ is complexed by a lucerne extract, but at pH 6.5, 79 per cent is complexed (Bloomfield and Sanders, 1977). Under these circumstances, even although gradients of sorption isotherms increase as pH rises, there may be a concurrent increase in the complexing ability of the soil solution, resulting in a greater equilibrium concentration of Cu in solution (Jarvis, 1981b). Even although concentration in solution may increase, absorption by roots may then be reduced by the changes in the form of the soluble Cu.

   2. *Mineral Ions*. A further means by which pH may affect the Cu contents of plants is indirectly through the competitive effect of Al on the absorption of Cu by roots. Studies with plants grown in solution culture (Hiatt *et al.*, 1963), and in soils (Blevins and Massey, 1959; Brown and Foy, 1964) have shown that absorption of Cu is reduced as Al concentrations are raised, as may occur when soils become acidic. There is, however, an unexplained promotion of absorption when low concentrations of Al are added.
   There are also a number of other interactions involving Cu and the supply of plant nutrients. Examples of the occurrence and nature of some of these have been reviewed by Olsen (1972), and Table V gives some further examples. These interactions may also involve transport processes within the plant, but reduced absorption by roots resulting in reduced plant concentrations have resulted from increasing additions of Zn to soil (e.g. Gilbey *et al.*, 1970) and of Fe to peat (Cheshire *et al.*, 1967). In contrast, increasing additions of N to both soils (e.g. Hill *et al.*, 1978) and solutions (Jarvis, 1981b) have increased absorption. It has been suggested (Timmer and Leyden, 1980) that the reduction in plant Cu contents at high P results from a reduced exploitation of the soil by mycorrhiza, and hence reduced absorption. There are other reports of reduced Cu contents in plants when supplies of nutrients are increased, but these often result from enhanced rates of growth rather than any direct effect on Cu.

IV.   PROPERTIES AFFECTING TRANSPORT FROM ROOTS TO SHOOTS

   Transport of Cu to plant shoots, as well as absorption by roots, may be influenced by soil properties. The transport of Cu in many species, especially those of the Gramineae is restricted (Jarvis, 1978, 1981b). This is an important phenomenon resulting in large proportions of the absorbed Cu being immobilised even when plants are suffering from acute Cu deficiency (Jarvis and Robson, 1981).   Such restricted transport may be partly responsible (along with the immobility of Cu in soils) for the low recovery in the shoots of plants of Cu added to soils as fertilizers (Gilkes and

TABLE V.   Effects of Increased Supply of Nutrients on Cu Concentration in Plants.

Nutrient	Effect	Comments	Reference
N	Decrease	Deficiency resulting from increased growth	Chaudhry and Loneragan, 1970; Hill et al.,1978.
	Increase	Increased absorption/transport at luxury supplies of Cu from soil and solution culture	Cheshire et al., 1967; Hill et al., 1978; Jarvis, 1981b.
P	Decrease	(i) Reduced absorption/transport (ii)Reduced mycorrhizal absorption of Cu at high P	see review by Olsen, 1972 Timmer and Leyden, 1980.
Fe	Decrease	Reduced absorption from solution Reduced absorption from peat	Moore et al., 1957 Cheshire et al., 1967.
Zn	Decrease	Deficiency resulting from Zn fertilizer application. Reduced absorption from soil	Millikan, 1953; Gilbey et al., 1970 Chaudhry and Loneragan, 1970.

Sadlier, 1978).

The effect of restricted transport in plants may, in some circumstances, be as important in determining Cu concentration in shoots as the effects of supply to, and absorption, from the soil solution. When ryegrass was grown under controlled environment conditions on 21 soils with wide-ranging properties, from 49 to 79 (median 63) per cent of the total Cu absorbed was retained in the roots at the last of six harvests of the shoots. Variation of Cu concentration in the shoots of plants grown on a particular soil between harvests was almost as great as that between plants grown on all soils (Jarvis and Whitehead, 1981). In this, and in other studies (see Table V) the transport of Cu has been related to the supply and transport of N, i.e. increasing transport of Cu with increasing supply of N from soil or solution. It has also been suggested (Table V) that an increasing supply of P from the soil may restrict transport of Cu from roots to shoots.

TABLE VI   *Percentage of Variance in Cu Concentration ($\mu g \ g^{-1}$) in Ryegrass grown on 21 Soils Accounted for by Variation in Soil and Plant Factors[a]*

	A	B	C	D	E
	EDTA extractable Cu in soils	A + Soil pH	B + Soil organic matter (%)	C + Plant dry weight (g pot^{-1})	D + Plant N content (%)
Shoots:					
harvest 1	10	7	16	28	34
harvest 6	-	35	34	33	57
Roots:					
harvest 6	58	57	64	63	65

[a] *Data taken from Jarvis and Whitehead, 1981*

V. CONCLUSIONS

This review has indicated that the relationships between soil properties and the concentration of Cu in plants are extremely variable. This is demonstrated in Table VI which shows the percentage of the variance in Cu concentration in ryegrass grown on 21 soils that could be accounted for by variation in some of the soil properties which most consistently influence concentration in plants, namely, total extractable Cu, pH, and organic matter. The only major contribution to the variance in concentration in roots came from extractable Cu contents, but this contributed significantly to variance in shoots only at the first harvest. The influence of soil pH and organic matter varied with time, but major proportions of the variance in concentration in shoots could often be accounted for by variation in plant dry weight and N content.

Relationships between soil and plant Cu will be difficult to formulate until more is known of:

(1) the factors influencing desorption from, rather than sorption by, soils containing appropriately low concentrations,

(2) the forms in the soil solution and the extent to which they may be utilised by plants, and

(3) the factors influencing absorption and transport in plants.

Even when such information is available, the effect of other plant factors may make the prediction of the concentration of Cu in plant tissues from soil data difficult.

REFERENCES

Agricultural Research Council. (1965). Nutrient Requirements of Farm Livestock No. 2. Ruminants. HMSO, London.

Archer, F.C. (1971). *In:* Trace Elements in Soils and Crops, p. 150. MAFF, Tech. Bul. 21. HMSO, London.

Archer, F.C. (1980). *In:* Proc. ADAS Conference on Inorganic Pollution and Agriculture, HMSO, London.

Beckwith, R.S., Tiller, K.G. and Suwadji, E. (1975). *In:* Trace Elements in Soil-Plant-Animal Systems. (D. Nicholas, and A. Egan, eds.), p.135. Academic Press Inc., New York.

Benians, G., Scullion, P., and Fitzhugh, G.R. (1977). *J. Soil Sci.* 28, 454.

Blevins, R.L., and Massey, H.F. (1959). *Soil Sci. Soc. Amer. Proc.* 23, 296.

Bloomfield, C., and Sanders, J.R. (1977). *J. Soil Sci.* 28, 435.

Borggaard, O.K.   (1976).   *Act. Agric. Scand.* 26, 144.
Bowen, H.J.M.   (1979).   Environmental Chemistry of the Metals, Academic Press, London.
Bradford, G.R., Bair, F.L., and Hunsaker, V.   (1971).   *Soil Sci.* 112, 225.
Brooks, R.R., Morrison, R.S., and Reeves, R.D.   (1978).   *Science,* 199, 887.
Brown, J.C., and Foy, C.D.   (1964).   *Soil Sci.* 98, 362.
Caldwell, T.H.   (1971a).   *In:* Trace Elements in Soils and Crops, p.62. MAFF Tech. Bul. 21, HMSO, London.
Caldwell, T.H.   (1971b).   *In:* Trace Elements in Soils and Crops, p.73. MAFF Tech. Bul. 21, HMSO, London.
Caldwell, T.H.   (1976).   *In:* Trace Element Deficiencies in Crops. A.D.A.S. Advisory Paper No. 17. MAFF, Pinner.
Cavallaro, N., and McBride, M.B. (1978).   *Soil Sci. Soc. Amer. J.* 42, 550.
Cavallaro, N., and McBride, M.B. (1980).   *Soil Sci. Soc. Amer. J.* 44, 729.
Chaudhry, F.M., and Loneragan, J.F.   (1970).   *Aust. J. agric. Res.* 21, 865.
Cheshire, M.V., De Kock, P.C., and Inkson, R.H.E.   (1967).   *J. Sci. Fd Agric.* 8, 156.
Coombes, A.J., Phipps, D.A., and Lepp, N.W.   (1977).   *Z. Pflanzenphysiol. Bd.* 82, 435.
Davies, R.I., Cheshire, M.V., and Graham-Bryce, I.J. (1969).   *J. Soil Sci.* 20, 65.
Davies, D.B., Hooper, L.J., Charlesworth, R.R., Little, R.C., Evans, C., and Wilkinson, B.   (1971).   *In:* Trace Elements in Soils and Crops, p.88. MAFF Tech. Bul. 21, HMSO, London.
Davies, J.A., and Leckie, J.O.   (1978).   *Environ. Sci. Technol.* 12, 1309.
Dolar, S.G., and Keeney, D.R.   (1971).   *J. Sci. Fd Agric.* 22, 273.
Donald, C.M., and Prescott, J.A.   (1975).   *In:* Trace Elements in Soil-Plant-Animal Systems. (D. Nicholas and A. Egan, eds.), p.7. Academic Press, New York.
Elgala, A.M. and Schlichting, E.   (1976).   *Landwirtsch. Forsch.* 29, 34.
Forbes, E.A., Posner, A.M., and Quirk, J.P.   (1976).   *J. Soil Sci.* 27, 154.
Gilbey, D.J., Greathead, K.D., and Gartrell, J.W.   (1970).   *J. Agric. W. Aust.* 11, 70.
Gilkes, R.J., and Sadlier, S.B.   (1978).   *Aust. J. Soil Res.* 16, 113.
Gilkes, R.J., and Sadlier, S.B.   (1979).   *Aust. J. Soil Res.* 17, 121.

Goodman, B.A., and Cheshire, M.V. (1976). *J. Soil Sci.* 27, 337.

Goodman, B.A., and Linehan, D.J. (1979). *In:* The Soil-Root Interface. (J.L. Harley, and R. Scott Russell, eds.). p.67. Academic Press, London.

Gupta, U.C. (1971). *Plant and Soil,* 34, 249.

Gupta, U.C. (1972). *Soil Sci.* 114, 131.

Gupta, U.C. (1979). *In:* Copper in the Environment. I. Ecological Cycling. (J.O. Nriagu, ed.). p.255. John Wiley and Sons Inc., New York.

Haq, A.U., Bates, T.E., and Soon, Y.K. (1980). *Soil Sci. Soc. Amer. J.* 44, 772.

Henkens, Ch. H. (1957) *Landbouwvoorl.* 14, 581.

Hiatt, A.J., Amos, D.F., and Massey, H.F. (1963). *Agron. J.* 55, 284.

Hill, J., Robson, A.D., and Loneragan, J.F. (1978). *Aust. J. agric. Res.* 29, 925.

Hodgson, J.F. (1963). *Adv. Agron.* 15, 119.

Hodgson, J.F., Lindsay, W.L., and Trierweiler, J.F. (1966). *Soil Sci. Soc. Amer. Proc.* 30, 723.

Jarvis, S.C. (1978). *J. Sci. Fd Agric.* 29, 12.

Jarvis, S.C. (1981a). *J. Soil Sci.* 32 (in press)

Jarvis, S.C. (1981b). *Annal. Bot.* (in press)

Jarvis, S.C., and Robson, A.D. (1981). Submitted for publication in *Annal. Bot.*

Jarvis, S.C., and Whitehead, D.W. (1981). *Plant and Soil.* (in press).

Kline, J.R., and Rust, R.H. (1966). *Soil Sci. Soc. Amer. Proc.* 30, 188.

Knezek, B.D., and Ellis, B.G. (1980). *In:* Applied Soil Trace Elements. (B.E. Davies, ed.). p.259. John Wiley and Sons Inc., Chichester.

Kubota, J., Lemon, E.R., and Allway, W.H. (1963). *Soil Sci. Soc. Amer. Proc.* 27, 679.

Lal, R., and Taylor, G.S. (1970). *Soil Sci. Soc. Amer. Proc.* 34, 245.

Le Riche, H.H., and Weir, A.H. (1963). *J. Soil Sci.* 14, 225.

Lindsay, W.L. (1972). *In:* Micronutrients in Agriculture. (J.J. Mortvedt., P.M. Giordano, and W.L. Lindsay, eds). p.41. Soil Sci. Soc. Amer. Inc., Madison.

Lindsay, W.L. (1979). Chemical Equilibria in Soils. John Wiley and Sons, New York.

Lindsay, W.L., and Norvell, W.A. (1978). *Soil Sci. Soc. Amer. J.* 42, 421.

Loneragan, J.F. (1975). *In:* Trace Elements in Soil-Plant-Animal Systems. (D. Nicholas and A. Egan, eds.) p.109. Academic Press, New York.

van Luit, B., and Henkens, Ch.H.   (1967).   Versl. Landbouwk,
    Onderz. 695, Wageningen.
Lundblad, K., Svanberg, O., and Ekman, P.   (1949).   *Plant
    and Soil* 1, 277.
Martens, D.C.   (1968).   *Soil Sci.* 106, 23.
McBride, M.B., and Blasiak, J.J.   (1979).   *Soil Sci. Soc.
    Amer. J.*   43, 866.
McKenzie, R.M.   (1978).   *Aust. J. Soil Res.* 16, 209.
McKenzie, R.M.   (1980).   *Aust. J. Soil Res.* 18, 61.
McLaren, R.G., and Crawford, D.V.   (1973a).   *J. Soil Sci.*
    24, 172.
McLaren, R.G., and Crawford, D.V.   (1973b).   *J. Soil Sci.*
    24, 443.
Mercer, E.R., and Richmond, J.L.   (1972).   Agricultural
    Research Council, Letcombe Laboratory Annual Report
    1971, 20.
Millikan, C.R.   (1953).   *Aust. J. Biol. Sci.* 6, 164.
Mills, C.F.   (1975).   *In:* Copper in Farming, p.55. Copper
    Development Association, Potters Bay, England.
Mitchell, R.L.   (1964).   *In:* Chemistry of the Soil (F.E.
    Bear, ed.).   p.320.   Reinhold Publ. Corp., New York.
Mitchell, R.L.   (1971).   *In:* Trace Elements in Soils and
    Crops.   p.8.   MAFF Techn. Bul. 21.   HMSO, London.
Mitchell, R.L.   (1974).   *Neth. J. agric. Sci.* 22, 295.
Mitchell, R.L., Reith, J.W.S., and Johnston, I.M. (1957).
    J. Sci. Fd Agric. 8, S51.
Misra, S.G., and Tiwari, R.C.   (1966).   *Soil Sci.* 101, 465.
Moore, D.PO, Harward, M.E., Mason, D.D., Hader, R.J., Lott,
    W.L., and Jackson, W.A.   (1957).   *Soil Sci. Soc. Amer.
    Proc.* 21, 65.
Mortvedt, J.J.   (1977).   *In:* Soil Testing: Correlating
    and Interpreting the Analytical Results, ASA Special
    Publ. No. 29.   (M Stelly, ed.).   p.100.   Amer. Agron.
    Soc., Madison, USA.
Nambiar, E.K.S.   (1975).   *In:* Trace Elements in Soil-Plant-
    Animal Systems.   (D. Nicholas, and A. Egan, eds.).
    p.151.   Academic Press Inc., New York.
Nambiar, E.K.S.   (1977a).   *Plant and Soil*, 46, 175.
Nambiar, E.K.S.   (1977b).   *Plant and Soil* 46, 185.
Navrot, J., and Singer, A.   (1976).   *Soil Sci.*, 121, 337.
Nielson, N.E.   (1976a).   *Plant and Soil, 45, 659.*
Nielson, N.E.   (1976b).   *Plant and Soil, 45, 679.*
Nriagu, J.O.   (1979).   *In:* Copper in the Environment.
    I. Ecological Cycling.   (J.O. Nriagu, ed.).   p.1.
    John Wiley and Sons, Inc., New York.
Oliver, S., and Barber, S.A.   (1966).   *Soil Sci. Soc. Amer.
    Proc.*, 30, 468.

Olsen, S.R. (1972). *In:* Micronutrients in Agriculture.
(J.J. Mortvedt, P.M. Giordano, and W.L. Lindsay, eds.)
p. 243. Soil Sci. Soc. Amer. Inc., Madison.

Osiname, O.A., Schulte, E.E., and Corey, R.B. (1973).
J. Sci. Fd Agric. 24, 1341.

Petruzzelli, G., and Guidi, G. (1976). *Z. Pflanzenernaehr.
Bodenkd.* 6, 679.

Piper, C.S., and Beckwith, R.S. (1949). Brit. Commonw. Spec.
Conf. Agric., Australia.

Purves, D., and Ragg, J.M. (1962). *J. Soil Sci.* 13, 241.

Reddy, C.N., and Patrick, W.H. (1977). *Soil Sci. Soc. Amer.
J.* 41, 729.

Reilly, C. (1969). *New Phytol.* 68, 1081.

Reith, J.W.S. (1968). *J. Agric. Sci. Camb.* 70, 39.

Reuther, W., and Labanauskas, C.K. (1966). *In:* Diagnostic
Criteria for Plants and Soils (H.D. Chapman, ed.),
p. 157. University of California.

Schwab, G.O., Taylor, G.S., Fouss, J.L., and Stibbe, E.
(1966). *Soil Sci. Soc. Amer. Proc.* 30, 634.

Shuman, L.M. (1979). *Soil Sci.* 127, 10.

Suttle, N.F. (1978). *In:* Sulphur in Forage. Proc. Symp.
Wexford. 1978. p.197. An Foras Taluntais, Dublin.

Thornton, I. (1979). *In:* Copper in the Environment.
I. Ecological Cycling. (J.O. Nriagu, ed.). p.171.
John Wiley and Sons Inc., New York.

Tiller, K.G., Honeysett, J.L., and DeVries, M.P.C. (1972).
Aust. J. Soil Res., 10, 165.

Timmer, L.W., and Leyden, R.F. (1980). *New Phytol.* 85, 15.

Veltrup, W. (1976). *Physiol. Plant,* 36, 217.

Whitehead, D.C. (1966). Commonw. Bur. Past. Fld Crops
Mimeo Pub. 1. Farnham Royal

Younts, S.E., and Patterson, R.P. (1964). *Agron. J.* 56,229.

13

DIAGNOSIS OF COPPER DEFICIENCY
AND TOXICITY

A.D. Robson

Department of Soil Science and Plant Nutrition
University of Western Australia
Nedlands, Western Australia

D.J. Reuter

Department of Agriculture
Adelaide, South Australia

I. INTRODUCTION

Copper deficiency is widespread, occurring in many areas
and in many different plant species (see reviews by Stephens
and Donald, 1958; Caldwell, 1971a, b; Kubota and Allaway,
1972; Reuter, 1975; Gartrell, this Symposium). Copper
toxicity is less common occurring either in the limited areas
of high natural availability of Cu or after the addition of
Cu as a pollutant in sewage sludge, in mine wastes (Page,
1974; Purves, 1977) and in Cu containing pesticides (Delas,
1963).
Correction of both Cu deficiency and toxicity requires
that accurate and sensitive procedures be available to
diagnose these conditions. This review considers the use of
symptoms, soil analysis, plant analysis, and biochemical
assays in the diagnosis of Cu deficiency and toxicity. Most
emphasis will be directed towards an assessment of procedures
for diagnosing Cu deficiency in agricultural plants.

COPPER IN SOILS AND PLANTS

II. DIAGNOSIS BY FIELD OBSERVATIONS

A. Symptoms

   1. Copper Deficiency. Since the discovery 50 years ago
that Cu was an essential plant nutrient, numerous publications
and reviews have described characteristic symptoms of severe
Cu deficiency in many plant species (e.g. Piper, 1942;
Riceman and Anderson, 1943; Millikan, 1953; Hewitt, 1963;
Andrew, 1963; Reuther and Labanauskas, 1966; Caldwell, 1971a;
King, 1974; Gartrell *et al.*, 1979; Reuter *et al.*, 1981a).
These specific symptoms have proved useful for the recognition
of Cu deficiency in the field and for the delineation of
potentially Cu-responsive situations. Also it is now well
known (see Lucas and Knezek, 1972) that certain plant species
(e.g. oats, wheat, lucerne, lettuce, carrots, onions, tomatoes, tobacco and citrus), and even specific cultivars
(Smilde and Henkens, 1967; Caldwell, 1971a; Graham and Pearce,
1979) are very sensitive to Cu deficiency. As such they are
especially suited as "indicator" species for detecting Cu
deficient soils. Even so, the occurrence and severity of Cu
deficiency is typically variable within a given landscape
(King and Alston, 1975) and even within a field (Henriksen
and Jensen, 1958; Davies *et al.*, 1971), and this is caused
both by natural soil variability and other soil and environmental factors.
   In most species, symptoms of chlorosis, necrosis, leaf
distortion and terminal dieback are observed first in young
shoot tissues (Hewitt, 1963), and at least in some species
this can now be associated with the poor redistribution of Cu
in the phloem of Cu deficient plants (Nualsri, 1977; Hill
et al., 1979a, b; Loneragan *et al.*, 1980; Reuter *et al.*,
1981b; Loneragan, this Symposium). Wilting is also a commonly
observed symptom: in wheat this symptom is caused by structural weaknesses (reduced lignification) within the xylem
elements (Graham, 1976a; Bussler, this Symposium).
   Furthermore, in most annual and perennial sensitive
species, the most spectacular effects of Cu deficiency are
observed during reproductive development: seed and fruit
yields are dramatically reduced either by reductions in the
number of florets or seeds produced (e.g. Riceman, 1948) or
by the production of sterile pollen (Graham, 1975). As a
result, the most obvious effects of Cu deficiency in annual
leguminous pastures are observed in the year following pasture
establishment on newly cleared land (Rossiter, 1951; Carter
and Day, 1977). Moreover in annuals, the effects of Cu

deficiency on seed yield may be further aggravated when soil moisture becomes limiting during seed development because of effects of Cu deficiency in both decreasing root growth and delaying flowering (Reuter *et al.*, 1981a).

However, effective diagnosis can not always be made by plant symptoms alone. Symptoms vary between species and with the degree of Cu stress. Indeed, some species (e.g. celery, potatoes, sugar beet (Caldwell, 1971a); subterranean clover (Reuter *et al.*, 1981a)) do not exhibit characteristic plant symptoms. Additionally symptoms of "marginal" Cu deficiency (sometimes described as "hidden hunger"), even in Cu sensitive species, may not be recognised or their appearance is delayed beyond the stage when remedial Cu treatment can be effectively applied.

Moreover, experience has shown that symptoms of Cu deficiency may be masked or modified in plants damaged by insects or pathogens (Reuther and Labanauskas, 1966). They may also be affected by other nutritional stresses: e.g. Reuter *et al.*, (1981c) observed that a *potential* deficiency in soil Cu was not recognised in subterranean clover plants stressed by severe P deficiency, but symptoms were displayed in plants of higher P status.

2. Copper Toxicity. Plant species differ appreciably in their tolerance to high levels of soil Cu. Indeed, certain species of *Caryophyllaceae, Labiatae* and some mosses are able to tolerate high accumulations of Cu and have been used by mineral exploration companies to locate Cu ore bodies. Moreover, in soils where high levels of Cu have been concentrated in the surface horizon (e.g. following prolonged use of Cu containing fungicides), deep rooted perennial species are rarely affected, but shallow rooted annuals often fail to establish (Delas, 1963).

The more common symptoms of Cu toxicity include reduced shoot vigor, poorly developed and discoloured root systems, and leaf chlorosis (Smith and Specht, 1953; Delas, 1963; Reuther and Labanauskas, 1966; Daniels *et al.*, 1972). Often the chlorotic symptoms in shoots superficially resemble those of Fe deficiency, although the mechanisms responsible for this symptom in different plant species has not yet been eludicated (Mengel and Kirkby, 1979). Nevertheless, because the symptoms of Cu toxicity can be confused with other factors, diagnosis by symptoms alone is not satisfactory.

B. *Additional Observations*

Copper deficiency and toxicity may be associated with particular soils and with particular management practices. Awareness of these associations can assist in the assessment of the likelihood of Cu disorders. For example, certain soil types are more prone to Cu deficiency than others (Gilbert, 1952; Caldwell, 1971a, b; Donald and Prescott, 1975) and the availability of soil Cu declines when wet lands are drained (Mitchell *et al.*, 1957b). Similarly Cu deficiency in sensitive crops can be induced by prolonged or excessive use of N (Chaudhry and Loneragan, 1970); P (Bingham, 1963) and lime fertilizers (Henkens, 1957; Caldwell, 1971b), and can often be exacerbated by low soil moisture after flowering.

A knowledge of previous Cu applications (fertilizers and pesticides) should also aid field diagnosis because the availability of applied Cu in mineral soils persists for a long time (e.g. Gartrell, 1980, this Symposium). However on highly organic soils repeated large applications of Cu are required to prevent Cu deficiency in plants (see Murphy and Walsh, 1972).

Copper toxicity commonly occurs on very acid (pH<5) soils with low cation exchange capacities (Reuther and Smith, 1954; Leeper, 1978). On these soils, unlike highly organic soils, high levels of Cu application sometimes result in Cu toxicity being induced in sensitive species (Delas, 1963; Page, 1974; Purves, 1977).

III. DIAGNOSIS BY SOIL AND PLANT ANALYSIS

Reliable diagnosis of Cu disorders are, however, not always possible from field observations. Much research has concentrated on devising soil and plant tests for confirming the existence of Cu deficiency and toxicity. These laboratory tests need to be appropriate, reproducible and rapid. The development of improved instrumentation has simplified the analysis of copper in soil and plant extracts (e.g. Simmons and Loneragan, 1975) and has greatly increased the rate of analysis possible.

The objectives of any diagnostic test is to separate nutritional adequacy from deficiency and toxicity. In developing these procedures two important stages are involved:

* selecting the most appropriate test, and
* calibrating the test against plant response (i.e. plant yield, nutrient uptake or concentration) and thereby estimating the "critical" concentration which separates normal from nutritionally stressed plants.

The review will now consider the advances made in diagnosing Cu disorders by soil and plant analysis.

A. *Soil Analysis*

Soil analysis has been used for predicting the need to apply Cu fertilizer to annual crops at sowing time, but for deep-rooted tree crops, plant analysis is normally preferred for diagnosing Cu and other nutritional disorders.

Total levels of soil Cu are of limited diagnostic value, except to identify extremely deficient soils with very low total Cu concentrations (Lucas and Knezek, 1972). Similarly, Reuther and Smith (1954) concluded that slight toxic effects in orchard trees are likely to occur when the total soil Cu exceeded about 1.6 mg for each meq. of exchange capacity per 100 g soil. They also found toxicity in citrus seedlings grown on very sandy acid soils (pH<5) when total soil Cu levels exceeded 150 µg/g (Reuther and Smith, 1953).

A host of chemical soil tests (and also biological assay techniques using *Aspergillus niger.*: see Gerretsen, 1952; Henriksen and Jensen, 1958) have been devised for estimating the ability of soils to supply Cu for plant growth (see Table 1). Normally the soils for these tests are taken from the surface horizon (e.g. 0-10, 0-15 cm etc.) and are ground and mixed before being extracted.

These tests are empirical since there is no certainty that the Cu solubilized during extraction is necessarily the soil Cu which the plant roots can absorb (see later). In addition, the Cu from the resulting solid phases of the soil is extracted, often under equilibrium conditions, to varying degrees into the extracting medium, and these reactions occur when the soil particles are in intimate contact with the extractant. However, plants distribute their roots through only a small fraction of the surface horizon. Furthermore, most soil tests extract considerably more Cu (20-80 times) than that which is accumulated by the shoots and roots of annual crops (Davies *et al.*, 1971; Viets and Lindsay, 1973). Thus the tests only crudely estimate the "available" Cu status of soils, given that no account has been taken of available Cu at lower depths in the soil nor of the environmental

TABLE I. Some examples of soil Cu tests used for deficiency diagnosis

Extractant	Soln/soil Ratio	Extn time (hr)	Cu concentration ($\mu g/g$) Range	Crit. level	Test plant[+]	Source[§]
(a) Water and dilute salts						
Cold H_2O	40	16	0.01-0.24	NA	Oats (F)	1
0.01M Ca(NO_3)$_2$-NH_4OH(pH6.0)	10	18	0.001-0.018	0.005-0.015	Wheat (F)	3
N NH_4OAc (pH7)	12	16	0.02-0.33	0.03	Sub-clover (F)	4
N Mg(NO_3)$_2$(pH5.9)	10	1	0.2 - 20	NA	Oats (G)	5
(b) Acids						
2.5% CH_3COOH(pH2.5)	40	88	0.01-1.1	0.1	Oats (F)	1,2
0.1N HCl	10	1	0.3 - 5	NA	Corn (G)	6
1N HCl	10	1	0.7 - 11	NA	"	6
HCl (pH2.0)	4	NA	<2.5->10	3.1	Barley(F)	7
HCl (pH2.0)	4	1	<0.1->0.5	0.2 - 0.3	Oats, Barley (F)	8
HNO_3(dilute)	10	2	0.2-24	5	Pastures (G,F)	9
(c) Chelating agents						
0.01M EDTA + N NH_4OAc(pH7)	10	1	0.3-66	NA	Oats (G)	5
0.02M Na_2EDTA + 0.5% NH_4Cl (pH4.6)	10	2	<0.5 - >3	1-3	Barley, Oats (F)	8
"	10	2	0.2 - 32	NA	Pastures (F)	10
0.02M Na_2 EDTA (pH7)	12	16	0.1 - 7.3	c 0.5	Subclover (F)	4
0.05M NH_4 EDTA (pH4)	5	1	0.5 - 3.3	1.3 - 2.0	Cereals (F)	11,12
"	5	2	0.3 - 54	2.5 - 6.0	"	11,12
0.05M NH_4 EDTA (pH7)	5	1	0.3 - 6.8	0.8 - 1.1	"	2,13
0.05M Na_2EDTA+1M NH_4OAc (pH6)	5	16	0.1 - 1.3	0.4 - 0.8	Wheat (F)	3
0.05M EDTA + NH_4OH (pH9)	5	0.7	1.9 - 11	NA	Pine (F)	14
0.005M DTPA + $CaCl_2$+ TEA (pH7.3)	2	2	0.26-2.6	<0.2	Corn, Oats (G)	15
"	2	2	NA	<0.1	Tropical pastures (F)	16
"	2	2	0.01 - 0.34	0.1 - 0.2	Wheat (F)	3
0.005M DTPA + $CaCl_2$ + TEA (pH10.9)	10	1	0.2 - 63	NA	Oats (G)	5
0.2 M (NH_4)$_2$ C_2O_4(pH3)	10	16	0.3 - 2.2	NA	Wheat (F)	17

[+] F = calibrated in field experiments; G = calibrated in glasshouse experiments.   NA = not available

[§] Source = 1 = Mitchell et al. (1957a); 2 = Mitchell et al. (1957b); 3 = King and Alston (1975); 4 = McKenzie (1966); 5 = Dolar et al. (1971); 6 = Martens (1968); 7 = Steenbjerg and Boken (1950); 8 = Henriksen and Jensen (1958); 9 = van Luit and Henkens (1967); 10 = Beavington and Wright (1977); 11 = Davies et al. (1971); 12 = Pizer et al. (1966); 13 = Reith (1968); 14 = Viro (1955); 15 = Lindsay and Norvell (1978); 16 = Bruce (1978); 17 = Gartrell (1980).

variables which influence both Cu supply to plant roots and
the demand for Cu by the plants (e.g. growth potential).
Nevertheless some chemical soil tests have proved reasonably
useful for separating very deficient from adequately supplied
soils, but procedures for predicting Cu toxicity are less
adequately calibrated (Baker, 1974; Page, 1974; Tiller and
Merry, this Symposium).

1. The Search for an Appropriate Extraction Procedure.
Soil chemists have characterized soil nutrients into defined
chemical pools, each of which have attributes of concentration, size, turn over rate and equilibrium status with other
pools of the same nutrient (e.g. Viets, 1962). Moreover, it
is now accepted (Viets and Lindsay, 1973) that soil tests
should extract nutrients from the same pools from which plant
roots derive their supply. Thus, extracting procedures
should reflect both the present activity (intensity) and the
ability of soil to replenish the soil solution (capacity).
In addition the pH of the extracting medium should be selected
and buffered to avoid undesirable side reactions occurring
during extraction (e.g. Lopez and Graham, 1972; Kishk *et al.*,
1973; Lindsay and Norvell, 1978).
    McLaren and Crawford (1973a, b; 1974) have distinguished
five pools for soil Cu:

* soil solution and exchangeable Cu (soluble in 0.05M
$CaCl_2$)
* Cu weakly bound to specific sites (soluble in 2.5%
acetic acid)
* organically bound Cu (soluble in K-pyrophosphate)
* Cu occluded in oxide materials, (soluble in acidic NH_4
oxalate), and
* residual Cu, mainly in clay lattices (soluble in HF).

Their evidence agrees with other studies (e.g. Kline and
Rust, 1966; Martens, 1968; Dolar *et al.*, 1971; Baker, 1974)
and indicates strongly that Cu in the soil solution is
replenished by Cu weakly, but specifically adsorbed on *soil
organic matter.*
    Given the overriding importance of organically-bound Cu
in regulating Cu supply to plant roots, most soil testing
laboratories now use, with some measure of success, organic
chelating agents, such as EDTA and DTPA as empirical extractants of "available" soil Cu (see Table 1). These tests have
largely superseded the dilute salt and acid extractants used
initially, which probably underestimated (e.g. $Ca(NO_3)_2$) or
overestimated (e.g. HCl) the available Cu status of soils.

Moreover, acidic solvents, such as HCl, appear to extract Cu from pools which are not available to plants (Martens, 1968; Dolar *et al.*, 1971), are less reproducible, (Henriksen and Jensen, 1958) and are clearly unsuitable for extracting calcareous soils.

Because of the very complex nature and diverse properties of soil organic matter, there is no guarantee that a simple extraction procedure will *consistently* solubilize Cu from the same pools or parts of pools in soils of widely different characteristics. As a consequence, equal amounts of extract- able Cu do not necessarily infer equal availability of Cu in different soils (Baker, 1974). Thus King and Alston (1975) found that "critical" soil test values for 3 extraction tech- niques were more definitive when wheatland soils of similar properties were grouped, rather than when all soils were examined together. Others too, have found less variability in their calibration data and better separation of deficient and adequate soils when additional soil factors such as pH (McKenzie, 1966), and pH, organic matter content and plant yield potential (van Luit and Henkens, 1967) were concurrently examined with extractable soil Cu levels.

Moreover in regions where soils vary greatly, precision in defining diagnostic indices for soil Cu may be complicated by large differences in bulk densities between soils, i.e. when soil test values are expressed on a weight/weight basis. Indeed, Pizer *et al.*, (1966) recognized this, and adjusted soil test values to a common soil bulk density level. A wide range in bulk densities can also be expected in mineral soils: e.g. when comparing soil test values from siliceous sands, calcareous aeolian sands and lateritic podzolic soils.

2. Concluding Remarks on Soil Testing. An examination of Table 1 highlights some of the diverse number of Cu soil tests that have been advocated in the past. Clearly the extracting capacity and diagnostic indices vary appreciably between the tests. The array of tests can be related with the empirical nature of the tests, but it also indicates the advances which have been made in the art of soil testing for Cu. Given that the tests are empirical it seems now desirable to standardize soil testing procedures for Cu, and we propse that three tests may be most useful for delineating poten- tially Cu-responsive situations:

* For acid and near neutral soils (containing no free
lime): 0.02M EDTA in 1M NH_4OAC (pH 7) as reviewed and
recommended by Borggaard (1976).
* For near neutral and calcareous soils: 0.005 M DTPA
containing 0.01M $CaCl_2$ and 0.1M triethanol-amine (pH 7.3) as
recommended by Lindsay and Norvell (1978).
* For polluted or potentially toxic soils: 0.1M EDTA
(disodium salt) adjusted to pH 6 with NH_4OH as recommended
by Clayton and Tiller (1979).

Further calibration of each of these extractants is
required to develop critical levels. For example, while the
DTPA procedure was specifically designed to simultaneously
identify deficient levels of Cu, Zn, Mn and Fe in neutral and
calcareous soils, this test has yet to be adequately calibrated (Lindsay and Norvell,1978; Bruce, 1978), although it
appears to be useful for monitoring the residual effects of
previous applications of Cu fertilizer (Follett and Lindsay,
1971).
Because plant species differ widely in their ability to
extract soil Cu (e.g. Graham and Pearce, 1979), it is to be
expected that "critical" soil test values will also vary with
the type of crop being grown (see examples for DTPA Zn and
Fe; Lindsay and Norvell, 1978). There is therefore a need to
calibrate soil tests for each of the major species being grown
in a region, unless the soil test is only being used to
separate normal from very affected soils. When calibrating
these tests it would also now seem advisable to relate soil
test values to Cu concentrations in plants or sensitive plant
parts, as well as to yield response. Although this type of
calibration is rarely undertaken it has potential since both
the soil and plant are involved in the analysis, and often
soil and plant samples are collected by field agronomists
when diagnosing unknown disorders in plants.

B. Plant Analysis

1. *Deficiency*. The use of plant analysis to diagnose Cu
deficiency depends upon an understanding of the relationship
between Cu concentrations within the plant and growth. In
this section of the review the nature of this relationship
and factors affecting it, will be discussed. We shall
consider the use of Cu concentrations in whole shoots, in
selected tissues within shoots, and in seeds as indicators of
the Cu status of plants.

Many of the published values for critical concentrations of Cu in plants (particularly for horticultural crops and for intensive forestry crops) have been established by comparisons of the concentrations in "deficient" (usually defined as showing symptoms) or low producing trees with those in "adequate" or high producing trees in field surveys (see Reuther and Labanauskas, 1966; Leece, 1976). For other plants critical concentrations in plants have often been established in experiments where there are only 2 or 3 levels of Cu supply (deficient and adequate). There are relatively few plants for which critical concentrations have been established in experiments with at least 5 levels of Cu supply and with several harvests. This method of experimentation is essential to systematically estimate critical concentrations within plants (Ulrich, 1950, 1961; Ulrich and Hills, 1967). Values obtained using this approach are more accurate than those obtained in field surveys where the concentrations in visibly "deficient" and sufficient plants often overlap (Teakle and Turton, 1943; Beck, 1962).

Unlike soil analysis, plant analysis can only be used to indicate whether the plant was deficient when it was sampled. Nutrient supply can either decrease (e.g. by depletion) or increase (e.g. by the growth of roots into a nutrient rich zone) after sampling. Hence when using plant analysis to define deficient situations some knowledge of the likely pattern of supply of the nutrient is required. Alternatively crops with low but adequate concentrations of Cu at early stages of growth can be resampled at flowering as a guide to the fertilizer requirements of subsequent crops (as suggested by Gartrell *et al.*, 1979 for wheat).

a. *Whole shoots*. Copper concentrations in whole shoots have been used as indicators of the Cu status of the plant (Reuther and Labanauskas, 1966) but in many situations they have not proved to be reliable (Piper, 1942; Steenbjerg, 1951; Caldwell, 1971b; Thiel and Finck, 1973; King and Alston, 1976).

There are at least four difficulties with the use of Cu concentrations in whole shoots to assess the Cu status of plants. Firstly critical concentrations in whole shoots decline considerably as plants age (Nualsri, 1977; Loneragan *et al.*, 1980; Reuter *et al.*, 1981b). For example, critical concentrations of Cu in whole shoots of subterranean clover, at maximum yield, declined from 3.9 µg/g at 26 days from sowing to 1.6 µg/g at 98 days from sowing (Fig. 1, Reuter *et al.*, 1981b).

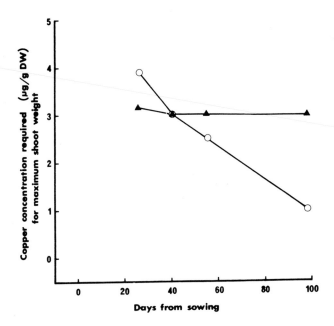

FIGURE 1. *The effect of plant age on the Cu concentration in whole shoots (○) and young leaf blades (▲) required for maximum yield of Trifolium subterraneum (drawn from data from Reuter et al., 1981b).*

The reason for this decline may be related to the changing distribution of both dry weight and copper contents among plant parts during late vegetative and reproductive phases of plant development (see Loneragan *et al.*, 1976, 1980; Hill *et al.*, 1978; Reuter *et al.*, 1981b, c). For cereals and other crops, adjustments in critical concentrations for plant age can be made (see Ulrich and Hills, 1967). However for pasture legumes and grasses which are grazed, it is very difficult to make any adjustment except to sample at one of the relatively few well-defined stages of development (for example jointing or flowering).

The second difficulty in the use of Cu concentrations in whole shoots for diagnosis is that for both cereals (Piper, 1942; Steenbjerg, 1943, 1951, 1954; Steenbjerg and Jakobsen, 1963) and legumes (Andrew and Thorne, 1962; Nualsri, 1977; Reuter *et al.*, 1981b, c), the concentration of Cu in the plant increases as the degree of deficiency increases from marginal to severe (see Fig. 2). This C-shaped curve has been referred

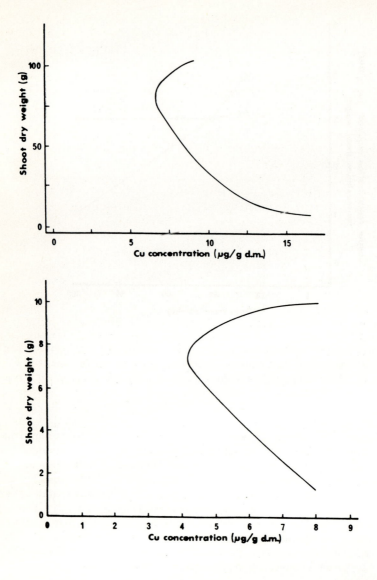

FIGURE 2. *The relationship between Cu concentrations in whole shoots and dry weight of shoots for (a) barley (Steenbjerg, 1951) and (b) Desmodium uncinatum (Andrew and Thorne, 1962).*

to as the Piper-Steenbjerg effect or curve (Rosell and Ulrich, 1964) and as the Steenbjerg effect (Bates, 1971). There are a number of possible explanations for such curves (Bates, 1971; Loneragan, 1978). For both wheat and subterranean clover the Piper-Steenbjerg curvature in the relationship between Cu concentrations in whole shoots and yield occurs because of effects of Cu supply on the distribution of Cu within the plant.

In wheat, Cu deficiency delays the senescence of old leaves and delays the loss of Cu from these leaves (Loneragan *et al.*, 1976, 1980; Hill *et al.*, 1978, 1979a, b). Copper concentrations in the old leaves of Cu-deficient wheat plants can thus exceed concentrations within the old leaves of Cu-adequate plants (Loneragan *et al.*, 1976, 1980) giving a strong Piper-Steenbjerg effect. More marked Piper-Steenbjerg effects would be expected at luxury N supply because the effects of N in delaying leaf senescence (Ballard and Petrie, 1936) may also further delay the movement of Cu out of old leaves.

In subterranean clover, the Piper-Steenbjerg curvature was not due to higher Cu concentrations in the blades of old leaves of Cu-deficient plants than in those of Cu-adequate plants (Reuter *et al.*, 1981b, c). Copper concentrations in the blades of all leaves including those that were completely senesced closely reflected Cu supply. However, Cu concentrations in the petioles of both young and old leaves of severely Cu-deficient plants exceeded those in moderately Cu-deficient plants producing a Piper-Steenbjerg curve. Additionally, Cu deficiency depressed the dry weight of the stems plus petioles relative to the dry weight of leaf blades (Reuter *et al.*, 1981a). As the proportion of stems plus petioles increases with internode elongation the degree of Piper-Steenbjerg curvature would also be expected to increase.

Piper-Steenbjerg curvature is particularly troublesome when severe deficiency is not accompanied by specific symptoms (e.g. as in subterranean clover, Reuter *et al.*, 1981a). However, because in both wheat and subterranean clover the curvature occurs because of effects of Cu supply on the distribution of copper within the plant, analysis of a particular plant part such as a young leaf blade may avoid this problem.

A third difficulty with using Cu concentrations in whole shoots for diagnosis is that the concentrations of Cu in whole shoots of deficient plants (i.e. plants responsive to Cu application) are frequently not very different from those in Cu-adequate plants (for example Caldwell, 1971b). The low sensitivity of Cu concentrations in whole shoots to Cu supply may at least partly reflect uneven distribution of Cu within the plant. For example, in Cu-deficient wheat plants Cu concentrations in old leaves exceeded those in young leaves whereas in Cu-adequate plants the reverse applied (Loneragan *et al.*, 1976, 1980; Robson, Loneragan, Gartrell and Snowball, unpublished data).

A further difficulty with using Cu concentrations in whole shoots to diagnose Cu deficiency is that critical concentrations in whole shoots are affected by N supply. Increasing the N supply from sub-optimal to luxury levels increased the Cu concentration in whole shoots of oats required for maximum *grain* yield (Thiel and Finck, 1973). This effect may partly arise from effects of increasing the N supply in delaying the senescence of old leaves and hence the movement of Cu from these leaves. However, at the beginning of tillering, the copper concentrations in young leaves of plants producing maximum grain yield was also greater in plants supplied with an optimal N supply than in those with a sub-optimal N. Unfortunately Thiel and Finck (1973) did not examine the effect of N supply on the relationship between Cu concentrations in young leaves and dry weight of shoots at the beginning of tilering.

There is little known on the effect of the supply of other nutrients on critical concentrations for Cu in whole shoots. Increasing P supply from sub-optimal to optimal levels did not markedly affect the concentrations of Cu in whole shoots required for maximum growth of subterranean clover (Reuter *et al.*, 1981c).

b. Plant parts within shoots. Some of the difficulties associated with the use of Cu concentrations in whole shoots to assess Cu status of plants can be overcome by selecting a specific plant part. Many procedures for the diagnosis of Cu deficiency, particularly in horticultural crops, are based on concentrations in a particular leaf rather than in whole shoots (see reviews by Reuther and Labanauskas, 1966; Leece, 1976). The choice of the particular leaf to be analysed appears to be largely related to the ease with which it can be sampled. However, a better basis for selecting the most appropriate tissue is an understanding of the mobility of the nutrient within the plant (Loneragan *et al.*, 1976). The mobility of Cu is variable depending on Cu status and other environmental conditions (see Loneragan this Symposium). The immobility of Cu in Cu-deficient plants has led to the conclusion that Cu concentrations in young leaves are the most sensitive and accurate indicators of Cu status in many diverse species - wheat (Loneragan, 1975; Loneragan *et al.*, 1976, 1980; Nambiar, 1976; Gartrell *et al.*, 1979; Brennan

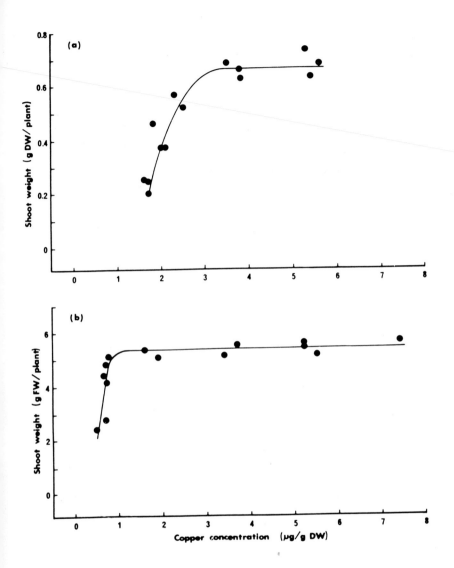

FIGURE 3. The relationship between Cu concentrations in young leaves and shoot weight for (a) subterranean clover (Reuter et al., 1981b) and (b) wheat (Robson, Loneragan, Gartrell and Snowball, unpublished data). Plants were harvested 37 and 40 days after sowing for wheat and subterranean clover respectively.

et al., 1980); peanuts (Nualsri, 1977; Nualsri *et al.*, 1977);
chrysanthemum (Graves, 1978) and subterranean clover (Reuter
et al., 1981b, c, Fig. 3).

Critical concentrations of Cu within young leaves vary
with species (Fig. 3). For example for peanuts (Nualsri
et al., 1977) and wheat (Nambiar, 1976; Brennan *et al.*, 1980;
Loneragan *et al.*, 1980), growth was reduced by Cu deficiency,
whenever Cu concentrations in young leaves were less than
1 μg/g, whereas in subterranean clover the critical concentration of Cu in young leaves was 3 μg/g (Reuter *et al.*,
1981b, c). The higher critical concentration of Cu in young
leaves of subterranean clover may reflect the involvement of
Cu in nitrogen fixation in this species (Greenwood and
Hallsworth, 1960; Snowball *et al.*, 1980). Copper does not
appear to be involved in nitrogen fixation by peanuts
(Nualsri, 1977).

Among cultivars of subterranean clover variation in
critical concentrations of Cu in young leaves does not appear
to be large (Reuter, Robson, Loneragan and Trantham-Fryer,
unpublished data). Similarly, in wheat, there is also no
clear evidence that critical Cu concentrations in young
leaves differed among cultivars, even though the cultivars
varied in the growth response to applied Cu (Nambiar, 1976).

The critical concentration of Cu in young leaves did not
decline or declined much less with plant age than did the
critical concentration in whole shoots for the species so
far examined (peanuts, Nualsri *et al.*, 1977; subterranean
clover Reuter *et al.*, 1981b, see Fig. 2; and wheat, Robson,
Loneragan, Gartrell and Snowball, unpublished data). In
wheat, Cu concentrations in the flag leaf are not a satisfactory index of Cu status once grain filling has begun,
presumably because of N and Cu export from the flag leaf to
developing grains (Hill *et al.*, 1978; Loneragan *et al.*,
1980).

The supply of other nutrients and other environmental
conditions does not appear to change the relationship between
Cu concentrations in young leaves and plant growth. Increasing P supply from sub-optimal to luxury levels for plant
growth did not affect the concentration of Cu in young leaves
required for maximum growth of subterranean clover (Reuter
et al., 1981c). In peanuts varying the N and Zn status of
plants did not affect the critical concentration of Cu in
young leaves (Nualsri, 1977). The critical concentration of
Cu in young wheat leaves remained constant in plants grown
on a wide range of soils (Brennan *et al.*, 1980). Imposing a
drought stress after anthesis did not affect the relationship between Cu concentrations in young leaves and grain

yield (Robson, Loneragan, Gartrell and Snowball, unpublished
data).

Copper concentrations in young leaves of wheat are now
being used to provide fertilizer recommendations for farmers
(Gartrell *et al.*, 1979). If Cu deficiency is diagnosed in
young plants it can be corrected within the same season by
foliar application of Cu without much loss of yield (Graham,
1976b).

c. Grain. Copper concentrations in grain are generally
of less value in assessing the adequacy of Cu supply than
concentrations in young leaves. Copper concentrations in
grain showed very small increases with increasing Cu supply
even though yield was increased substantially (Caldwell,
1971b). Nevertheless in some environments, Cu concentrations
in grain can be used to identify sites where grain yield will
respond to applied Cu. In South Australia, where Cu concentrations in grain from untreated plots were <2.5 µg/g, grain
yield responded to Cu application (Fig. 4, King and Alston,
1975). Similarly, in south-east England, a Cu concentration
in ears or grain <2 µg/g was considered to indicate deficiency and a concentration >3 µg/g to indicate adequacy
(Caldwell, 1971a; Davies *et al.*, 1971). However, in Western
Australia Cu concentrations in cereal grains are frequently
less than 2 µg/g even where yield did not respond to
increased Cu supply (Robson, Loneragan, Gartrell and Snowball
unpublished data). Similarly Cu concentrations of grain of
several wheat genotypes not deficient in Cu were less   than
2 µg/g (Nambiar, 1976).

There appears to be little value in using Cu concentrations in embryos dissected from mature grains to diagnose Cu
deficiency. While Cu concentrations in embryos are greater
than those in endosperms, increasing the Cu supply from
deficient to adequate levels increased Cu concentrations to
similar extents in both embryos and endosperms (Hill *et al.*,
1979c).

2. Toxicity. Plant analysis can be used to identify
situations where high levels of Cu are depressing plant
growth. For barley (Beckett and Davis, 1977; Davis *et al.*,
1978), snapbeans (Walsh *et al.*, 1972) and for tobacco
(Struckmeyer *et al.*, 1969) growth was depressed when Cu
concentrations in whole shoots exceeded 20 µg/g. Increasing
Cu supply may also depress plant growth by interfering with
the uptake of other nutrients (e.g. iron, Reuther and Smith,
1953). In these situations Cu concentrations in the plant
may be of less value in identifying the nature of the

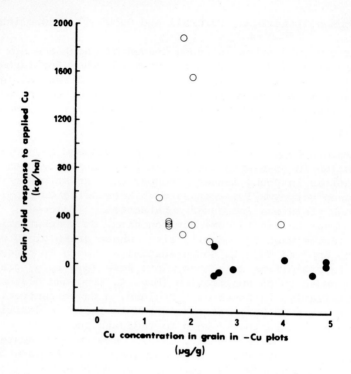

FIGURE 4. *Grain yield response in wheat to applied Cu in relation to Cu concentrations in the grain of plants not supplied with Cu (drawn from data of King and Alston, 1975). -○ Significant (p<0.05) response to applied Cu,● Non-significant (p>0.05) response to applied Cu.*

disorder than concentrations of the other nutrients.

C. *Biochemical Assays*

Nutrient deficiencies in plants can be diagnosed by measuring either enzyme activity or metabolite concentrations (see review by Bar-Akiva, 1971). If biochemical assays are to be sensitive and accurate indices of the adequacy of nutrient supply for plant growth, the biochemical responses must be specific and must correlate closely with effects of nutrient supply on plant growth.

Copper deficiency decreases the activity of ascorbic acid oxidase, cytochrome oxidase, polyphenol oxidase and diamine oxidase (Walker and Webb, this Symposium). However, most emphasis has been placed on the use of the activity of ascorbic acid oxidase as an indicator of the adequacy of Cu supply (Brown and Hendricks, 1952; Bar-Akiva *et al.*, 1969; Perumal and Beattie, 1966). The effect of Cu deficiency in decreasing ascorbic acid oxidase activity appears to be specific: e.g. deficiencies of Fe, Mo, Zn and Mn did not affect the ascorbic acid oxidase activity of citrus leaves (Bar-Akiva *et al.*, 1969). However, adding Cu to the growing media increased the activity of ascorbic acid oxidase in some plants even though it did not increase plant growth (soy beans, lupins, - Brown and Hendricks, 1952; apples - Perumal and Beattie, 1966).

Because many factors affect the absolute levels of both enzyme activity and metabolites, Bar-Akiva (1971) suggested that reactivation of enzyme activity or restoration of certain metabolic products by supply of the nutrient *in vitro* may be a more useful indicator than absolute levels. Floating leaf citrus discs on 0.2% $CuSO_4$ for 10 hours in darkness at $30^{\circ}C$ before the enzyme activity was assayed restored the activity of ascorbic acid oxidase in Cu-deficient leaves to that of Cu-adequate leaves (Bar-Akiva *et al.*, 1969). The ratio of Cu-induced activity to initial activity closely reflected the effect of Cu supply on growth (Fig. 5 from Bar-Akiva *et al.*, 1969). Cytochrome oxidase activity of young leaves from Cu deficient plants of subterranean clover also doubled after incubation in Cu solutions: Cu treated plants did not respond (Walker and Loneragan, 1981). By contrast, the incubation of leaves in Cu solutions stimulated the activities of o-diphenol oxidase and inhibited the activities of diamine oxidase in both Cu-deficient and Cu-adequate plants (Walker and Loneragan, 1981).

The effect of Cu deficiency on enzyme activity may vary considerably within the shoots of plants. Copper deficiency depressed the activities of cytochrome oxidase, ascorbic acid oxidase, diamine oxidase and o-diphenol oxidase in shoot tips and young leaves of subterranean clover, but had no effect on the activity of these enzymes in mature or old leaves, stems, and petioles (Walker and Loneragan, 1981). In earlier studies with sunflowers and oats (Lyszcz *et al.*, 1977) and with citrus (Bar-Akiva *et al.*, 1969) Cu deficiency depressed the activity of ascorbic acid oxidase to the same extent in both old and young leaves. Nevertheless because of the marked effects of Cu deficiency on the distribution of Cu within plant shoots (Loneragan *et al.*, 1980;

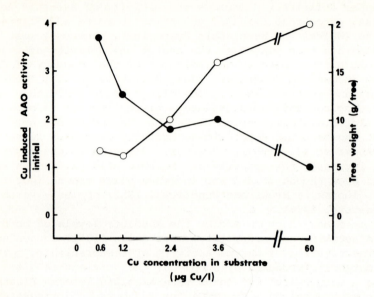

FIGURE 5.   *The effect of Cu supply on the growth of trees
of Eureka lemon (●) and on the effect of Cu infiltration on
the activity of ascorbic acid oxidase (AAO) relative to
initial activity (○) (after Bar-Akiva et al., 1969).*

Reuter *et al.*, 1981b) it would appear likely that the activity of Cu enzymes in young leaves would be more sensitive to
copper supply than that in old leaves.
     The concentration of a metabolite may also be suitable for
use as an indicator of Cu status.   Copper deficiency decreases lignification in plants (Schutte and Matthews, 1969;
Rahimi and Bussler, 1973) and also decreases polyphenol oxidase activity which may lead to the accumulation of phenols
in plants (Judel, 1972; Adams *et al.*, 1975).   However,
phenols also accumulate in B-deficient plants (Reed, 1947;
Lee and Aronoff, 1967; Watanabe *et al.*, 1964) and in Fe-
deficient plants (Romheld and Marschner, 1979).   Recent evidence suggests that Cu deficiency may exert a primary effect
on the composition of phenolic acids in cell walls of wheat
(Robson, Hartley and Jarvis, unpublished data).   Prior to
effects of Cu on growth, the ferulic acid content of cell
walls of young leaves of wheat was increased and the p-
coumaric acid content of cell walls was decreased by a low
copper supply.   This suggests that an assay based on the
composition of phenolic acids in cell walls may be a sensitive accurate indicator of Cu status.

IV. CONCLUSION

Copper deficiency in plants is difficult to diagnosis using symptoms, soil testing or biochemical assays when these approaches are used singly. There are two major difficulties with the use of symptoms. Firstly, yield may be reduced without any symptoms and secondly the symptoms of Cu deficiency can be confused with other disorders. Current procedures for soil testing are empirical and do not always discriminate between responsive and non-responsive situations. Soil testing appears to be most reliable when restricted to soils with similar soil properties and when used to identify extremely deficient sites. Biochemical assays appear to offer considerable promise as indicators of Cu status but further research is required to test their utility and general validity as diagnostic tools.

Recent developments in our understanding of the mobility of copper within plants has allowed the development of improved procedures for the diagnosis of Cu deficiency using plant analysis. Copper concentrations in young leaves are much more sensitive and accurate indicators of Cu status than are concentrations in whole shoots or grain. The critical concentrations of Cu in young leaves are unaffected by any of the environmental factors so far examined (drought stress after flowering, nitrogen and phosphorus supply, soil type). The critical concentrations of Cu in young leaves also shows much less change with plant age than does critical concentrations in whole shoots.

Diagnosis of Cu toxicity in plants by symptoms alone is not reliable, however Cu toxicity can be diagnosed by plant analysis. In several species under wide-ranging conditions, growth was depressed by Cu toxicity whenever concentrations in whole shoots exceeded 20 g/g. Copper toxicity may also be detected by soil analysis but insufficient calibrations have been conducted to define critical levels.

The most accurate assessment of the Cu status of a soil is gained when all the relevant information (field observations, soil type, previous Cu application, soil and plant analysis) is considered together.

REFERENCES

Adams, P., Graves, C.J., and Winsor, G.W. (1975). *J. Sc. Fd. Agric. 26,* 899-901.

Andrew, C.S. (1963). *Aust. J. Agric. Res. 14*, 654-9.
Andrew, C.S., and Thorne, P.M. (1962). *Aust. J. Agric. Res. 13*, 821-35.
Baker, D.E. (1974). *Fed. Proc. 33*, 1188-93.
Ballard, L.A.T., and Petrie, A.H.K. (1936). *Aust. J. exp. Biol. med. Sci. 14*, 135-63.
Bar-Akiva, A. (1971). *In* "Recent Advances in Plant Nutrition" (R.M. Samish, ed.) pp. 115-142. Gordon and Breach, New York.
Bar-Akiva, A., Lavon, R., and Sagiv, J. (1969). *Agrochimica 14*, 47-54.
Bates, T.E. (1971). *Soil Sci. 112*, 116-30.
Beavington, F., and Wright, W.A. (1977). *Aust. J. Agric. Res. 28*, 981-90.
Beck, A.B. (1962). *Aust. J. exp. Agric. Anim. Husb. 2*, 40-5.
Beckett, P.H.T., and Davis, R.D. (1977). *New Phytol. 79*, 95-106.
Bingham, F.T. (1963). *Soil Sci. Soc. Amer. Proc. 27*, 389-91.
Borggaard, O.K. (1976). *Acta Agric. Scand. 26*, 144-9.
Brennan, R.F., Gartrell, J.W., and Robson, A.D. (1980). *Aust. J. Soil Res. 18*, 447-59.
Brown, J.C., and Hendricks, S.B. (1952). *Plant Physiol. 27*, 651-60.
Bruce, R.C. (1978). *Trop. Grassl. 12*, 170-83.
Caldwell, T.H. (1971a). *In* "Trace Elements in Soils and Crops" pp. 62-72. Ministry of Agriculture, Fisheries and Food Tech. Bull. 21. HMSO London.
Caldwell, T.H. (1971b). *In* "Trace Elements in Soils and Crops" pp. 73-87. Ministry of Agriculture, Fisheries and Food. Tech. Bull. 21. HMSO London.
Carter, E.D., and Day, H.R. (1977). *Agric. Record. 4*, 9-15.
Chaudhry, F.M., and Loneragan, J.F. (1970). *Aust. J. Agric. Res. 21*, 865-79.
Clayton, P.M., and Tiller, K.G. (1979). Division of Soils Tech. Paper No. 41. CSIRO, Australia.
Daniels, R.R., Struckmeyer, B.E., and Peterson, L.A. (1972). *J. Amer. Soc. Hort. Sci. 9*, 249-54.
Davies, D.B., Hooper, L.J., Charlesworth, R.R., Little, R.C., Evans, C., and Wilkinson, B. (1971). *In* "Trace Elements in Soils and Crops" pp. 88-118. Ministry of Agriculture, Fisheries and Food Tech. Bull. 21. HMSO London.
Davis, R.D., Beckett, P.H.T., and Wollan, E. (1978). *Plant Soil 49*, 395-408.
Delas, J. (1963). *Agrochimica 7*, 258-88.
Dolar, S.G., Keeney, D.R., and Walsh, L.M. (1971). *J. Sci. Fd. Agric. 22*, 282-6.

Donald, C.M., and Prescott, J.A. (1975). *In* "Trace Elements in Soil-Plant-Animal Systems" (D.J.D. Nicholas and A.R. Egan, eds.) pp. 7-37. Academic Press, New York.

Follet, R.H., and Lindsay, W.L. (1971). *Soil Sci. Soc. Amer. Proc. 35,* 600-2.

Gartrell, J.W. (1980). *Aust. J. Exp. Agric. Anim. Husb. 20,* 370-6.

Gartrell, J.W., Brennan, R.F., and Robson, A.D. (1979). *J. Agric. W. Aust.* (4th Series) *20,* 18-20.

Gartrell, J.W., Robson, A.D., and Loneragan, J.F. (1979). *J. Agric. W. Aust.* (4th Series) *20,* 86-8.

Gerretsen, F.C. (1952). *Trans. Int. Soil Sci. Soc.* (Dublin) *1,* 151-66.

Gilbert, F.A. (1952). *Adv. Agron. 4,* 147-73.

Graham, R.D. (1975). *Nature 254,* 514-5.

Graham, R.D. (1976a). *Aust. J. Plant Physiol. 3,* 229-36.

Graham, R.D. (1976b). *J. Exp. Bot. 27,* 717-24.

Graham, R.D., and Pearce, D.T. (1979). *Aust. J. Agric. Res. 30,* 791-9.

Graves, C.J. (1978). *Ann. Bot. 42,* 117-25.

Greenwood, E.A.N., and Hallsworth, E.G. (1960). *Plant Soil 12,* 97-127.

Henkens, Ch.H. (1957). *Landbouwvoorlichting 14,* 629.

Henriksen, A., and Jensen, H.L. (1958). *Acta Agric. Scand. 8,* 441-69.

Hewitt, E.J. (1963). *In* "Plant Physiology, A Treatise Vol. III" (F.C. Steward, ed.) pp. 137-360. Academic Press, New York.

Hill, J., Robson, A.D., and Loneragan, J.F. (1978). *Aust. J. Agric. Res. 29,* 925-39.

Hill, J., Robson, A.D., and Loneragan, J.F. (1979a). *Ann. Bot. 43,* 449-57.

Hill, J., Robson, A.D., and Loneragan, J.F. (1979b). *Ann. Bot. 44,* 279-87.

Hill, J., Robson, A.D., and Loneragan, J.F. (1979c). *Aust. J. Agric. Res. 30,* 233-7.

Judel, G.K. (1972). *Z. Pflanz Bodenk 131,* 159-70.

King, P.M. (1974). *J. Agric. Sth. Aust. 77,* 96-9.

King, P.M., and Alston, A.M. (1975). *In* "Trace Elements in Soil-Plant-Animal Systems" (D.J.D. Nicholas and A.R. Egan, eds.) pp. 339-52. Academic Press, New York.

Kishk, F.M., Hassan, M.N., Ghanem, I., and El-Sissy, L. (1973). *Plant Soil 39,* 487-96.

Kline, J.R., and Rust, R.H. (1966). *Soil Sci. Soc. Amer. Proc. 30,* 188-90.

Kubota, J., and Allaway, W.H. (1972). *In* "Micronutrients in Agriculture" (J.J. Mortvedt, P.M. Giordano and W.L. Lindsay, eds.) pp. 525-54. Soil Sci. of Amer. Inc. Madison.

Lee, S., and Aronoff, S. (1967). *Science 158,* 798-9.

Leece, D.R. (1976). *J. Aust. Inst. Agric. Sci. 42,* 3-19.

Leeper, G.W. (1978). "Managing the heavy metals on the land". Marcel Dekker, New York.

Lindsay, W.L., and Norvell, W.A. (1978). *Soil Sci. Soc. Amer. J. 42,* 421-8.

Loneragan, J.F. (1975). *In* "Trace Elements in Soil-Plant-Animal Systems" (D.J.D. Nicholas and A.R. Egan, eds.) pp. 109-134. Academic Press, New York.

Loneragan, J.F. (1978). *In* "Plant Nutrition 1978". (A.R. Ferguson, R.L. Bieleski and I.B. Ferguson, eds.) pp. 283-98. Government Printer, Wellington.

Loneragan, J.F., Snowball, K., and Robson, A.D. (1976). *In* "Transport and Transfer Processes in Plants" (I.F. Wardlaw and J.B. Passioura, eds.) pp. 463-9. Academic Press, New York.

Loneragan, J.F., Snowball, K., and Robson, A.D. (1980). *Ann. Bot. 45,* 621-32.

Lopez, P.L., and Graham, E.R. (1972). *Soil Sci. 114,* 295-6.

Lucas, R.E., and Knezek, B.D. (1972). *In* "Micronutrients in Agriculture" (J.J. Mortvedt, P.M. Giordano and W.L. Lindsay, eds.) pp. 265-88. Soil Sci. Soc. Amer. Inc. Madison.

Lyszcz, S.T., Ruszkowska, M., Wejcieska, U., and Zinkiewicz, E. (1976). *Acta Agrobot. 29,* 99-105.

Martens, D.C. (1968). *Soil Sci. 106,* 23-28.

McKenzie, R.M. (1966). *Aust. J. Exp. Agric. Anim. Husb. 6,* 170-4.

McLaren, R.G., and Crawford, D.V. (1973a). *J. Soil Sci. 24,* 172-81.

McLaren, R.G., and Crawford, D.V. (1973b). *J. Soil Sci. 24,* 443-52.

McLaren, R.G., and Crawford, D.V. (1974). *J. Soil Sci. 25,* 111-19.

Mengel, K., and Kirkby, E.A. (1979). "Principles of Plant Nutrition". Int. Potash Inst., Berne.

Millikan, C.R. (1953). *Aust. J. Biol. Sci. 6,* 164-77.

Mitchell, R.L., Reith, J.W.S., and Johnson, I.M. (1957a). *In* "Plant Analysis and Fertilizer Problems" pp. 249-61. I.R.H.O. Paris.

Mitchell, R.L., Reith, J.W.S., and Johnson, I.M. (1975b). *J. Sc. Fd. Agric. 8,* 51-9.

Murphy, L.S., and Walsh, L.M. (1972). *In* "Micronutrients in Agriculture" (J.J. Mortvedt, P.M. Giordano and W.L. Lindsay, eds.) pp. 347-87. Soil Sci. Soc. Amer. Inc. Madison.

Nambiar, E.K.S. (1976). *Aust. J. Agric. Res. 27,* 465-77.

Nualsri, L. (1977). Ph.D. Thesis, Univ. of West. Aust.

Nualsri, L., Robson, A.D., and Loneragan, J.F. (1977). *Int. Soc. Soil Sci. Comm. IV, V. Kuala Lumpur.*

Page, A.L. (1974). Office of Research and Development U.S. Environmental Protection Agency 670/2-74-005.

Perumal, A., and Beattie, J.M. (1966). *Proc. Am. Soc. Hortic. Sci. 88,* 41-7.

Piper, C.S. (1942). *J. Agric. Sci. 32,* 143-83.

Pizer, N.H., Caldwell, T.H., Burgess, G.R., and Jones, J.L.O. (1966). *J. Agric. Sci. 66,* 303-14.

Purves, D. (1977). *In* "Fundamental Aspects of Pollution Control and Environmental Science I" pp. 130-3. Elsevier Scientific Publishing Company, Amsterdam.

Rahimi, A., and Bussler, W. (1973). *Z. Planz. Boden. 136,* 25-32.

Reed, H.S. (1947). *Hilgardia 17,* 377-411.

Reith, J.W.S. (1968). *J. Agric. Sci. 70,* 39-45.

Reuter, D.J. (1975). *In* "Trace Elements in Soil-Plant-Animal Systems" (D.J.D. Nicholas, A.R. Egan, eds.) pp. 291-323. Academic Press, New York.

Reuter, D.J., Robson, A.D., Loneragan, J.F., and Tranthim-Fryer, D.J. (1981a). *Aust. J. Agric. Res.* (in press).

Reuter, D.J., Robson, A.D., Loneragan, J.F., and Tranthim-Fryer, D.J. (1981b). *Aust. J. Agric. Res.* (in press).

Reuter, D.J., Robson, A.D., Loneragan, J.F., and Tranthim-Fryer, D.J. (1981c). *Aust. J. Agric. Res.* (in press).

Reuther, W., and Labanauskas, C.K. (1966). *In* "Diagnostic Criteria for Plants and Soils". (H.D. Chapman, ed.) pp. 157-79. Univ. of Calif. Division of Agricultural Science.

Reuther, W., and Smith, P.F. (1953). *Soil Sci. 75,* 219-24.

Reuther, W., and Smith, P.F. (1954). *Proc. Soil Sci. Soc. Fla. 14,* 17-23.

Riceman, D.S. (1948). *Bull. Comm. Sci. Ind. Res. Aust. No. 234.*

Riceman, D.S., and Anderson, A.J. (1943). *J. Agric. South Aust. 47,* 15-23.

Romheld, V., and Marschner, H. (1979). *In* "The Soil-Root Interface" (J.L. Harley and R. Scott Russell, eds.) pp. 405-17.

Rosell, R.A., and Ulrich, A. (1964). *Soil Sci. 97,* 152-67.

Rossiter, R.C. (1951). *Aust. J. Agric. Res. 2,* 14-23.

Schutte, K.H., and Matthews, M. (1968). *Trans. Roy. Soc. S. Afr. 38,* 183-200.

Simmons, W.J., and Loneragan, J.F. (1975). *Analyt. Chem. 47,* 566-8.

Smilde, K.W.', and Henkens, Ch.H. (1967). *Neth. J. Agric. Sci. 15,* 249-58.

Smith, P.F., and Specht, A.W. (1953). *Plant Physiol. 28,* 371-82.

Snowball, K., Robson, A.D., and Loneragan, J.F. (1980). *New Phytol. 85,* 63-72.

Steenbjerg, F. (1943). *Z. Tidsskrift Plantea 47,* 557-99.

Steenbjerg, F. (1951). *Plant Soil 3,* 97-109.

Steenbjerg, F. (1954). *Plant Soil 5,* 226-42.

Steenbjerg, F., and Boken, E. (1950). *Plant Soil 2,* 195-221.

Steenbjerg, F., and Jakobsen, S.T. (1963). *Soil Sci. 95,* 69-88.

Stephens, C.G., and Donald, C.M. (1958). *Adv. Agron. 10,* 167-256.

Struckmeyer, B.E., Peterson, L.A., and Hsi-Mu, T.F. (1969). *Agron. J. 61,* 932-6.

Teakle, L.J.H., and Turton, A.G. (1943). *J. Agric. West Aust. Ser. 2) 20,* 238-59.

Thiel, V.H., and Finck, A. (1973). *Z. Pflanz. Bodenk. 134,* 107-25.

Ulrich, A. (1950). *Soil Sci. 69,* 291-309.

Ulrich, A. (1961). *In* "Plant Analyses and Fertilizer Problems" pp. 190-211. Amer. Inst. Biol. Sci. Washington Pubn 8.

Ulrich, A., and Hills, F.J. (1967). *In* "Soil Testing and Plant Analysis Part II Plant Analysis" pp. 11-24. Soil Sci. Soc. Amer. Inc. Madison.

Van Luit, B., and Henkens, Ch.H. (1967). *Verslagen van Landbouwkundige Onderzoekingen 695.*

Viets, F.G. (1962). *Agric. Fd. Chem. 10,* 174-8.

Viets, F.G. and Lindsay, W.L. (1973). *In* "Soil Testing and Plant Analysis Vol. 1" pp. 153-72. Soil Sci. Soc. Amer. Inc. Madison.

Viro, P.J. (1955). *Soil Sci. 80,* 69-74.

Walker, C.D., and Loneragan, J.F. (1981). *Ann. Bot.* (in press).

Walsh, L.M., Erhardt, W.H., and Seibel, H.D. (1972). *J. Environ. Qual. 1,* 197-200.

Watanabe, R., Chorney, W., Skok, J., and Wender, S. (1964). *Phytochemistry 3,* 391-3.

14

DISTRIBUTION AND CORRECTION OF COPPER DEFICIENCY IN CROPS AND PASTURES

J.W. Gartrell

Plant Nutrition Branch
Western Australian Department of Agriculture
South Perth, Western Australia

I. INTRODUCTION

Fifty years ago Sommer (1931) and Lipman and MacKinney (1931) published the first unequivocal evidence of the essentiality of Cu for plants. Benefits to plants of Cu fungicide sprays (Lutman 1911; Floyd 1913 and 1917) and soil applications (Densch and Hunnius 1924; Hudig et al. 1927; Felix 1927) in the absence of pathogenic disease had led earlier workers to suspect the essential nature of Cu, but the earliest attempts to obtain proof failed due to traces of Cu in the nutrient salts and in the water from metal stills used in the glasshouse experiments (Stout and Arnon, 1939). The development and application of new techniques combined with a questioning attitude towards conventional wisdom are still the prerequisites for notable progress in our knowledge of Cu nutrition.

By 1931, Cu deficiency had been produced in housed animals (Hart et al. 1928; Cunningham 1931; Elvehjem and Hart 1932; Sheldon 1932). In this same year a field response of cattle to Cu was reported by Neal et al. (1931). In the following 20 years Cu deficiency was shown to be the cause of many severe disorders of animals and plants hampering agriculture in localities throughout the world (Gilbert 1952).

TABLE 1.   *Occurrences of Cu Deficiencies on Coarse
Textured Soils Formed on Igneous or Non-
calcareous Sedimentary Parent Materials.*

Country		Parent material	Weathering	Ref
Australia	*(i)*	Granite	Extreme	*(a)*
	(ii)	Dissected lateritic profiles on granite	Deeply in past;) variable recent)	*(a)*
	(iii)	Sedimentary rocks	Extreme to mod)	
	(iv)	Dissected lateritic profiles on sedimentary rocks	Deeply in past;) variable recent)	*(a)*
	(v)	Weathered detritus	Extreme in past)	
U.S.A.	*(i)*		Very high	*(a,b, c,d)*
	(ii)		High	*(e)*
	(iii)		Moderate	*(f)*
U.K.	*(i)*	Granite	High	*(g,h)*
	(ii)	Coarse siliceous sediments	Extreme to mod	*(g,h)*
Sweden	*(i)*		High	*(j)*
France	*(i)*	Granite	High	*(k)*
New Zealand	*(i)*	Pumice	High	*(l,m)*
South Africa	*(i)*		High	*(n)*
Kenya	*(i)*	Pumice and ash		*(o)*
India	*(i)*			*(p,q, r)*
	(ii)		High	*(s)*

References: *(a) Donald and Prescott 1975; (b) Camp and
Fudge 1939; (c) Dickey <u>et al</u> 1948;
(d) Gilbert 1952; (e) Berger and Truog 1948;
(f) Gilbert 1957; (g) Caldwell 1971;
(h) Mitchell 1971; (j) Lundblad 1949;
(k) Coppenet and Calvez 1955;
(l) Cunningham 1956; (m) Wells 1957;
(n) Beyers and Roach 1965; (o) Pinkerton 1967;
(p) Lal <u>et al</u>. 1971; (q) Takhar <u>et al</u>. 1972;
(r) Grewal et al. 1969; (s) Joshi and Joshi
1957.*

This review aims to: (1) identify the causes of Cu deficiencies in field crops and pastures, and to review knowledge relevant to understanding these field situations; (2) outline methods of correcting Cu deficiencies of crops and pastures, and mechanisms governing the effectiveness of corrective measures in the field; (3) identify aspects of the behaviour of Cu in soils and plants relevant to animal nutrition.

In earlier chapters of these proceedings the individual processes governing the behaviour of Cu in soils and plants have been discussed in terms of the particular scientific discipline to which they belong. Knowledge of Cu in field situations is fragmentary: the relative importance in the field, of the many processes shown in laboratory studies to be capable of influencing the behaviour of Cu, is poorly defined. For these reasons much of what follows is, of necessity, unsatisfactorily tenuous, revealing needs for further research.

II. GLOBAL DISTRIBUTION OF Cu DEFICIENCIES IN CROPS AND PASTURES

Examination of Cu deficiency in world agriculture reveals recurring associations with soil conditions, management practices and plant species.

A. *Soils*

1. *Coarse textured soils formed on igneous or non-calcareous sedimentary materials and varying in organic content.* This group of soils includes a large proportion of the Cu deficient areas of western and southern Australia. Cu deficiency on these soils is also common throughout northern and western Europe and has been reported from other countries (Table I). Organic matter contents range from less than one per cent (e.g. Brennan et al. 1980) to over 15 per cent (e.g. Caldwell 1971), tending to be least in regions with arid summers such as southern Australia where the extensive areas of Cu deficient soils contain little organic matter.

Cu deficiency occurs on soils formed on sedimentary or acid igneous rocks but not basic igneous rocks. A high degree of weathering in either past or recent times is invariably associated with those Cu deficient soils developed on acid igneous material. The soils vary in texture from sandy earths to coarse sands and may contain substantial quantities of 'free' Fe and Al oxides.

2.   *Soils formed on calcareous marine and littoral deposits.* Extensive areas of Cu deficient soils all formed from calcareous dune sands or limestones from littoral zones of varying ages occur in the south east of South Australia (Donald and Prescott 1975) and fringe the Western Australian coast. The soil types developed on these calcareous materials include calcareous sands, groundwater rendzinas, terra rossas, solodised solenetzs and deep podsolised siliceous sands (Blackburn 1964; Tiver 1955).

In England Cu deficiency is common on the shallow chalks varying in organic content from high to low and often very friable and loose ('black puffy soils') (Caldwell 1971).

In South Africa, coastal sands have produced Cu deficiency in crops and livestock (Perold 1949).

3.   *Organic soils.* Peats and mucks have commonly produced Cu deficient crops throughout northern and western Europe, the eastern States of the United States and in New Zealand (see Table II).

Acid, neutral and alkaline peats have produced equally severe Cu deficiency (Pizer et al 1966).

4.   *Fine textured soils formed on igneous or non-calcareous sedimentary materials.* Cu deficiencies in plants are rare on fine textured mineral soils developed on igneous material, being confined to soils developed from fine textured acid igneous rocks. Limited areas of Cu deficient fine textured soils developed on fine non-calcareous sediment occur in Western Australia (Gartrell 1979; Teakle et al. 1940). Cu deficient fine textured soils of uncertain origin have been reported in southern Australia (King and Alston 1975) in the United States (Oserhowsky and Thomas 1938; Gilbert 1957) and in India (Dwivedi and Shanker 1977).

TABLE II. Occurrences of Cu Deficiencies on Peats and Mucks

Country	References
Sweden	Lundblad 1939, 1940
Holland	Mulder 1926, 1950; Henkens 1951
Germany	Sjollema 1933
Denmark	Steenbjerg and Boken 1950
Poland	Liwski 1963
Norway	Caldwell 1971
Russia	Caldwell 1971
England and Wales	Pizer *et al.* 1966; Caldwell 1971
Ireland	Walsh and Berry 1958; Fleming and Delaney 1961; Ennis and Brogan 1961
New Zealand	Cunningham 1946; Cunningham *et al.* 1956
United States	Davies 1950; Gilbert 1952 and 1957
South Africa	Anderssen 1932; Isaac 1934
Australia	Gartrell and Glencross 1969

B. Management Practices

1. Supply of Other Nutrients. Applications of N fertiliser have markedly accentuated Cu deficiencies in Europe (Henkens 1951), England (Davies et al. 1971), Ireland (Fleming and Delaney 1961) and Australia (W.J. Toms and J.W. Gartrell, unpublished data). In Western Australia Cu deficiency has appeared on extensive areas of initially infertile soils after soil nitrogen supply has been elevated by leguminous pastures (J.W. Gartrell, unpublished). Increasing use of NPK fertilisers has been implicated in the increasing appearance of Cu deficiency in India (Grewal et al. 1969).

The application of Zn fertiliser has accentuated Cu deficiencies in Australia (Dunne 1956; Gartrell 1969) and Pakistan (Kausar et al. 1976).

2. Liming and Drainage. Increasing soil alkalinity by liming has had small and variable effects on the Cu status of plants (Caldwell 1971).

Nicholson and Firth (1958) reported that lowering the watertable to below 71 cm increased the severity of Cu deficiency on peaty soil in England.

    3. Incorporation of Plant Material into Soil: In the U.K. ploughing in unharvested brassica root crops appears to increase the severity of Cu deficiency in the following crop (Davies et al. 1971).   Severe Cu deficiency has appeared in crops planted in haystack residues in Western Australia.

C. Plant Species

    Cu deficiencies occur much more frequently in some species than in others. Table III lists cultivated species in which Cu deficiency has been recorded, according to their apparent sensitivity to low Cu supply.
    These differences between crop and pasture species can be very important.
    Among cereal species, rye shows exceptional tolerance to low levels of soil Cu supply (Riceman and Donald 1938; Smilde and Henkens 1967) to the extent that healthy rye can be grown where wheats fail completely without the application of Cu. The Cu efficiency of cereal rye appears transferable to *Triticale* (Graham 1978 and this Symposium).
    Among the other small grained cereals the order of sensitivity to Cu deficiency usually observed in the field has been wheat > barley > oats but Nambiar (1976) has recently shown that intraspecific variation can be as large as interspecific variation. In Nambiar's study, Halberd wheat and Clipper barley were less sensitive to Cu deficiency than Avon oats and Gabo, Glaive, Pinnacle, Chatilerma, UP-301 and Duramba wheats. Brown and McDaniel (1978) found marked differences in the Cu efficiency of two varieties of oats *(Avena byzantina)*.
    Cu concentrations in cereals and pasture grasses are insensitive to increasing soil Cu supply compared with legume and other herbaceous species (Riceman et al. 1940; Teakle et al. 1941; Piper 1941; Beck 1962). In Western Australia *Lotus* spp. thrive on soils where *T. subterraneum* completely fails to set seed from Cu deficiency, while in less severely Cu deficient situations *T. subterraneum* has been far less sensitive than wheat. Narrow-leafed lupins *L. augustifolius* have been little affected by Cu deficiency on sites severely Cu deficient for wheat in Western Australia (Gartrell, unpublished).

TABLE III. Relative Responsiveness of Crops to Cu [a]

Responsiveness to Cu		
Low	Medium	High
Beans	Barley	Wheat
Peas	Oats	Rice
Potatoes	Broccoli	Lucerne
Asparagus	Cabbage	Carrots
Cereal rye	Cauliflower	Lettuce
Pasture grasses	Celery	Spinach
Lotus spp.	Clovers	Table beets
Soybeans	Maize	Sudangrass
Lupins	Parsnips	Citrus
Rape	Radishes	Onions
Pines	Sugar beats	
	Turnips	
	Pome and	
	stone fruits	
	Vines	
	Pineapples	

[a] *Sources: Anon. (1966), supplemented from various reports*

In pastures, the variation in Cu concentrations between species can be very high (Piper and Walkley 1943; Beeson et al. 1947; Mitchell 1954; Adams and Elphick 1956; Beck 1962; Fleming 1965; Gladstones et al. 1975).

III. CAUSES OF Cu DEFICIENCIES IN FIELD GROWN PLANTS

Cu deficiency mostly affects young shoot tissues and meristems (Robson and Reuter, Chapter 13). It occurs when the rate of Cu supply to the young tissues is insufficient to maintain the minimum concentrations needed for normal cell division and development in these tissues. Low concentrations in young tissues may result from:
(1) Restricted contact between absorbing plant roots and plant available Cu in the soil due to (a) low levels of plant available Cu in the soil (b) limited root development in soil zones where Cu is present.
(2) Restricted ability of plant roots to absorb Cu.

(3) Restricted movement of Cu within the plant.
(4) Rapid growth rate and proliferation of plant tissue,
i.e. high demand.

These factors are the bases for the recurring
association between Cu deficiencies and soil conditions,
management practices and plant species noted in the previous
section. Because information linking field occurrences of
Cu deficiencies to the causes of Cu deficiency is fragmentary
and incomplete, explanations for the occurrences of Cu
deficiencies are open to question.

A. Restricted Contact Between Roots and Cu in the Soil

1. Low Levels of Non-occluded Cu in the Soil: A large
proportion of the Cu in soils may be occluded within
mineral particles or clay lattice structures (Mitchell 1971),
in Fe-Al nodules (McLaren and Crawford 1974), in organic
matter particles and large organic molecules (Petruzelli and
Guidi 1976). Because this occluded Cu cannot be absorbed by
plant roots, total soil Cu concentration is a poor indicator
of the ability of different soils to supply Cu to plants.

Various procedures have been used in attempts to measure
the pool of non-occluded Cu from which plants draw their
supplies. Generally the correlations between plant uptake
and soil test level have not been high (Robson and Reuter,
this Symposium). Lundblad (1949) and Lucas and Knezek (1972)
have pointed out that concentrations of soil Cu associated
with plant deficiencies are similar for both mineral and
organic soils provided Cu concentration is expressed in terms
of soil volume rather than soil mass. Cu deficiencies have
most commonly been reported on mineral and organic soils
containing less than $0.7\mu g$ Cu ml^{-1} soil determined by acid
ammonium oxalate or EDTA extraction (McKenzie 1966; Davies
et al. 1971; Mitchell 1971; Cox and Kamprath 1972; Brennan
et al. 1980).

The frequent association of Cu deficiency with organic
soils appears due to the low levels of non-occluded Cu
commonly found in organic soils rather than the strength of
bonding between non-occluded Cu and organic radicles.
Indeed Lundblad et al. (1949) found Cu was slightly more
available in peat soils than mineral soils possibly due to
increased Cu concentrations in soil solution through the
formation of soluble organic-Cu complexes (Hodgson et al.
1965). Lindsay (1972) suggests organic-Cu complexes may
explain the insensitivity of Cu uptake by plants to
increasing alkalinity. The Cu in peat accumulations comes

almost entirely in groundwater from the surrounding rock
formations (Hodgson 1963). Peats formed in drainage water
from materials low in diffusable Cu have low total Cu
contents (Hodgson 1963). A large proportion of the Cu in
low-Cu peats is likely to have been incorporated into the
accumulated vegetable matter. The Cu content of peats tend
to decrease with increasing distance from the shoreline
(Knott 1938). Cu deficiency is therefore likely to be most
severe in areas of peat furthermost from the shore.

The Cu contents of soil parent materials are reflected in
the levels of non-occluded Cu in soils that have all formed
under similar conditions. Table IV shows generalised values
for the comparative Cu concentrations in different classes
of rock material.

The common occurrence of Cu deficiencies on calcareous
marine and coastal deposits is largely due to the low Cu
content of these materials. Weathering may increase the Cu
concentration by removing proportionately more Ca and Mg than
Cu from calcareous soils (Caldwell 1971).

Both fine and coarse textured Cu deficient soils may be
produced from these calcareous parent materials.

Soils formed on acid igneous rocks and sandstones
commonly contain low levels of non-occluded Cu, particularly
after prolonged intensive weathering. Intensive weathering
also produces coarse textured soils from the coarse grained
acid igneous rocks (such as granites) and from the sandstones,
explaining the common association between coarse texture and
Cu deficiency.

Shales tend to have a higher Cu content than the coarse
textured materials from which they have separated during
sedimentation (Hodgson 1963). Cu deficiency on fine textured
soils developed on shales is uncommon although it has been
recorded, for example, on Eocene clay in Western Australia
(Gartrell 1979).

The high Cu content of basic igneous rocks is reflected
in the absence of any reports of Cu deficiencies on soils
developed on them. In Western Australia where Cu deficiency
invariably occurs on lateritic soils derived from granites,
it has not been recorded on similar soils derived from basic
igneous rocks.

Excellent examples of how parent materials low in Cu
together with prolonged intensive weathering combine to
produce Cu deficient soils are given by Donald and Prescott
1975 (Australia), Pizer et al. 1966 (England), Lundblad
et al. 1949 (Sweden), Mulder 1950 (Holland), Mitchell 1971
(Scotland) and Wells 1957 (New Zealand),

Whereas prolonged intensive leaching has been an obvious
factor in producing Cu deficient sands, the small quantities

TABLE IV. Typical Cu Concentrations in Rocks and Soils (Ranges in parentheses)[a]

Ug Cu g^{-1}

Igneous rocks		Sedimentary rocks			Soils
Basic	Acidic	Shales	Sandstones	Limestones	
120	10	45	20	4	15
(20-350)	(5-15)	(10-150)	(5-60)	(1-10)	(1-300)

[a]*Sources: Wells 1957; Turton et al. 1962; Mitchell 1971; Krauskopf 1972; Valaderes 1975.*

of Cu leached in, say 50 years are unlikely to be agriculturally important. For example, it takes 3,200 years to reduce the Cu concentration of 20 mm soil of bulk density 1.5 by 1 µg Cu g^{-1} soil with 500 mm of 10-^{8}MCu^{2+} leachate per annum. Where old land surfaces have been exposed to intensive leaching in earlier geological times, Cu deficient soils may occur in arid present-day climates as in much of southern Australia. For this reason the geographical occurrence of Cu deficiencies is not necessarily closely related to present day climatic patterns.

Donald and Prescott (1975) suggest that the absence of extensive areas of Cu deficiency in North America and much of Europe is due to replenishment of Cu lost from old land surfaces by deposition of fresh rock material from the Pleistocene ice sheet. Youthful vulcanism is another means by which the Cu levels of old land surfaces have been replenished over extensive areas of the globe.

The remarkable extent of Cu deficiency in Australia has been summed up by Donald and Prescott as being due to "the low relief and lack of uplift within the continent, the slow and shallow dissection of Tertiary surfaces, the prolonged leaching of ancient profiles and formation of "new soils from old", the loss of finer fractions due to wind and water sorting, and the limited development of new surfaces or new materials through glaciation or vulcanism".

2. Limited Root Development: Due to the immobility of Cu in soils, the position and extent of roots capable of absorbing Cu in relation to the position of non-occluded Cu in soil profiles governs Cu absorption into roots. There is little information on the role of root geometry in the development and occurrence of Cu deficiencies. However, it seems likely that differences in the rate and pattern of root development at least partly explain differences in susceptibility between species (Table II) and between cultivars (Graham, this Symposium) to Cu deficiency.

In addition, restrictions to root development caused by root pathogens, toxicities and excessively wet or dry soil conditions could be expected to aggravate Cu deficiencies. Toms (private communication), on a site which had not previously produced Cu deficiency in wheat crops apparently free of "take-all" (*Gaeumannomyces graminis*), recorded Cu deficiency and response to fertiliser Cu by wheat severely infected by take-all in a subsequent season. Naturally occurring organic compounds can stimulate or inhibit root growth (Mylonas and Mccants 1980), introducing the likelihood of contradictory effects on Cu uptake following

additions of such compounds. Reductions in plant Cu uptake
following additions of large quantities of fresh plant
material (e.g. Table IIX) may be partly due to the
production of root inhibitors.

Al toxicity, by inhibiting root development, also appears
capable of reducing Cu uptake. The appearance of Cu
deficiencies following the use of acid forming N fertilisers
may in part be due to increased Al levels in the soil
solution.

In studies aimed at understanding the effects of
varying treatments on Cu uptake by plants the influence of
treatments on root development and distribution must be
taken into account. In the past the possible influence of
this part of the system has been largely ignored.

B. *Restricted Ability of Plant Roots to Absorb Cu*

High concentrations of Zn and Al in soil solution
depress absorption of Cu into plant roots and may aggravate
Cu deficiency. The application of Zn fertilisers in the
absence of Cu fertilisers has induced Cu deficiencies in
cereals grown on soils low in Cu in Western Australia
(Dunne 1956) and Pakistan (Kausar et al.). Zn^{2+} appears to
compete in physiological absorption of Cu^{2+} (Hawf and Schmid
1967; Loneragan 1975).

Al in solution has been shown capable of both increasing
and decreasing Cu absorption, depending on the Al
concentration (Blevins and Massey 1959; Hiatt et al. 1963).
The importance of the effects of Al have not been
demonstrated in field situations.

It seems possible that other factors as yet not defined
may affect the ability of plant roots to absorb Cu.

C. *Restricted Movement of Cu within the Plant*

A high N level in plants markedly reduces the rate of
retranslocation of Cu from older leaves to the growing
points where it is most needed (Hill et al. 1978),
particularly at low plant Cu concentrations. Increasing
the N supply to crops has markedly accentuated Cu
deficiencies in many field situations (Henkens 1951; Davies
et al. 1971; Grewal et al. 1969; Gartrell, unpublished)
partly by reducing the mobility of Cu in the plant. With
increasing use of N fertilisers, or after the introduction
of a legume pasture phase into the farming system, Cu
deficiencies may appear in areas not previously affected.

This has been the case in large areas of Western Australia where soil Cu supply was adequate for wheat crops with a potential yield of 1 t ha^{-1} grain but not for those with potential yields of 2 to 3 t ha^{-1}.

D. Rapid Growth Rate

Cu concentrations in plant growing points decline with increasing growth rate unless there is a concomitant increase in the rate of absorption into the plant roots through increased root growth or absorption per unit length of root, or increased retranslocation within the plant (Williams 1948; Loneragan 1968). The insensitivity of cereal rye to Cu compared with wheat, oats and barley may, in part, be due to its lower growth rate. Low growth rates and shoot:root ratios may explain the lower susceptibility to Cu deficiency of some plant species.

The introduction of faster growing and higher yielding cereal varieties has been associated with the increased incidence of Cu deficiency in the Indian sub-continent (Grewal et al. 1969).

Chaudry and Loneragan (1970) concluded that the application of NH_4NO_3 reduced the Cu concentration of wheat plants by increasing shoot growth more than Cu uptake. This dilution effect partly explains the observations of accentuated Cu deficiency associated with the use of N fertilisers referred to in previous sections. Similarly the alleviation of a growth restriction due to phosphate deficiency can reduce plant Cu concentration (Robson and Reuter, this Symposium) and the same effect would be expected to follow the correction of other nutrient deficiencies.

E. Conclusions on Causes of Cu Deficiencies in Crops and Pastures

Cu deficiencies occur in agricultural systems where susceptible plant varieties are grown on either mineral or organic soils containing low levels of non-occluded Cu (< about 0.7 µg Cu ml^{-1} soil determined by acid ammonium oxalate extraction). Most commonly it is accentuated by high levels of N supply by mechanisms which have only recently been made clear (Loneragan, this Symposium).

The almost total absence of reports of Cu deficiencies from most of the South American and African continents is noteworthy. It seems highly likely that potentially Cu

deficient soils remain unrecognised within these
continents due to (i) they have not been used for the
cultivation of Cu deficiency susceptible plants, (ii) low
levels of nitrogen supply, (iii) other growth factors are
more limiting, or (iv) Cu deficiency exists but has not
been recognised as such.

There is insufficient information to allow differences
in plants' Cu efficiencies to be attributed specifically
to differences in absorption, translocation or utilisation
efficiencies, or to differences in top growth relative to
root growth and distribution.

Similarly there is little information on how other
factors which may alter root growth and development, rates
of Cu absorption by roots, Cu mobility within the plant and
plant shoot:root ratios, operate to produce Cu deficiencies
in field situations.

The relevance in field situations of much of our
knowledge of the behaviour of Cu is obscure and therefore
of doubtful value towards developing a clear understanding
of the occurrences of Cu deficiencies in agriculture at
this time.

IV. CORRECTION OF Cu DEFICIENCY

A. *Soil Application*

Rates of fertiliser Cu reported to be needed to correct
Cu deficiencies range widely from 0.7 to 23 kg Cu ha-[1]
(Reuter 1975). The reasons for this 300-fold difference in
the rate of fertiliser Cu needed in different field
situations has not been clearly defined although fragmentary
evidence suggests some of the possibilities.

Differences in the effectiveness of Cu fertilisers and
the rate required to correct a deficiency are likely to
depend on:

(1) The placement geometry of the fertiliser Cu in
relation to the distribution of plant roots capable of
absorbing Cu.

(2) The ability of the Cu material to produce Cu^{2+} ions
either directly or after reaction with the soil or
rhizosphere.

(3) Reactions of fertiliser Cu with soil constituents
that affect absorption of Cu by plants.

(4) Differential requirements of plants due to
differences in growth rate, or in the efficiency of

absorption, translocation and utilisation.

Placement geometry: Due to extremely rapid immobilisation of Cu in soils (Jones and Belling 1967; Gilkes et al. 1975) utilisation of fertiliser Cu is highly dependent on plant roots entering the immediate vicinity of fertiliser granules (Reuter 1975; Gilkes and Sadleir 1979; Gilkes, this Symposium).
The effectiveness of fertilisers in supplying elements which are immobile in soils is greatly influenced by the number of granules per unit soil volume, and the position in which they are placed (Sorensen et al. 1979; Gilkes, this Symposium) and topdressing is usually an inefficient method of correcting Cu deficiencies unless the Cu is subsequently incorporated into the plant root zone.
Banding Cu fertiliser with the seed may be more efficient than topdressing without subsequent incorporation (Gilkes, this Symposium) while pre-seeding cultivations of banded Cu fertiliser increase its effectiveness (Table V).

TABLE V. *Effect of Pre-seeding Cultivations on the Effectiveness of 1.4 kg Cu ha^{-1} in Granulated Superphosphate (Cu-OSP) Placed in a Band on the Cu concentration of Barley Leaves Sampled Nine Weeks after Germination.*[a]

| | Cu concentration in barley leaves | |
| | Diameter of Cu-OSP granules | |
Number of cultivations after banding Cu-OSP	< 1 mm μg Cu g^{-1}	> 1 mm <3 mm μg Cu g^{-1}
0 (drilled with seed)	2.0	1.4
1	n.d.	1.7
5	3.7	2.9
9	4.1	3.7

n.d. = not determined

[a]*Source:* Unpublished data of J.W. Gartrell, R.F. Brennan and R.J. Jarvis.

Cu fertiliser may be relatively ineffective in parts of the profile containing few actively absorbing roots, for example in dry surface soils where plants are grown on sub-soil moisture (Nicholson and Firth 1958; Hodgson 1963). The results of Grundon (1980) dramatically demonstrate the unavailability of Cu in dry topsoil.

Applied at the same rate, large granuled Cu fertiliser is much less effective than fine granuled Cu fertiliser probably mainly because doubling granule diameter reduces the number of granules to one-eighth (for spherical particles). Figure 1 shows the results of a field experiment where Cu superphosphate (Cu-OSP) of different granule size was drilled with wheat seed (Gartrell, Brennan and Jarvis, unpublished data). In the year of application five times as much Cu had to be applied in Cu-OSP granules > 3 mm diameter to equal the effectiveness of granules < 1 mm diameter. In this case the rate of Cu needed for maximum wheat grain yield varied from 1.32 kg Cu ha^{-1} for Cu-OSP < 1 mm to 6.6 kg Cu ha^{-1} for Cu-OSP > 3 mm. The difference due to granule size was greatly diminished in the second year cereal crop probably due to the disintegration of the larger granules into smaller particles and redistribution by cultivation (Figure 1b).

Large variations in Cu content between Cu fertiliser granules increase the rate of Cu fertiliser needed to ensure that every plant contacts sufficient Cu-containing granules (Gilkes 1975). Uniform application may more easily be obtained using fluid rather than solid nutrient carriers (Giordano and Mortvedt 1972).

It seems likely that the extremely high rates of soil applied Cu found to be needed to correct Cu deficiencies of cereals in England (Caldwell 1971) are largely due to limited contact between fertiliser-Cu and plant roots with the methods used in, and under the conditions of, the English investigations.

The Form of Cu: Applied as fine powders a wide variety of Cu materials have been found similarly effective in supplying Cu to plants (Gilkes, this Symposium) presumably because they can supply dissociated Cu^{2+} which is the form of Cu absorbed by plant roots (Goodman and Linehan 1979; Graham, this Symposium). With increasing granule size, highly water soluble materials may be more effective than insoluble materials (e.g. Sharpee et al. 1969) for a number of possible reasons.

Glencross (private communication) found CuO mixed dry with superphosphate was only 50 per cent as effective as CuSO$_4$.H$_2$O mixed with superphosphate although they have been

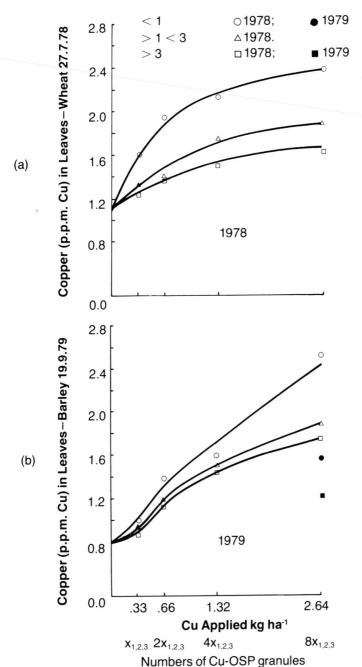

Fig. 1 The effect of granule diameter of Cu-OSP on Cu concentration in the youngest fully expanded leaves of (a) wheat in the year of application (1978) and (b) barley in the year after application and the year of application (1979).

found equally effective when granulated with superphosphate (Gartrell 1978) presumably having formed similar compounds within the superphosphate granules.

Slow release sources of Cu have not appeared to confer any real advantage over $CuSO_4.5H_2O$ in maintaining the Cu supply to successive crops and pastures, contrary to the suggestion of Sharpee et al. (1969). This is not surprising since $CuSO_4.5H_2O$ rapidly reacts with soil to become, in effect, a slow release form of Cu, and further fixation to forms of Cu of lower plant availability has not appeared to exert a dominant influence under field conditions.

Cu chelates and Cu polyflavonoids, by virtue of their higher mobility and solubility in the soil solution, may appear to offer increased Cu efficiency, but usually cost more than inorganic Cu sources. The efficiency of Cu chelates may be partially lost through Fe^{3+} and Ca^{2+} in soil partially replacing the chelated Cu (Lindsay and Norvell 1969; Norvell and Lindsay 1969). Alkalinity does not severely limit the plant availability of inorganic forms of Cu, as it does for Fe, Mn and Zn for which the use of chelates in some situations, may be justifiable.

A similar range of Cu compounds is used in Cu fertiliser throughout the world. The difference in rates of Cu fertiliser found needed to correct deficiencies in different countries are not usually due to differences in the chemical nature of Cu sources.

Reactions of Fertiliser Cu with Soil Constituents:
Reactions of applied Cu with soil may influence Cu availability in two ways:
  (1) By reducing dispersal of Cu^{2+} through the soil root zone (Gilkes, this Symposium).
  (2) After the initial rapid immobilisation of soluble fertiliser Cu^{2+}, by slow conversion to forms unavailable, or of reduced availability to plants.

Surprisingly, there is little data showing direct effects of reactions of fertiliser with soil in reducing Cu uptake by plants.

Differences in stability constants of various soil-Cu complexes have been estimated but their significance in terms of availability to plants have rarely been measured. Because "plants are the final arbiter" (Loneragan 1975) little meaning can be ascribed to these r eported stability constants.

Hoffman (1939) demonstrated Cu-humic acid complexes have lower plant availability than free Cu^{2+} ions, but Cu-humic acids are probably formed almost immediately after the application of fertiliser Cu in moist soil and it seems wrong to classify them as a form in which Cu is slowly "fixed". Indeed there is strong evidence that in many soils organically bound Cu in solution is the main source of Cu absorbed by plants after dissociation of the Cu^{2+} (Hodgson et al. 1966; Martens 1967; Petruzelli and Guidi 1976).

Brennan et al. (1980) in glasshouse experiments found that $CuSO_4 5H_2O$ (mixed with the soil immediately before planting) was less effective in increasing Cu uptake by wheat plants in soils which most rapidly reduced the effectiveness of $CuSO_4 5H_2O$ applied several days earlier and incubated in the presence of moisture. This suggests higher initial rates of Cu fertiliser may be needed to correct Cu deficiencies on soils with a greater capacity for reducing the effectiveness of fertiliser Cu by slow reactions.

Different Requirements of Crop Plants: Although interspecific differences in sensitivity to Cu deficiency have been noted (Graham, this Symposium; Section IIC, this review), the differential requirements of different crop plants under field conditions are poorly defined and poorly understood. Recommendations for crops in Michigan (Anon 1966) vary from 2.2 to 13 kg Cu ha^{-1} per year until total Cu applied reaches 22 to 45 kg Cu ha^{-1} depending on the crop species grown.

With increasing rates of N application higher rates of Cu fertiliser are needed to avoid Cu deficiency in wheat (Gartrell, unpublished data) but this has not been specifically taken into account in the formulation of Western Australia recommendations. Relatively high rates of N are usually applied to crop species grown in situations where high Cu rates are recommended. Generally, regional recommendations for Cu fertiliser rates for different crop species appear to be based on minimum rates which have corrected Cu deficiencies in particular species, by a narrow range of materials and application methods, over a wide range of conditions encountered in that region. Often recommendations are adopted from elsewhere and if they "seem to work" are prescribed for years, without question or investigation. Thus, from the literature available at this time, it is not possible to accurately assess the extent of the influence of the individual factors governing the effectiveness of Cu fertilisers in field situations.

B. *Foliar Sprays*

Foliar sprays are highly effective in correcting Cu deficiencies in crops and pastures. In most countries soil applications are preferred for general treatment while the use of foliar sprays is mainly confined to emergency treatment where the deficiency is not recognised until after planting.

Soil applications of sufficiently high Cu rates have the advantage of long-lasting residual effects eliminating one managerial and operational task for several years. Additionally, soil applications can eliminate the risk of production losses which may occur if spray applications are delayed until after the appearance of deficiency symptoms (Nambiar 1976). In the absence of deficiency symptoms spraying at late tillering has been found more effective than spraying earlier (Davies et al. 1971; Graham 1976).

In England foliar spraying is the most common method of correcting Cu deficiency in crops, partly because it can be included in other sprays applied as normal practice, and partly due to doubts about the residual effectiveness of soil applications.

Another reason in favour of foliar sprays in England is the much higher capital cost of the 33 to 66 kg ha-[1] $CuSO_4.5H_2O$ needed for topdressed applications compared with 0.85 kg ha-[1] $CuSo_4.5H_2O$ sprayed. It seems likely that lower rates of soil applied Cu could be adequate were it incorporated into the soil rather than broadcast as seems to be the commonly reported method.

High nitrogen fertiliser applications and highly productive crops doubtless also contribute to the need for high Cu fertiliser rates in England. Nevertheless, Davies et al. (1971) were surprised to find that in two out of three trials less than 1 kg Cu ha-[1] sprayed a year previously produced 85 per cent of the increase in barley yields obtained from a current spray. This probably reflects the effectiveness of the residues from 1 kg Cu ha-[1] mixed through the soil by cultivations before the second crop.

$CuCl_2.2CuO.4H_2O$ and Cu_2O at 2.5 kg ha-[1] are preferred to $CuSO_4.5H_2O$ which may scorch leaves, corrode equipment and easily wash off foliage after rain (Caldwell 1971). In Western Australia $CuSO_4.5H_2O$ sprayed at 1 to 2 kg ha-[1] has been highly effective with negligible spray scorch. Spray equipment designed to apply corrosive materials is now available.

Other Cu materials effective in sprays include $CuNH_4PO_4.H_2O$ (Bridger et al. 1962), Cu chelates and polyflavonoidlignosulphonates (Murphy and Walsh 1972). Cu fungicide sprays have fortuitously corrected Cu deficiencies through foliar absorption or soil accumulation.

C. Seed Treatment

Seed treatments, dusts or Cu solutions, tend to be less effective than soil or foliar applications, due mainly to the problem of separation in the case of dusts, and the sensitivity of germinating seeds to Cu at rates high enough to prevent deficiency appearing in the plant at later stages of growth (e.g. Beyers and Roach 1965). In addition, as with foliar sprays, they have the disadvantage of requiring re-application each year, for several years. Russian work cited by Murphy and Walsh (1972) indicates seed treatments can be at least partially effective in correcting Cu deficiencies.

V. RESIDUAL EFFECTIVENESS OF Cu ADDITIVES FOR PLANTS

Long-lasting residual effects of Cu applications are recognised (Teakle 1942; Steenbjerg and Boken 1950; Gilbert 1952; Toms 1958; Riceman 1961; Berger 1965; Pizer et al. 1966; Reith 1968; Davies et al. 1971).

Davies and associates (1971) reported an apparent increase in the second year in the effectiveness of 33 kg ha^{-1} CuSO$_4$.5H$_2$O topdressed. Gartrell (1980) found Cu applied four years earlier was two to three times more effective in increasing the Cu content of wheat plants than current applications banded with the seed. Later results from this continuing trial show Cu applied 12 years earlier is still two to three times more effective in increasing Cu uptake by cereals than current applications banded with the seed (author's unpublished data). In the other trial for which plant uptake data were available, Cu applied 12 years earlier was more effective than that banded with the seed (Gartrell 1980).

Long-lasting residual effects of fertiliser Cu for pastures and grazing animals have also been reported (Underwood et al. 1943; Reuter et al. 1977; Gartrell 1979; Hannam 1979).

On Western Australian farms we have no evidence of declining residual effectiveness of single applications of Cu fertiliser at 1.5 to 3 kg Cu ha^{-1} applied to initially deficient soils up to 35 years ago. Declines may yet appear; we hope to detect them using the tissue technique described by Robson et al. (1977) and Gartrell et al. (1979).

Table VI shows an example of the residual effects of Cu on the Cu status of pasture and sheep, 13 and 17 years respectively after Cu application.

The bases for the remarkable residual effectiveness of Cu in soils appear to be:

(1) Amounts removed in farm products and leaching are small compared with amounts applied.

(2) The availability of fertiliser Cu to plants in the year of application is limited by the commonly used forms and methods of application. Subsequent activities may reduce these restrictions.

(3) Slow reactions reducing the plant availability of fertiliser Cu have not been found to exert a dominant influence on the supply of Cu to plants in the few field situations for which data is available.

A. *Removal in Plants, Livestock, Leaching, Erosion*

Farm products remove little Cu compared with amounts applied to correct deficiencies, and even less compared with the amounts in the profiles of soils (see Table VII).

From the table it can be seen that the small amount of Cu in impurities in the phosphatic fertiliser applied annually can exceed the amount removed in farm products under Australian conditions.

The excess of Cu applied compared with amounts removed in farm products and leaching allows for possible residual effects from both soil and spray applications of Cu. The figures in Table VII also show that "depletion of soil reserves" is an unlikely explanation for the appearance of Cu deficiency over a span of only a few years.

There may be situations in which significant quantities of Cu are removed by the downward movement of particulate matter bearing absorbed Cu, but such situations have not as yet been identified. The erosion of topsoil containing the applied Cu has occurred in localised patches in Western Australia necessitating re-applications of Cu, but such erosion is not a normal occurrence.

TABLE VI. *The Influence of Cu Fertiliser Applied in 1963 on the Cu Concentrations in Pasture in 1976, and in the Livers of Sheep in 1980 After Four Years Continuously on the Experimental Plots*[a]

$CuSO_4 \cdot 5H_2O$ Applied 1963 $Kg\ ha^{-1}$	Pasture Cu conc. October 2, 1976 $\mu g\ Cu\ g^{-1}$	Sheep liver Cu Conc. December 1, 1980	
		Mean $\mu g\ g^{-1}$	Range $\mu g\ g^{-1}$
0	3.0	15	4–35
2.75	4.3	82	9–135
8.25	6.3	180	63–310

[a] *Source: Unpublished data of J.W. Gartrell, B.R. Beetson and H.G. Masters*

TABLE VII.

Material	Cu concentration in material $\mu g\ g^{-1}$	Cu contained in material $g\ ha^{-1}$
		(b)
Soil profiles, depth 1.5 m		
Bulk density = 1.5		
"Deficient" soil	0.2	4,400
"Low normal"	1.0	22,000
"Normal"	2.0	44,000
Leachate 500 mm ha^{-1}	$10^{-8} M Cu^{2+}$	1.3

a (a) Total Cu

  (b) Acid ammonium oxalate extractable Cu

+   Unpublished data of D.J. Reuter, R.J. Hannam, G.J. Judson and M.E. Dodson quoted by Reuter (1975). Other figures calculated by J.W. Gartrell on basis of levels commonly found in Western Australian conditions.

*   See also Williams (1977).

TABLE VII. *Amounts of Cu in Farm Produce, Cu Treatments and Soil Profiles*[a]

Material	Cu concentration in material $\mu g \cdot g^{-1}$	Cu contained in material $g \cdot ha^{-1}$
Farm Produce	(a)	
1. Lucerne hay, 10,000 kg ha^{-1}	8	80
2. Cereal grain, 2,000 kg ha^{-1}	4	8
3. Wool from non-deficient wethers stocked at 12 ha^{-1}	n.a.	0.3^+
Cu Treatments		
1. Superphosphate 100 kg ha^{-1} (West Australian)*	50	5
2. $CuSO_4 \cdot 5H_2O$		
0.85 kg ha^{-1} (e.g. spray)	250,000	210
6 kg ha^{-1} (e.g. fertiliser)	250,000	1,500
60 kg ha^{-1} (high rate fertiliser)	250,000	15,000

B. Management Factors Influencing the Residual Effectiveness of Cu Fertilisers

Increased effectiveness of Cu fertilisers in subsequent years would be expected to result from the mixing effects of cultivations as discussed in earlier sections. Additionally, Cu retained in the skeletal Cu-OSP grain described by Gilkes (1975) may be protected from "fixation" reactions and subsequently released by chemical and mechanical decomposition of the grains. In the experiments reported by Gartrell (1980) this was not a factor because dry physical mixtures of $CuSO_4.5H_2O$ and OSP were used in contrast to the wet incorporated Cu-OSP material in the experiments producing the data shown in Figure 1(a) and (b).
As noted earlier, increasing the N supply may increase the need for Cu but this is due to increased plant need rather than decreased soil supply.
Caution is needed in interpreting sudden appearances of Cu deficiencies in agricultural systems.

C. Reactions of Fertiliser Cu with Soil Constitutents

To date there is no clear evidence of chemical reactions between fertiliser Cu and soil situations. The studies of Brennan et al. (1980) suggest that such reactions may occur and that the incubation technique used in these studies provides a means of identifying those soils in which residual effectiveness is likely to decline most rapidly. By implication, occlusion of Cu in precipitated oxides has been referred to as a means by which applied Cu may be rendered unavailable to plants (e.g. Hodgson 1963; Le Riche and Weir 1963; McLaren and Crawford 1973) but the rate at which this may occur does not appear to have been measured.
Additions of large quantities of fresh plant material can markedly reduce the effectiveness of fertiliser Cu and the reduction is relatively greater where the Cu has been incubated in moist soil for several days before planting (e.g. R.F. Brennan, J.W. Gartrell and A.D. Robson, unpublished data, Table IIX).
The aggravation of Cu deficiency in cereals after ploughing in root crops (Davies et al. 1971) may be a field example of this effect. The mechanism by which fresh plant material depresses Cu availability has not been defined. Chemical reactions of Cu with organic compounds, competition for available Cu by stimulated microbial populations, and inhibition of root development and ability to absorb Cu may

TABLE IIX. *Effects of Adding Finely Ground Wheat Straw on the Availability and Relative Effectiveness of Incubated and Freshly Applied Cu for Wheat on Lancelin Sand*[a]

Wheat straw added g 100g-[1] soil	Cu uptake into wheat tops		Relative effectiveness of incubated to freshly applied Cu
	ug Cu mg Cu applied-[1]		
	Fresh Cu	Incubated Cu	
0	5.8	4.8	0.83
2.5	5.4	3.9	0.73
5.0	2.9	1.9	0.66
10.0	0.71	0.36	0.51

[a]*Source: Unpublished data of R.F. Brennan; J.W. Gartrell and A.D. Robson.*

all be involved.

VI. THE BEHAVIOUR OF Cu IN SOILS AND PLANTS IN RELATION TO SHEEP AND CATTLE NUTRITION

The Cu content of herbage is only one of several factors governing the Cu status of sheep and cattle. Cu absorption and retention by sheep and cattle are strongly modified by dietary components other than Cu, genetic and behavioural differences between animals, and gastro-intestinal conditions (Underwood 1977; Suttle and McLaughlan 1976; Wiener et al. 1978; Gartrell 1979). Thus at any given level of Cu in the diet, Cu responsive conditions may appear depending on the influence of the modifiers.

Aspects of the behaviour of Cu in soils and plants most relevant to animal nutrition are:

(1) Low levels of Cu occur in plants grown on soils with a low level of plant available Cu due to various combinations of low levels of Cu in soil parent materials prolonged intensive leaching and low soil bulk densities.

(2) Low levels of Cu are likely to occur in rapidly growing plants if the rate of top growth outstrips the rate at which the roots absorb Cu. Varieties capable of

faster growth, or a change in conditions allowing
faster growth, or the geometrically increasing nature of
plant growth rate under certain conditions, may produce
low Cu concentrations in plants. However, depending on
the relative levels of supply, the levels of the
dietary modifiers may also be lower in faster growing
plants.
(3) Some plant species tend to have lower concentrations
of Cu than others (Graham, this Symposium), but the
species with lowest Cu concentrations do not necessarily
supply least Cu to animals. Again, depending on the
relative levels of supply, and the undefined
physiological characteristics of the vast range of
pasture species, the levels of the modifiers are also
likely to be least in the species lowest in Cu. For
example, Stoszek et al. (1979) showed beef cattle
obtained more Cu from quackgrass with 4.6 ug g-^{1} Cu than
from tall fescue containing 6.6 ug g-^{1} Cu. The
quackgrass had lower Mo and S concentrations than the
tall fescue.
(4) Cu fertilisers, thoroughly mixed through the plant
root zone, are highly effective in increasing the amount
and concentration of Cu in plant tops, but may be
ineffective if applied in such a way that contact between
fertiliser Cu and plant roots is limited.
At increasing concentrations of Cu above deficiency
levels in plant tops, the increase in tops Cu
concentration per increment of applied fertiliser Cu
falls markedly, due apparently to saturation of the
pathway by which Cu is absorbed and transported from the
roots to the tops. Figure 1a demonstrates this effect.
The relationship varies among different plant species
but, due to this physiological block, increasing rates
of Cu fertiliser may have little effect on the Cu
concentrations of plants adequately supplied with Cu (e.g.
Brennan et al. 1980, for wheat on Cu adequate Merredin
soil compared with the deficient soils). For this reason,
soil application of Cu may be ineffective in increasing
the Cu supply to Cu responsive animals on normal to high
Cu pastures where the problem is due to abnormal levels
of modifiers (e.g. Hartmans 1969).
(5) Where Cu fertilisers are ineffective in increasing
Cu levels in plants, foliar sprays of 1 kg $CuSO_4.5H_2O$ or
less, or other forms of Cu supplementation, can be used.
There is a risk of Cu toxicity with Cu sprays, which can
lead to very high levels of intake soon after spraying
(e.g. 1 kg ha-^{1} Cu $SO_4.5H_2O$ on 5,000 kg ha-^{1} dry matter

gives a concentration of 60 µg Cu g^{-1} dry matter).There
may be an even greater risk of Cu toxicity where animals
graze pastures recently topdressed with Cu fertilisers
because of the higher rates usually applied.
(6) The sudden appearance of a Cu responsive condition
in animals in a particular area previously free of any
sign of Cu deficiency is not likely to be due to a
sudden "depletion of soil reserves" or a decline in the
residual effectiveness of fertiliser Cu applications.

Among the dietary factors found to modify the Cu status of
sheep and cattle, intakes of Mo and S have had the most
profound effects on inducing Cu responsive conditions.
Suttle and McLaughlan (1976) found :

$$\log y = -1.153 - 0.0019 \, Mo - 0.0755 \, S - 0.0131 \, Mo \times S$$

where, y is Cu availability to ruminants,
Mo is concentration of Mo, mg kg^{-1} feed,
and S is concentration of S, g kg^{-1} feed.

The calculated availabilities of Cu from various Mo and S
concentrations in feed, as derived from this equation are
shown in Table IX (after Purser 1979).

The factors controlling the levels of Mo and S in soils
and plants will not be reviewed here. Levels of Cu, Mo and
S in plants are differentially governed by numerous diverse,
and poorly understood, influences. As a result a wide
range of combinations of Cu, Mo and S in herbage do occur,
and generalisations become of doubtful value. Nevertheless,
a few generalisations tend to hold:
(1) Relative to Cu, Mo levels are usually higher in
granites, limestones and organic shales. Mo leaches
much more readily than Cu particularly in alkaline
conditions. As a result, groundwaters containing little
Cu often contain large quantities of Mo. Organic
deposits formed in these situations contain large
quantities of plant available Mo, but little Cu, giving
rise to molybdenosis, ("peat scours") in sheep and
cattle grazing the high Mo pastures.
(2) The availability of Mo to plants increases greatly
with increasing alkalinity whereas the availability of
Cu is only slightly, and variably affected (Piper and
Beckwith, 1949; Caldwell 1971).
Additionally, the physiological barrier to Mo uptake by
plants is very much less than that to Cu. Whereas Cu
deficiency in plants is associated with levels of 1 to 3
µg g^{-1} Cu, Mo deficiency is associated with levels of

TABLE IX. The Effect of the Concentration of S and Mo in Feed Upon the Availability of Cu (%)[a]

Mo concentration in pasture	Availability of Cu in diet		
	S concentration in feed		
$\mu g \ g^{-1}$	$g \ kg^{-1}$		
	1.0	2.0	3.0
	%	%	%
0.2	5.9	4.9	4.1
0.6	5.8	4.8	3.9
1.0	5.7	4.7	3.8
3.0	5.3	4.1	3.1
10.0	4.2	2.6	1.6

[a]*Source: Purser (1979) calculated from Suttle and McLaughlan (1976).*

0.03 to 0.15 $\mu g \, g^{-1}$, while Cu levels in plants rarely exceed 20 $\mu g \ g^{-1}$ but Mo levels may reach 300 $\mu g \ g^{-1}$ or more (Kubota and Allaway 1972).

Liming of acid soils, and the application of Mo to alkaline soils can greatly alter the Mo:Cu ratio in herbage, inducing molybdenosis in ruminants.

(3) S in soils and plants behaves quite differently to either Cu or Mo depending on the conditions. The sulphur incidentally applied as a constituent of common phosphatic, nitrogenous and potassic fertilisers markedly influences the S supply to pastures. S leaches very readily under high rainfall conditions particularly on coarse sands low in organic matter and colloids. In these conditions S deficiency can occur in the middle and end of the growing season despite S fertiliser having been applied earlier. The S intake of animals grazing pastures can rise to very high levels immediately after topdressing with superphosphate due to the ingestion of fertiliser particles. This may result in a sudden, but temporary, fall in the Cu status of ruminants in autumn, the time superphosphate is normally applied to pasture in warm winter rainfall areas.

An extremely wide range of combinations of Cu, Mo and S

levels in herbage are possible in nature. Man's practices greatly increase this range. In addition we confine our animals for long periods to relatively small areas which may be grossly abnormal, in contrast to the free ranging conditions in which sheep and cattle evolved. For these reasons Cu responsive conditions in animals may appear when new techniques are introduced into farming systems.

VII. CONCLUSION

The absence of relevant data and the need for a great deal of further work have been noted throughout this Chapter. Since Hodgson's (1963) review of the chemistry of micronutrients in soils, some progress has been made towards a better understanding of the agronomically important aspects of the behaviour of Cu under field conditions but much remains to be done.

There is a need for the soil chemist and plant nutritionist to prove the relative importance of existing concepts in field situations. A model of Cu in soil-plant systems is required to provide an objective basis for deciding priorities in Cu research. Until such a model is developed our knowledge of the behaviour of Cu in soils and plants will remain fragmentary and open to misinterpretation.

REFERENCES

Adams, A.F.R. and Elphick, B.L. (1956). *N.Z. J. Sci. Technol. 38:345.*

Allison, R.V., Bryan, O.C. and Hunter, J.H. (1927). *Fla. Sta. Bull. No. 190.*

Anderssen, F.G. (1932). *J. Pomol. Hort. Sci. 10:130.*

Andersson, A. (1977). *Swed. J. Agric. Res. 7:7.*

Anon. (1966). *Michigan Agr. Exp. Sta. Ext. Bull. E-550.*

Beck, A.B. (1962). *Aust. J. Exp. Agric. Anim. Husb. 2:40.*

Beeson, K.C., Gray, L. and Adams, M.B. (1947). *J. Amer. Soc. Agron. 39:356.*

Bennetts, H.W. and Chapman, F.E. (1937). *Aust. vet. J. 13:138.*

Berger, K.C. (1965). "Introductory Soils". The MacMillan Co., New York.

Berger, K.C. and Truog, E. (1948). *Proc. Soils. Sci. Soc. Am. 13:372.*

Beyers, C.P. de L., and Roach, W.A. (1965). *Landboukollege Stellenbosch - Elsenburg. Tech. Bull. No. 46.*

Blackburn, G. (1964). CSIRO Soils and Land Use Series *No. 45.*

Blevins, R.L. and Massey, H.F. (1959). *Proc. Soil Sci. Soc. Am. 23.*

Brar, M.S. and Sekhon, G.S. (1978). *J. Indian Soc. Soil Sci. 26:84.*

Brennan, R.F., Gartrell, J.W. and Robson, A.D. (1980). *Aust. J. Soil Res. 18:449.*

Bridger, G.L. Salutsky, M.L. and Starotska, R.W. (1962). *J. Agric. Food. Chem. 10:181.*

Brown, J.C. and Harmer, P.M. (1950). *Proc. Soil Sci. Soc. Am. 15:284.*

Caldwell, T.H. (1971). *N.A.A.S. Tech. Bull. No. 21, p. 62.*

Camp, A.F. and Fudge, B.R. (1939). *Fla. Sta. Bull. No. 335.*

Chaudry, F.M. and Loneragan, J.F. (1970). *Aust. J. Agric. Res. 21:865.*

Cipola, G. (1937). *Univ. nac. litoral (Corrientes Argentina), Inst. Exp. Agropecurias Pub. No. 4.*

Coppenett, M. and Calvez, J. (1955). *C.R. Acad. Sci. Paris 241.*

Coppenett, M. and Jolivet, E. (1952). *C.R. Acad. Agric. Fr. 38:691.*

Cox, F.R. (1973). *N. Carol. Agr. Exp. Sta. Tech. Bull. No. 219, p. 182.*

Cox, F.R. and Kamprath, E.J. (1972). In "Micronutrients in Agriculture" (J.J. Mortvedt, P.M. Giordano and W.L. Lindsay, eds), p. 289. Soil Science Society of America Inc., Wisconsin.

Cunningham, I.J. (1931). *Biochem. J. 25:1267.*

Cunningham, I.J. (1946). *N.Z. Dept. of Agric. Bull. No. 238.*

Cunningham, I.J., Hogan, K.J. and Green, J.N. (1956). *N.Z. J. Sci. Tech. A38:225.*

Davies, D.B., Hooper, L.J., Charleswroth, R.R., Little, R.C., Evans, C. and Wilkinson, B. (1971). *N.A.A.S. Tech. Bull. No. 21, p. 88.*

Davis, J.F. (1950). *Plant Food J. 4:5.*

Dickey, R.D., Drosdoff, M. and Hamilton, J. (1948). *Fla. Sta. Bull. No. 447.*

Densch, A. and Hunnius, W. (1924). *Z. PflErnahr. Dung. A3:369.*

Donald, C.M. and Prescott, J.A. (1975). In "Trace Elements in Soil-Plant-Animal Systems" (D.J.D. Nicholas and A.R. Egan, eds) p. 7. Academic Press, New York.

Du Toit, J.L. (1956). *S. afr. Sug. J. 40:797.*

Dunne, T.C. (1956) *Proc. Aust. Conf. Pl. Nutr., Melbourne 1;278* (CSIRO Aust, Melbourne).

Dwivedi, K.N. and Shanker, H. (1975). *Indian J. Agron. 20:131.*

Elvehjem, C.A. and Hart, E.B. (1932). *J. Biol. Chem. 95:363.*

Ennis, M.T. (1962). *Ir. J. agric. Res. 1:39.*

Ennis, M.T. and Brogan, J.C. (1961). *Ir. J. agric. Res. 1:35.*

Felix, E.L. (1927). *Phytopath. 17:49.*

Fleming, G.A. (1965). *Outlook on agriculture 4:270.*

Fleming, G.A. and Delaney, J. (1961). *Ir. J. agric. Res. 1:81.*

Floyd, B.F. (1913). *Fla. Sta. Rep. 1912/13,* p. 27.

Floyd, B.F. (1917). *Fla. Sta. Bull. No. 140.*

Gartrell, J.W. (1969). *West. Aust. Dept. Agric. Tech. Bull. No. 3.*

Gartrell, J.W. (1978). *West. Aust. Dept. Agric. Div. of Pl. Res. Rep. 1977/78,* p. 14.

Gartrell, J.W. (1979). *Proc. Seminar Aust. Soc. Anim. Prod.* (W.A.), Perth (D.B. Purser, ed.) p. 28.

Gartrell, J.W. (1980). *Aust. J. exp. Agric. Anim. Husb. 20:370.*

Gartrell, J.W., Brennan, R.F. and Robson, A.D. (1979). *J. Agric. West. Aust. (4th Ser.) 20:18.*

Gartrell, J.W., Robson, A.D. and Loneragan, J.F. (1979). *J. Agric. West. Aust. (4th Ser.) 20:86.*

Geering, H.R., Hodgson, J.F. and Sdano, C. (1969). *Proc. Soil Sci. Soc. Am. 33:81.*

Gilkes, R.J. (1975). *Aust. J. Soil Res. 13:203.*

Gilkes, R.J. and Sadleir, S.B. (1979). *Aust. J. Soil Res. 17:121.*

Gilkes, R.J. Young, R.C. and Quirk, J.P. (1975). *Aust. J. Soil Res. 13:33.*

Giordano, P.M. and Mortvedt, J.J. (1972). In "Micronutrients in Agriculture" (J.J. Mortvedt, P.M. Giordano and W.L. Lindsay, eds.), p. 505. Soil Science Society of America Inc., Wisconsin.

Gladstones, J.S., Loneragan, J.F. and Simmons, W.J. (1975). *Aust. J. Agric. Res. 26:113.*

Goodman, B.A. and Linehan, D.J. (1979). In "The Soil-Root Interface" (J.L. Harley and R. Scott Russell, eds.), p. 67. Academic Press, London.

Graham, R.D. (1978). Nature *271:542.*

Graham, R.D. (1976). *J. Exp. Bot. 27:717.*

Grewal, J.S., Bhumbla, D.R. and Randhawa (1969). *J. Indian Soc. Soil Sci. 17:27.*

Grundon, N.J. (1980). *Aust. J. Exp. Agric. Anim. Husb. 20:717.*

Hannam, R.J. (1979). *S. Aust. Dept. Agric. Soil Consv. Br. Rep. 511/79.*

Hart, E.B., Steenbock, H., Waddell, J. and Elvehjem, C.A. (1928). *J. Biol. Chem. 77:777.*

Hartmans, J. (1969). AGRI Digest, *18:42.*

Harris, H.C. (1947). *Proc. Soil Soc. Am. 12:278.*

Hawf, L.R. and Schmid, W.E. (1967). *Pl. Soil, 27:249.*

Henkens, C.H. (1957). Landbouwvoorlichting *14:629.*

Hem, J.D. (1960). *U.S. Geol. Survey Water Supply Paper No. 1459-B,* p. 33.

Heydeman, A. (1959). *Geochim. Cosmochim. Acta 15:305.*

Hill, J., Robson, A.D. and Loneragan, J.F. (1978). *Aust. J. agric. Res. 29:925.*

Hodgson, J.F. (1963). *Adv. Agron. 15:119.*

Hodgson, J.F., Geering, H.R. and Norvell, W.A. (1965). *Proc. Soil Sci. Soc. Am. 29:665.*

Hodgson, J.F., Lindsay, W.L. and Trierweiler, J.F. (1966). *Proc. Soil Sci. Soc. Am. 30:723.*

Hoffman, W. (1939). *Bodenk. Pflauzenern 13:139.*

Hudig, J., Meyer, C. and Goodyk, R.J. (1926). *Z. PflErnahr. Dung. 8A:14.*

Isaac, W.E. (1934). *Trans. roy. Soc. S. Africa 22:187.*

Jones, G.B. and Belling, G.B. (1967). *Aust. J. agric. Res. 18:733.*

Joshi, N.V. and Joshi, S.G. (1957). *J. Indian Soc. Soil Sci. 5:21.*

Kausar, M.A. Chaudhry, F.M., Rashid, A., Latif, A. and Alam, S.M. (1976). *Pl. Soil 45:397.*

King, P.M. and Alston, A.M. (1975). In "Trace Elements in Soil-Plant-Animal Systems" (D.J.D. Nicholas and A.R. Egan eds.), p. 339. Academic Press, New York.

Knott, J.E. (1938). *Ohio Veg. Growers' Assoc. Proc. 23:10.*

Krauskopf, K.B. (1972). In "Micronutrients in Agriculture" (J.J. Mortvedt, P.M. Giordano and W.L. Lindsay, eds.), p. 7. Soil Science Society of America Inc., Wisconsin.

Kubota, J. and Alloway, W.H. (1972). In "Micronutrients in Agriculture" (J.J. Mortvedt, P.M. Giordano and W.L. Lindsay, eds.), p. 525. Soil Science Society of America Inc., Wisconsin.

Lal, C., Grewal, J.S. and Randhawa, N.S. (1971). *J. Res. Ludhiana 8:52.*

Le Riche, H.H. and Weir, A.H. (1963). *J. Soil Sci. 14:225.*

Lindsay, W.L. and Norvell, W.A. (1969). *Proc. Soil. Sci. Soc. Am. 33:62.*

Lipman, C.B. and Mackinney, J. (1931). *Plant Physiol. 6:593.*

Liwski, S. (1963). *Polish Agr. Ann. 87, Ser. A., No. 3.*

Lucas, R.E. and Knezek, B.D. (1972). In "Micronutrients in Agriculture" (J.J. Mortvedt, P.M. Giordano and W.L. Lindsay, eds.), p. 265. Soil Science Society of America, Inc., Wisconsin.

Lundblad, K. (1939). *Svenska Vall-o MossForen. Medd.* 2:71.

Lundblad, K., Svanberg, O. and Ekman, P. (1949). *Pl. Soil* 1:277.

Lutman, B.F. (1916). *Verm. agri. Exp. Sta. Bull. No. 196.*

Loneragan, J.F. (1968). *Trans. 9th Int. Cong. Soil Sci.,* Adelaide, *II:173.*

Loneragan, J.F. (1975). In "Trace Elements in Soil-Plant-Animal Systems" (D.J.D. Nicholas and A.R. Egan, eds.), p. 109. Academic Press, New York.

Loneragan, J.F., Snowball, K. and Robson, A.D. (1980). *Ann. Bot.* 45:621.

McLaren, R.G. and Crawford, D.V. (1973A). *J. Soil Sci.* 24:172.

McLaren, R.G. and Crawford, D.V. (1973B). *J. Soil Sci.* 24:443.

Madsen, L.L. (1942). *USDA Yearbook Agri.* 323.

Marston, I.W., Moore, H.O. and Bull, L.B. (1938). *Bull. Coun. scient. ind. Res.,* Melbourne 113:1.

Martens, D.S. (1968). *Soil Sci.* 106:23.

Mitchell, R.L. (1954). In "Chemistry of the Soil" (F.E. Bear ed.), ACS Monograph *No. 126,* p. 253. Reinhold, New York.

Mitchell, R.L. (1971). *N.A.A.S. Tech. Bull. No. 21,* p. 8.

Mitchell, R.L., Reith, J.W.S. and Johnston, I.M. (1957). Analyse des Plantes et Problemes des Engrais Mineraux IRHO, Paris, p. 249.

Mulder, E.G. (1950). Lotsya *3:41.*

Mulder, E.G. (1954). *Pontif. Acad. Sci. Scripta Varia 14:215.*

Murphy, L.S. and Walsh, L.M. (1972). In "Micronutrients in Agriculture" (J.J. Mortvedt, P.M. Giordano and W.L. Lindsay, eds.), p. 347. Soil Science Society of America Inc., Wisconsin.

Mylonas, V.A. and Mccants, C.B. (1980). *Pl. Soil.* 54:485.

Nambiar, E.K.S. (1976). *Aust. J. agric. Res.* 27:453.

Neal, W.M., Becker, R.B. and Shealy, A.L. (1931). Science 74:418.

Nicholson, H.H. and Firth, D.H. (1958). *J. agric. Sci. Camb.* 50:243.

Norvell, W.A. and Lindsay, W.L. (1969). *Proc. Sci. Soc. Am.* 34:360.

Oserkowsky, J. and Thomas, H.E. (1938). *Plant Physiol.* 13:451.

Perold, I.S. (1949). *Specialist Conf. in Agric., Aust.*

Petruzelli, G. and Guidi, G. (1976). *Z. PflErnahr. Bodenkd.* 6:679.

Petruzelli, G., Guidi, G. and Sesqui, P. (1972). *Proc. Atti IX Simposis. Intrnazionale di Agrochimica, Punta Ala* (Italy), p. 329.

Pinkerton, A. (1967). *J. Soil Sci. 18:18.*

Piper, C.S. (1942). *J. Agric. Sci. 32:143.*

Piper, C.S. and Walkley, A. (1943). *J. Coun. sci. ind. Res. Aust. 16:217.*

Pizer, N.H., Caldwell, T.H., Burgess, G.R. and Jones, J.L.O. (1966). *J. agric. Sci., Camb. 66:303.*

Primaveri, A.M. and Primaveri, A. (1970). Agrochimica *14:490.*

Purser, D.B. (1979). *Proc. Seminar Aust. Soc. Anim. Prod.* (W.A.), Perth (D.B. Purser, ed.). p. 20.

Rademaker, B. (1937). *Fortschr. Landw. Chem. Forsch. 1937:149.*

Reith, J.W.S. (1968). *J. agric. Sci., Camb. 70:39.*

Reuter, D.J. (1975). In "Trace Elements in Soil-Plant-Animal Systems" (D.J.D. Nicholas and A.R. Egan, eds.), p. 291. Academic Press, New York.

Reuter, D.J., Hannam, R.J., Judson, G.J. and Dodson, M.E. (177). *Agric. Rec. 4:36.*

Reuther, W. and Labanauskas, C.K. (1966). In "Diagnostic Criteria for Plants and Soils" (H.D. Chapman, ed.), p. 157. *Univ. California, Div. Agri. Sci.,* Riverside.

Riceman, D.S. and Donald, C.M. (1938). CSIRO Pamphlet 78.

Riceman, D.S., Donald, C.M. and Evans, S.T. (1940). *Coun. sci. ind. Res. Aust. Pamphlet 96.*

Robson, A.D., Loneragan, J.F., Gartrell, J.W. and Snowball, K. (1977). *Aust. Cereals Agron. Conf.,* Perth 1977.

Sauchelli, V. (196). "Trace Elements in Agriculture", Van Nostrand Reinhold Co., New York.

Sharpee, K.W., Ludwick, A.E. and Attoe, O.J. (1969). *Agron. J. 61:746.*

Sheldon, J.H. (1932). *Brit. Med. J. No. 3749,* p. 869.

Sjollema, B. (1933). *Biochem. Z. 267:151.*

Smilde, K.W. and Henkens, C.H. (1967). *Neth. J. Agric. Sci. 15:249.*

Sommer, A.L. (1931). *Plant Physiol. 6:339.*

Sorensen, R.C., Penas, E.J. and Flowerday, A.D. (1970). *Nebraska Quart.,* Spring 1970, p. 9.

Stevenson, F.J. and Ardakani, M.S. (1972). In "Micronutrients in Agriculture" (J.J. Mortvedt, P.M. Giordano and W.L. Lindsay, eds.), P. 79. Soil Science Society of America, Inc., Wisconsin.

Stout, P.R. and Arnon, D.I. (1939). *Am. J. Botany 26:144.*

Steenbjerg, F. and Boken, E. (1950). *Pl. Soil 2:195.*

Stoszek, M.J., Oldfield, J.E., Carter, G.E. and Weswig, P.H. (1979). *J. Anim. Sci. 48:893.*

Suttle, N.F. and McLaughlan, M. (1976). *Proc. Nutr. Soc. 35:22A.*

Takhar, P.N., Mann, M.S. and Randhawa, N.S. (1972). *Indian Fmg. 23:5.*

Teakle, L.J.H. (1942). *J. Agric. West. Aust. 19 (2nd Ser.) 242.*

Teakle, L.J.H., Thomas, I. and Turton, A.G. (1941). *J. Agric. West. Aust. 18 (2nd Ser.):70.*

Teakle, L.J.H., Turton, A.G. and Throssell, G.L. (1940). *J. Agric. West. Aust. 17 (2nd Ser.):161.*

Tiver, N.S. (1955). *South Aust. Dept. Agric. Bull. No. 441.*

Toms, W.J. (1958). *J. Agric. West. Aust. (3rd Ser.) 7:197.*

Turton, A.J., Marsh, N.L., McKenzie, R.M. and Mulcahy, M.J. (1962). *CSIRO Soil Publn No. 20.*

Underwood, E.J. (1977). In "Trace Elements in Human and Animal Nutrition" 4th Edn. Academic Press, New York, San Francisco, London.

Underwood, E.J., Robinson, T.J. and Curnow, D.H. (1943). *J. Agric. West. Aust. 20:80.*

Valadares, J.M. (1975). Bragantia *34:125.*

Viets, F.G.(1966). In "Zinc Metabolism" (A.S. Prasad, ed.), p. 90. Charles G. Thomas, Springfield.

Walsh, T. and Barry, T.A. (1958). *Proc. roy. Inst. Acad. 59 (Sec. B):305.*

Wells, N. (1957). *N.Z. J. Sci. Technol. 38B:884.*

Wild, A.S. and Teakle, L.J.H. (1942) *J. Agric. West. Aust. 19 (2nd Ser.):71.*

Wiener, G., Suttle, N.F. Field, A.C., Herbert, J.G. and Wooliams, J.A. (1978). *J. agric. Sci. Camb. 91:433.*

Williams, C.H. (1977). *J. Aust. Inst. Agric. Sci. 43:99.*

Williams, R.F. (1948). *Aust. J. scient. Res. B1:333.*

15

CONCLUSION: COPPER IN SOILS AND PLANTS — AN OVERVIEW

A.D. Robson

Department of Soil Science and Plant Nutrition
University of Western Australia
Nedlands, Western Australia

In the fifty years since the discovery of the essentiality of copper for higher plants, significant advances in our understanding of the behaviour of Cu in soils and plants have been made. However, there is still much we do not understand. Hence in many instances we are still unable to predict the rates and extents of reactions of Cu either in soil or within the plant. This inability to predict the behaviour of Cu limits our ability to answer important practical questions relating to the alleviation of Cu deficiency and toxicity in agricultural and horticultural crops. For example, we cannot predict the rate of decline in the availability of Cu to plants that occurs with time after Cu fertilization in both mineral and organic soils. In this overview I will attempt to state briefly what we know and what we need to know to be able to predict the behaviour of Cu in soils and plants.

In the very few studies of Cu in soil solutions very low concentrations (10^{-6}-10^{-8}M) have been found. These concentrations are much lower than those used in many studies of both the reactions of Cu in soils and the absorption of Cu from solution by plants. Further, in the soil solution Cu exists complexed to organic ligands rather than as Cu^{2+}.

The concentration of Cu in soil solutions are far too low for them to be controlled by the solubility of Cu minerals found in soil. Instead the concentrations of Cu in soil solutions are controlled by both reactions of Cu with atoms on the surface of the solid phase (adsorption) and by reactions of Cu with organic matter. Although both forms will be present the concentration in solution will be determined by

COPPER IN SOILS AND PLANTS

the solid component that is in equilibrium with the lowest solution concentration.

Copper is specifically adsorbed by Fe, Al and Mn oxides. Studies using such model substances have helped our understanding of the mechanisms of adsorption but these studies cannot be easily translated into the behaviour of Cu in soil. One reason is that soil materials are inevitably contaminated with other materials which may compete with Cu for adsorption or may co-operatively increase adsorption of Cu. Moreover many studies of the adsorption of Cu by soil constituents have used Cu^{2+} rather than anionic complexes of Cu with organic ligands. Little is known about the nature of these organic Cu complexes present in the soil solution. Hence little is known about their effects in the adsorption of Cu in soil. Some organic Cu compounds are known to be strongly adsorbed, others are hardly adsorbed at all.

Little attention has been paid to studying the rates of adsorption and desorption either for pure oxides or in soils. The rate which Cu reacts with soil may be important in determining the long-term effectiveness of Cu fertilizers, the rate at which Cu may be desorbed could be important in determining the supply of Cu to plants.

For many soils one-fifth to one-half of the Cu occurs in organically bound forms. Copper is more strongly bound to organic matter than is any other micronutrient. Organic constituents in soil form both insoluble and soluble complexes with Cu. The quantitative relevance of these reactions to the availability of Cu to plants has been little explored. We do not know which insoluble organic constituents form complexes which restrict the availability of Cu to plants. Similarly we do not know whether Cu is absorbed at the same rate by plants from all soluble complexes of Cu with different organic ligands.

Various schemes have been proposed to fractionate soil Cu by using a sequence of extractants of increasing vigour (for example $CaCl_2$, acetic acid, $Na_4P_2O_7$, oxalate, HF). However, it has not been possible to relate the availability of Cu to plants in a particular soil to the quantity of Cu in any particular fraction. Similarly it has not been generally possible to discriminate between Cu deficient and Cu adequate situations for plant growth by soil analysis.

Because Cu reacts strongly with both organic and inorganic components of soil it is very immobile in soils. This has several important consequences. Firstly little Cu is leached into the ground water. Secondly the uptake of Cu from fertilizer applications may be limited by the activity of roots in the fertilized zone. Cultivation may thus increase the

availability of Cu to plants by increasing the amount of
fertilized soil intercepted by roots. We do not know whether
the roots of Cu deficient plants proliferate in zones fertilized with Cu. However, the most efficient use of fertilizer
Cu requires the establishment of a large number of fertilized
sites within the soil. This can be achieved by adding large
numbers of small granules and by cultivating after fertilizer
application.

In many situations, particularly in mineral soils, applied
Cu has a high residual value even though reactions of Cu with
soil components decrease its availability to plants. The high
residual value of Cu may be associated with (i) the absence
of leaching, (ii) increased positional availability caused by
cultivation, and (iii) high rates of Cu application which are
many times greater than the amount of Cu removed in plant and
animal products.

The absorption, transport and function of Cu in plants
has been extensively studied. In many of these studies, the
absence of a convenient radioisotope has limited the development of an understanding of the nature and rates of the
processes. While it appears likely that the absorption of Cu
by plant roots is dependent on the metabolism of the plant,
the evidence is equivocal. The value of almost all studies
of copper absorption by plants is limited by inadequate
desorption of copper adsorbed onto the cell walls. Not all
the Cu adsorbed onto cell walls can be desorbed by high concentrations of other cations. While it seems likely that Cu
dissociates from organic complexes prior to absorption as the
Cu^{2+} ion, this is not certain.

Copper behaves as a variably mobile nutrient within
plants; in Cu adequate plants Cu is readily retranslocated
from old leaves to new growth, whereas in Cu deficient plants
Cu is retained in old leaves. Delayed senescence of the old
leaves of Cu deficient plants delays the movement of Cu from
these leaves. Copper movement from old leaves parallels
nitrogen movement leading to the suggestion of co-transport
in Cu-amino acid complexes. In many other biological systems
Cu is transported in association with sulphur containing
compounds. Perhaps the effect of senescence on Cu retranslocation is associated with increased concentrations of these
compounds. Recently a Cu-metallothionen has been isolated in
plants. What is the role of this and other similar compounds
in the regulation of the distribution of Cu within plants?

Although the mechanisms may still not be understood, a
greater understanding of the distribution of Cu within plants
in relation to its external supply has led to the development
of improved procedures for the diagnosis of Cu deficiency by

plant analysis. The concentrations of Cu in young leaves are
sensitive and accurate indicators of Cu status in many diverse
plant species.

Copper is involved in many enzymatic reactions within
plants and is a constituent of many enzymes. Two marked and
relatively specific symptoms of Cu deficiency are reduced
lignification and pollen sterility. Reduced lignification
leading to floppy plants and weak straw appears to be due to
the involvement of Cu in enzymes which control phenol metabolism and lignin synthesis. The involvement of Cu in phenol
metabolism may lead to effects of Cu deficiency in enhancing
disease susceptibility. Pollen sterility leading to a failure
to set grain has not yet been associated with impaired functioning of a particular enzyme.

Interactions between Cu and other nutrients are poorly
understood. There is however no evidence that interactions
between Cu and other nutrients occur within the plant at the
sites of function. Most interactions between Cu and other
nutrients occur at sites of absorption (for example Zn, P) or
at sites of retranslocation (for example N). Critical concentrations of Cu in young leaves appear to be independent of
Zn, P and N supply. Although other transition elements can
substitute at least partly for Cu in enzyme function *in vitro,*
it is not known whether such substitution can occur *in vivo*.

Genotypic differences in response to applied Cu are well
documented. Differences among species in tolerance to Cu
deficiency appears to be due to differences in ability to
take up Cu from soil rather than differences in internal
requirements. Differences among and within species in tolerance to Cu toxicity do not appear to rely on a single
physiological difference. The ability to exclude Cu from the
plant, to immobilize it within the plant and to tolerate high
Cu concentrations within the cell may all be important either
singly or together in a particular species tolerant of Cu
toxicity.

From this brief overview it is clear that despite considerable study we do not understand the reactions of Cu in soils
and plants sufficiently well to predict its behaviour.
Further research is required to fill this gap between the
process orientated research and its field application. In
particular it is essential that attempts be made to integrate
our understanding of particular processes occurring both in
soil and in plants into an overall quantitative model.

ACKNOWLEDGMENTS   I am grateful for helpful comments from
other members of the organizing committee of this symposium,
particularly Dr. N.J. Barrow and Dr. R.D. Graham.

ABSTRACTS OF POSTERS

INFLUENCE OF TEMPERATURE ON READILY-EXTRACTABLE COPPER IN SOILS

A. M. Alston
S. P. Harry
G. D. Reddy

Waite Agricultural Research Institute
University of Adelaide
Glen Osmond, South Australia

The influence of temperature on readily extractable copper in soils was measured in a pot experiment, and in incubation studies.

Copper extracted by 0.05 M $CaCl_2$ from a lateritic podzolic soil (pH 5.5) and a calcareous sand (pH 8.2) in which subterranean clover had grown for 11 weeks, was significantly higher where the soil had been maintained at a constant 22°C, than where the soil temperature was 12°C.

When a lateritic podzolic soil (pH 5.5), with and without added copper, was subjected to aerobic incubation for 4 weeks at constant temperature in the absence of plants, the amounts of copper extracted by 0.05 M EDTA in 1.0 M CH_3COONH_4 at pH 6 increased as the incubation temperature increased from 10 to 30°C. The amount of copper extracted from a rendzina (pH 7) similarly treated was significantly affected by temperature only where copper was added, in which case extractable copper decreased with increasing temperature.

It is clear that temperature did not influence extractable copper consistently: the effects varied with the soil, and on whether or not copper was added. Moreover, when the soils were incubated under a fluctuating temperature regime (10-30, mean 20°C), 20 to 40% less copper was extracted by EDTA than where the incubation temperature was constant at 20°C. This result emphasises the need for caution in applying the results of studies conducted at constant temperature to the field situation where the temperature continually fluctuates.

GENETICS OF THE COPPER EFFICIENCY FACTOR IN RYE

Robin D. Graham

Department of Agronomy
Waite Agricultural Research Institute
Glen Osmond, South Australia

Rye has long been known for its copper efficiency, defined here as a tolerance of soils too deficient in copper for other crops, notably wheat, oats and barley. At the time this study began, a copper-inefficient rye was not known and the first step was to study the hybrid of rye and wheat, triticale. Triticales have proven generally to inherit the copper efficiency factor from rye.

A series of substituted triticales containing various numbers and combinations of rye and wheat chromosomes were screened. All triticales which contained the rye chromosome 5R and had a hairy peduncle (Hp - a gene known to be located on 5R) were copper-efficient and all genotypes without 5R were copper-inefficient. Confirmation of the conclusion that the copper-efficiency factor is located on the 5R chromosome has come from a study of the rye addition lines of wheat. Only the 5R addition line, that is, a wheat genotype with an extra pair (5R) of chromosomes, was copper efficient; all other addition lines were copper-inefficient (sterile).

Subsequently we have tested two 5RL translocation lines, that is, wheat genotypes in which a part of the long arm of the 5R chromosome of rye is attached to a wheat chromosome. These two wheats have also performed better in copper-deficient soil than either the pure wheat genotype (minus the translocation) or a similar translocation line involving the 2R rye chromosome instead of the 5R. We are currently studying progeny of a cross of copper-efficient and inefficient ryes to see whether this factor segregates in a manner typical of a single dominant gene.

RAINFALL PATTERN AND RESPONSE OF WHEAT TO SOIL DRESSINGS OF COPPER SULPHATE

Noel J. Grundon
Errol K. Best

Queensland Wheat Research Institute
Toowoomba, Queensland

Unusual responses by wheat to soil dressings of copper sulphate have been obtained in field trials on neutral to alkaline clay soils in Queensland. When copper sulphate was applied with the seed at planting, vegetative growth improved in all trials, but foliar symptoms of copper deficiency appeared in four trials just prior to the boot stage of growth and little or no grain was produced. In the other two trials, deficiency symptoms did not appear and grain yields increased significantly.

In glasshouse trials, wheat grew normally and produced grain when soil dressings of copper sulphate were applied to surface soil from the field sites. There is therefore no inherent characteristics of the soils which makes fertilizer-copper unavailable for plant uptake.

The pattern of rainfall at each field site was compared with the rate of phenological development of the wheat. When symptoms of copper deficiency appeared and negligible grain yields occurred despite fertilizer application, no effective rain (less than 10 mm) fell in the 20 days before the plants reached the boot stage of growth. When soil dressings improved grain yield, effective rainfalls occurred within 14 days prior to booting.

It is postulated that wheat cannot absorb sufficient copper for normal growth and development from a dry fertilizer zone. When the zone is dry just before booting, wheat suffers an acute deficiency of copper during a critical phase of pollen formation, resulting in the production of infertile pollen. Hence, lack of rain to rewet the fertilizer zone at this stage leads to failure of the wheat to produce grain.

TOLERANCE OF SOME WINTER AND SUMMER CROPS
TO COPPER DEFICIENCY

Noel J. Grundon
Errol K. Best

Queensland Wheat Research Institute
Toowoomba, Queensland

Four winter and four summer crops were screened for their tolerance of copper deficiency in glasshouse trials using a neutral clay soil known to be severely deficient in copper. The winter crops were: wheat (Duratti, Cook, Songlen and Gatcher), barley (Clipper), triticale (Satu) and canary seed (Local), while the summer crops were sorghum (E57 and Goldrush), sunflowers (Sunfola-68 and Hysun-31), millet (White French) and panicum (Panorama).

Copper sulphate was applied as a soil dressing to 5 kg air dry soil in a 200 mm diameter pot. Basal dressings of nitrogen, phosphorus and potassium were applied before the experiment began and at three times during crop growth to maintain adequate supplies of these nutrients. Four rates of copper, 0, 0.06, 0.6 and 6 mg Cu pot^{-1}, were applied to the winter crops, and the same plus a higher rate (60 mg Cu pot^{-1}) were applied to the summer crops.

A rate of 6 mg Cu pot^{-1} was not sufficient to maximise grain yield from the winter crops, but the summer crops did produce maximum yields when 60 mg Cu pot^{-1} was applied. Significant differences were recorded among crops and among varieties within crops in tolerance of copper deficiency.

All crops and varieties responded significantly to applications of copper but only the summer crops and Satu triticale produced grain when nil or the lowest rates of copper were applied. On the basis of their responses in grain yield, the winter crops could be ranked as follows in decreasing order of tolerance to copper deficiency: Satu triticale>>>Clipper barley>Cook wheat>>Songlen wheat>Gatcher wheat>Duratti wheat = Local canary seed. The summer crops ranked as follows: Panorama panicum>Hysun-31 sunflower>> Sunfola-68 sunflower>White French millet>E57 sorghum>Goldrush sorghum.

COPPER-ZINC INTERACTIONS IN WHEAT, RYE AND TRITICALE GROWING ON SOILS OF CONTRASTING pH

Susan P. Harry
Robin D. Graham

Department of Agronomy
Waite Agricultural Research Institute
Glen Osmond, South Australia

Three South Australian field soils of pH 5.0, 7.1 and 8.8, each deficient in Cu and Zn, were chosen for a pot experiment in which growth responses of wheat (cv. Halberd and Gatcher), rye (cv. S.A. Commercial) and triticale (Armadillo type) were compared at low and high levels of the limiting trace elements. The texture of the three soils was sandy, the first two silicious, the third calcareous. All basal nutrients were supplied and Cu and Zn added at 4 mg/pot as sulphates.

A basic difference in the physiological effects of Cu and Zn deficiency was on pollen viability. Addition of Cu alone greatly increased pollen viability whereas addition of Zn alone either decreased pollen viability or had no effect.

The results have further established the tolerance of rye to extremes of soil pH and to both Cu and Zn deficiency separately and together, the relatively much greater sensitivity of wheat, and the generally intermediate performance of triticale. In one environment, pH 5.0 high Zn and Cu, triticale outyielded both parental types. Typically positive interactions between Zn and Cu were observed in both vegetative and grain yield and most strikingly in pollen viability on which the patterns in grain yield were based. In all environments and genotypes, adding Zn alone aggravated Cu deficiency and decreased pollen viability and yield, where there was any. However, adding Cu alone generally increased yield. Thus, genotypic differences in the Cu-Zn interaction showed up most strongly in soils of higher pH (7.0 and 8.8) where some grain was produced by rye and triticale in the unfertilised treatment. We conclude that although there were marked differences among the genotypes in their sensitivity to a single deficiency of Cu or of Zn, the Cu-Zn interaction was physiologically similar for all genotypes in each soil.

COPPER DEFICIENCY OF *PINUS* SP. IN EASTERN AUSTRALIA

J. Hill

CSIRO Division of Forest Research
St. Lucia, Queensland

N.D. Turvey

A.P.M. Forests Pty Ltd
Traralgon, Victoria

D.I. Bevege

Department of Forestry
Indooroopilly, Queensland

Pinus elliottii and *Pinus radiata* develop severe symptoms of copper deficiency during the first growing season on podzol soils on the coastal lowlands of south-east Queensland and Gippsland (Victoria) respectively. In both states the affected soils possess a very acid (pH 4) humic A_1 horizon. Applications of major element fertilizers (phosphorus and nitrogen) accentuate the deficiency.

Initial trials suggest that applying copper fertilizer to the soil surface or spraying copper on to foliage in the field causes an increase in growth but only a temporary increase in copper concentration in young foliage. Clearly, these practices do not result in a constant supply of copper to young growth, unless repeated regularly.

Trials planned and/or in progress are aimed at studying the interaction of copper with nitrogen and phosphorus, the application of copper as a spray in the nursery and the field and as a root dip prior to out-planting, the form and placement of copper fertilizers in the field, and the development of genetically resistant material.

Severe copper deficiency of *Pinus* in Queensland and Victoria is restricted in area at present. Future use of more intense management practices, e.g. nitrogen fertilizers, and the establishment of second rotation stands on soils known to be copper deficient for pastures, could extend the problem.

COPPER STATUS OF U.S. SOILS AND PLANTS

Joe Kubota

Soil Conservation Service,   USDA
U.S. Plant, Soil, and Nutrition Laboratory
Ithaca, N.Y.

The copper concentration in soils of the U.S. ranges from about 1 to over 40 ppm. The average concentration is about 9 ppm, with the highly weathered soils of the southeastern coastal plain having the least copper (about 5 ppm). Geographical patterns of copper concentration in soils are more strongly reflected by copper concentration in legumes than in grasses. The copper concentration is most uniformly distributed in plants from the broad glaciated plains of northern U.S., and the southeastern coastal plains. It is most variable in the western and Rocky Mountain states where soils on which the plants grow often vary widely from one valley to another in soil parent materials and the changes they have undergone with soil weathering. A map of the U.S. depicting broad geographical patterns of copper in grasses and legumes will be presented.

THE ACCUMULATION OF COPPER IN AUSTRALIAN ORCHARD SOILS

R.H. Merry
K.G. Tiller

Division of Soils, CSIRO
Glen Osmond, South Australia

A.M. Alston

Waite Agricultural Research Institute
University of Adelaide
Glen Osmond, South Australia

Elements such as copper, lead and arsenic accumulate in orchard soils when pesticides are used over many years. Because there has been a net decline in the area in which pome fruits (apples and pears) are grown, at least 30,000 ha of soils which have been affected by copper are now put to other, usually intensive, uses such as to pastures, vegetable growing or urban development. Similarly there has been a decrease in the area in which stone fruits (plums, peaches and apricots) are grown. Copper-containing fungicides are also used in stone fruit production.

This work was carried out to investigate the extent of copper accumulations in orchard soils from South Australia and Tasmania, the mobility of copper in these soils, uptake by pasture plants and some of the soil factors affecting uptake of copper by vegetables.

Copper accumulated in the surface soils of orchards at concentrations up to 320 μg g^{-1}. There was no evidence of the accumulation copper at depths below about 25 cm in soil profiles. Uptake by pasture plants was not related to concentration in the soil. In some situations copper in pasture plants increased to concentrations which may affect the nutrition of sheep. Soil temperature and pH were found to be the most important factors in determining copper uptake by plants. Fertilizers had little effect on copper uptake, probably because of the high rates of fertilizers which are normally applied to orchard soils.

THE IMPORTANCE OF BACKGROUND CORRECTION
IN FLAME ATOMIC ABSORPTION ANALYSIS OF PLANTS
WHEN DIAGNOSING COPPER DEFICIENCY

W.J. Simmons

Department of Soil Science and Plant Nutrition
University of Western Australia
Nedlands, Western Australia

The use of plant analysis to identify Cu deficiency in plants requires accurate chemical analysis if the correct diagnosis is to be obtained. Flame atomic absorption spectrometry is the most widely used method of measuring Cu in plants. Background absorption can lead to serious positive errors if it is not taken into account. Large negative errors can also be obtained if overcorrection occurs due to the use of the 333.8 nm line. Both errors are much more serious with a stoichiometric flame than a lean flame. The increasing use of automatic safety devices on flame atomic absorption spectrometers, while highly desirable for safe operation of the instrument, tend to use richer flames. This could lead to serious errors if the operator is not aware of the problem.

Analyses of digests of wheat dosed with different levels of Ca, showed that for a lean flame the addition of 4000 μg Ca produced a positive error of +0.3 μg/g. Use of the 333.8 nm line gave a negative error of -0.3 μg/g. A stoichiometric flame gave errors of +1.4 μg/g (uncorrected) and -1.0 (333.8 nm). When it is remembered that the critical value for Cu in the youngest fully emerged leaves of wheat, barley and oats is about 1 μg/g, the serious nature of such errors is evident. Either appropriate background correction with a continuum source or extraction of the Cu into a suitable solvent is essential if accurate results are to be obtained.

COMPARISONS OF THE CRITICAL CONCENTRATIONS OF COPPER IN YOUNG LEAVES FOR THE GROWTH OF WHEAT, OATS AND BARLEY

K. Snowball
A.D. Robson

Department of Soil Science and Plant Nutrition
University of Western Australia
Nedlands, Western Australia

Wheat, oats and barley were grown with several levels of Cu application on a Cu-deficient soil. Differences among cereals in growth responses to Cu appeared to be due to differences in copper uptake rather than to differences in Cu requirements within the plant. Critical concentrations of Cu in the youngest fully emerged leaf were approximately $1\ \mu g\ g^{-1}$ for all three cereals.

The effect of copper supply on the growth of tops and the copper concentration in the youngest fully emerged leaf (YFEL) of wheat, oats and barley, 61 days after sowing in a copper deficient soil.

Copper level (μg/3Kg soil)	Yield of tops* (% max. fresh weight)			Copper conc^{n}+ in YFEL ($\mu g\ g^{-1}$d.w.)		
	Wheat	Oats	Barley	Wheat	Oats	Barley
0	21	78	73	0.5	0.3	0.4
50	65	82	83	0.4	0.4	0.6
100	77	89	92	0.5	0.6	0.6
300	97	100	95	1.2	1.4	0.9
900	100	98	100	3.1	2.5	3.5

* Values are mean of 4 reps.
+ Values are mean of 2 reps.

COPPER DEFICIENCY AND TOXICITY IN SWEET POTATO (*IPOMOEA BATATAS* L.)

K.S. Velayutham
K. Simon Pondrilei

Department of Primary Industry
Konedobu, Papua New Guinea

Sweet potato plants were grown in perlite culture and treated with Cu concentrations of 0, 2, 4, 8 and 16 ppm. Three to four week old plants grown in 0 ppm Cu showed symptoms of Cu deficiency. Most of the young leaves were chlorotic and cupped upwards. Normal healthy plant growth was observed in the 2 ppm treatment but plants grown in 4, 8 and 16 ppm Cu were also chlorotic. The degree of chlorosis increased with increasing Cu concentration in the perlite. Chlorotic symptoms due to deficiency as well as toxicity spread from the top towards the base of leaves. After 5 weeks the intensity of chlorosis increased in all the treatments except 2 ppm treatment where the growth was vigorous. In addition to the chlorotic symptoms, the plants appeared stunted in the 0 ppm treatment (deficiency) and 8 and 16 ppm treatments (toxic). After 7 weeks the leaves in the 8 and 16 ppm treatments were nearly white possibly by Fe deficiency as Cu antagonises the uptake of Fe. In addition to this some pink pigmentation also developed in the leaves which could be due to phosphorus deficiency as Cu also antagonises the uptake of phosphorus. Apart from the deficient and toxic symptoms due to various treatments, the total yield obtained on a dry weight basis was 4.66, 8.14, 5.63, 5.27 and 3.79 g/culture at 0, 2, 4, 8 and 16 ppm treatments respectively.

A TRACER TECHNIQUE FOR COPPER-STABLE ISOTOPES IN ARTIFICIAL DISTRIBUTION

C.D. Walker
R.M. Welch

US Plant, Soil and Nutrition Laboratory
Tower Road
Ithaca,   New York

The longest-lived Cu radioisotopes, Cu-64 and Cu-67 both have relatively short half-lives of 12.8 and 58.5 hours respectively. Consequently long-term studies on the movement of Cu using radioisotopes in biological organisms or soil systems have been impractical.

Stable isotope tracer research is a rapidly developing field, where researchers employ the rare stable isotopes of an element (typically having a relative abundance of no more than a few percent) in relatively large doses. These isotopes may be safely handled, and analysed in samples of tissues or wastes at the leisure of the researcher. Trace metal isotopes may be examined by recovery of the isotope as a metal chelate capable of volatilisation under vacuum, and introduction of the volatile chelate into a mass spectrometer, producing direct comparisons of isotope abundance in adjacent peaks of the mass spectra.

Two stable isotopes of Cu exist, Cu-63 and Cu-65, having natural abundances of 69 percent and 31 percent respectively. Small deviations from the natural distribution as a result of experimental enrichment would be difficult to detect; however an artificial distribution may be obtained by growing an organism on nutrient media highly enriched in one of these isotopes, giving a low background on which a trace amount of the second isotope can then be detected.

This poster describes an experiment supplying stable Cu isotopes to lettuce seedlings, the methods of metal chelate preparation, mass spectrometer operation and spectra interpretation, and examines the uptake of Cu into individual lettuce tissues and its biochemical fractions.

FORMATION OF COPPER COMPLEXES BY AERATION OF METALLIC COPPER

W.R. Walker[1]
S.J. Beveridge[2]

[1]Department of Chemistry
[2] Faculty of Medicine
University of Newcastle, New South Wales

The role of copper in health and disease includes the copper I.U.D. and the 'copper bracelet' and inflammation.

The corrosion of copper is also relevant to copper contamination of drinking water and the dissolution of copper by biological fluids. Concerning the latter Oster and Salgo have stated that the following reactions may be involved:

$$Cu^O \xrightarrow{O_2} Cu_2O \xrightarrow{NaCl} Na[CuCl_2]$$

$$Cu^+ + O_2 \xleftrightarrow{H_2O} Cu^{2+} + R\cdot$$

where R· is the free radical ·OH or O_2^- (hydroxyl and superoxide).

Walker and Griffin have discussed the reaction:

$$2Cu^O + O_2 + 4HL \longrightarrow 2CuL_2 + 2H_2O \qquad \cdots \quad (1)$$

where HL could be one of the many acids present in human sweat.

The following reaction:

$$Cu^O + \tfrac{1}{2}O_2 + H_2O \longrightarrow Cu^{2+} + 2OH^- \qquad \cdots \quad (2)$$

suggests that in the presence of a bidentate chelate such as 1,10-phenanthroline (phen) and 2,2'-bipyridyl (bpy), binuclear hydroxy-bridged species might be prepared as follows:

$$2Cu^{O} + O_2 + 2H_2O + 2 \text{ chelate} \longrightarrow$$

$$[\text{chelate Cu} \underset{OH}{\overset{OH}{<>}} \text{Cu chelate}]^{2+} + 2OH^- \quad \ldots \quad (3)$$

This communication reports the testing of this hypothesis which is relevant to the role of both oxygen and copper in biochemistry.

When air is bubbled into an aqueous suspension of metallic copper in the presence of a chelate such as 1,10-phenanthroline (phen) and 2,2'-bipyridyl (bpy), the copper dissolves and binuclear hydroxy-bridged complexes such as

$$[\text{chelate Cu} \underset{OH}{\overset{OH}{<>}} \text{Cu chelate}] \ X_2 \cdot xH_2O$$

are formed. ($2X = CO_3^{2-}$, $X = ClO_4^-$ and Cl^-).

The dichlorides of such binuclear complexes have been prepared previously by aeration of freshly precipitated suspensions of CuCl. In the case of glycine however, only the bis(glycinato) copper (II) complex is formed from metallic copper. This complex has been labelled with Cu^{64} and has been shown to perfuse intact skin.

These observations afford some evidence for the production of free radicals such as ·OH and may be relevant to the mechanism of action of both the copper I.U.D. and the 'copper bracelet'.

Studies on the nature of the 'copper deposits' in drinking water have also been carried out and the implication of fluoride has been investigated. The complex $Cu[SiF_6]$ appears to be involved in the corrosion of copper pipes.

INDEX

Velayutham, K.S., 367
Vermicilite
  Cu retention, 34

Walker, C.D., 189, 368
Walker, S., 235
Walker, W.R., 369
Webb, J., 189
Welch, R.M., 368
Wheat
  Cu and N in, 173
  Cu distribution in leaves,
    171
  Cu leaf relationship, 177
  response to soil dressing,
    359
Wheat-rye hybrid, 358
Wheat straw
  availability of Cu in, 339
White tip of cereals, 228
Wilting, 223, 227, 288
Withertip of cereals, 228
Woolhouse, H.S., 235

Xylem sap, 152, 165-7, 173
  amino acids, 166-7
  Cu, 165-7

Yellow tip of cereals, 228

Zeta potential, 50
Zinc/Cu enzyme, 199

1 2 3 4 5 6 7 8 9 0
A B C D E F G H I J